C0-AWN-934

COMPARING BEHAVIOR:
Studying Man Studying Animals

COMPARING BEHAVIOR:
Studying Man Studying Animals

Edited by
D. W. Rajecki

Indiana University—Purdue University at Indianapolis

LAWRENCE ERLBAUM ASSOCIATES, PUBLISHERS
1983 Hillsdale, New Jersey London

Lawrence Erlbaum Associates, Inc., Publishers
365 Broadway
Hillsdale, New Jersey 07642

Library of Congress Cataloging in Publication Data
Main entry under title:

Comparing behavior.

Bibliography: p.
Includes index.
1. Psychology, Comparative. I. Rajecki, D. W., 1939– . [DNLM: 1. Ethology—Methods. QL 751 C737]
BF671.C62 1983 156 83-1550
ISBN 0-89859-259-3

Printed in the United States of America
10 9 8 7 6 5 4 3 2 1

Contents

List of Contributors

Robert K. Colwell, Department of Zoology, University of California at Berkeley
Jack Demarest, Department of Psychology, Monmouth College
Phyllis C. Dolhinow, Department of Anthropology, University of California at Berkeley
Irenäus Eibl-Eibesfeldt, Forschungsstelle für Humanethologie, Max-Planck-Institut für Verhaltensphysiologie, D-8131 Seewiesen, Federal Republic of Germany
Gordon G. Gallup, Jr., Department of Psychology, State University of New York at Albany
Jerry Hirsch, Department of Psychology, University of Illinois at Urbana-Champaign
Mary-Claire King, Department of Biomedical and Environmental Health Sciences, University of California at Berkeley
D. W. Rajecki, Department of Psychology, Purdue University School of Science, Indiana University-Purdue University at Indianapolis
Elliott Sober, Department of Philosophy, University of Wisconsin at Madison
Susan D. Suarez, Department of Psychology, State University of New York at Albany
Stephen J. Suomi, University of Wisconsin Primate Laboratory at Madison
S. L. Washburn, Department of Anthropology, University of California at Berkeley
Stephen Zawistowski, Department of Psychology, University of Illinois at Urbana-Champaign
Gail Zivin, Departments of Family Medicine, and Psychiatry and Human Behavior, Jefferson Medical College

List of Contributors

Editor's Preface

A major stimulus for this book came from my experiences with a recent article I helped write: namely, Rajecki, Lamb, and Obmascher (1978). In that paper we reviewed literature on social attachment in precocial hatchlings, puppies, infant monkeys, and human babies, all in order to evaluate certain theories of such phenomena. Part of the critical reaction to our review—published along with the review itself, incidentally—was that our arguments were flawed by our seemingly arbitrary collection of species. After all, some commentators wondered, what is the behavioral connection between birds, dogs, monkeys, and children? In this, the critics may have been right, for we had failed in that paper (in our preoccupation with other matters) to give adequate attention to formal methods or rules for comparing behavior.

Having had this concern brought forcefully to mind I began to look around for such guidelines, but with rather small success. Whereas there seemed to be a great deal of comparing going on at the time (for one thing, Wilson's *Sociobiology* (1975) had been published not long before), and whereas there was considerable controversy attending the claims or conclusions that followed from those comparisons, I did not find much in the way of systematic, formalized statements on just how comparative work should or should not be carried out. In reaction to this paradox, I decided to ask various people who were active in comparative work what their thoughts were on this contemporary issue, and proposed that we use their replies to create a book.

Since the aim of the book was to get a sense of how scientists viewed their own comparative domain, I tried to convey this perspective in the title: *Comparing behavior: Studying man studying animals*. It should be noted that the title is not intentionally sexist; the use of the word *man* simply underscores the traditional dichotomy drawn by many people between human beings, and other members of the animal kingdom (cf. Gould, 1981, p. 16).

As it happened, I sought out the ideas of writers whose work I admired. This led me to people in a variety of fields including anthropology, ethology, genetics, philosophy, psychology, and zoology. The diversity of approaches on the part of the contributors struck me as an interesting and exciting prospect for a discussion of how to compare behavior, but for a period before the submissions arrived I became uneasy. How could one organize a collection of papers that ranged so widely across so many academic disciplines? Indeed, would any chapter be relevant to any other? As the material came in, however, it did cohere, and suggested its own organization. It seemed to me that the papers fell naturally along a general-to-specific dimension. The four that I assigned to Section 1 struck me as the most broad, the four in Section 2 next so, while the two in Section 3 seemed the most focussed. A general-to-specific trend is also evident to me within sections.

Collecting particular papers together in certain sections was perhaps a bit arbitrary, as they might be made to fit in some other place as well. But I think that no harm has been done to any of them as a result of this arrangement. Most assuredly, the numerical ordering of the chapters on my part does *not* reflect or presume any judgment of relative merit or importance. Each of the sections begins with a reader's guide to its chapters. Attention to the guides should alert the reader to the organization and content therein.

As editor of this book I feel indebted to many people for its existence and calibre. First, there are certain contributors who worked especially long and hard, and who persisted in finishing their chapters. Second, there are other contributors who, having submitted their material early on, waited patiently while the former were persisting. The publisher should also be recognized for his faith in the project. I am further grateful to the publisher's anonymous reviewer for his or her valuable observations and recommendations, and to Stanley R. Aeschleman of my department for his help as a reviewer.

REFERENCES

Gould, S. J. *The mismeasure of man*. New York: W. W. Norton, 1981.

Rajecki, D. W., Lamb, M. E., & Obmascher, P. Toward a general theory of infantile attachment: A comparative review of aspects of the social bond. *Behavioral and Brain Sciences*, 1978, *1*, 417–464.

Wilson, E. O. *Sociobiology*. Cambridge, Mass.: Belknap Press, 1975.

COMPARING BEHAVIOR:
Studying Man Studying Animals

Reader's guide to Section 1
ISSUES AND ESSAYS: POINTS OF VIEW IN A COMPARATIVE APPROACH

In Chapter 1, Gallup and Suarez raise the issue of a long-standing human reluctance to be compared with nonhumans in the areas of psychology and behavior. They show the implications of such reluctance in classic notions of dualism in which humans are assumed to be somehow truly different from the remainder of the animal kingdom. Such dualism is said to be based on the *illusion of central position*—humans' alleged uniqueness regarding tools, culture, self-awareness, and other features—and the *romantic fallacy*—that humans are intrinsically good. Further, the authors probe more modern tendencies to set humans apart in terms of our capacity to learn, and to use language. Given their analysis the authors argue that there may be chinks in the arguments for a special human nature. Therefore, if human psychology is not completely removed from a comparison with that of other animals, this opens the door for an "intellectual framework from which to conceptualize and interpret human behavior and human evolution."

If we can take human-nonhuman comparisons seriously, Chapter 2 by Washburn and Dolhinow instructs us where to begin such an endeavor. It is their thesis that if human social behavior is to be part of any comparison, one must begin with an appreciation of the complexity and diversity of the contemporary human condition lest the richness and uniqueness of human life be lost in the process. That is, the authors argue that the most characteristic features of the social life of our species may not be comparable to other animals on a strictly biological basis. This is so because, for one thing, some of those features emerged as a product of evolution only after *Homo sapiens* diverged from the rest of the primate line. For example, it is known that the temporal order of aspects of human evolution was: first, bipedal locomotion; second, the use of stone tools; and third, the relatively large, internally reorganized brain. Thus, the human brain upon which our social behavior is predicated is seen as a product of the latter part of human evolution. For another thing, much of our human life is shaped by cultural, and not by biological influences, and a great deal of our culture is of very recent origin. Therefore, the Washburn–Dolhinow chapter alerts us to the danger of losing track of the distinction between "long-term evolutionary problems of the last 4 million years, and short-term historical problems of the last few thousand years." In other words, "if the study of human behavior is approached by evolutionary sequence or comparison of contemporary forms, the late events in human evolution are greatly underemphasized and reductionism seems almost inevitable."

Chapter 3 by Eibl-Eibesfeldt offers further instruction on the possibilities and promises of a comparative approach. He identifies two classes of observed resemblances between humans and nonhumans: *analogies* and *homologies*. Analogies are behavioral similarities that are due to convergent evolution. That is, biologically disparate organisms confronted with the same environmental demands may independently evolve a common morphological or behavioral "solution" to those problems. As an example of such an analogy, Eibl-Eibesfeldt cites evidence from a number of biologically distant species that points to a conclusion that development of parental care patterns is an important preadaption for the evolution of complex social interaction among adults. Homologies, in turn, are resemblances based on genetic similarities that are tracable to a common ancestor. Following a statement of the criteria for the establishment of homologies, he goes on to make a case that the adult human behavior of kissing is homologous with the behavior of kissfeeding (the passage of food or other material from mouth to mouth between individuals) seen in other primates. According to this analysis, "similarity of the movement pattern, the situational context, linkage by intermediate forms within the species, and intra-species comparison all suggest that the pattern of kissing is derived from kissfeeding." Hence the power of analogies and homologies to help explain the origins of contemporary human behaviors.

The first three chapters in this volume have a marked philosophical air. However, Chapter 4 by Rajecki adopts a more pragmatic tone concerning the study of nonhuman behavior in relation to human concerns. He inquires whether knowledge from animal research has made any real difference in how scientists, scholars, or practitioners deal with human problems or phenomena. In other words, has comparative psychology made any material impact on research, theory, or application in the human domain? The answer to this question is: Yes. In some detail, Rajecki traces the role of knowledge about nonhuman behavior in the development of four lines of thinking about human behavior. The first two are from the area of child psychology and include the formation of social bonds (attachment) in babies, and social dominance and conflict in children. The remaining two stem from an intersection of clinical and experimental psychology and include aversion therapy for alcoholism, and a depression-like reaction known as learned helplessness. All of these human fields have been clearly influenced by research on nonhumans, and Rajecki terms his treatment, "successful comparative psychology: four case histories." But what about *un*successful comparative psychology? Is it not possible to make improper or incorrect comparisons? To this possibility the author replies: "In the long run it is probably better to have compared and erred, than never to have compared at all. If one knows why a particular comparison is a failure, then one knows something important about the organisms to be compared."

1 Overcoming Our Resistance to Animal Research: Man in Comparative Perspective

Gordon G. Gallup, Jr.
Susan D. Suarez
State University of New York at Albany

MAN AS AN ANIMAL

A book devoted to the topic of comparative research on animal behavior is obliged to include some discussion of human behavior. However, it is important to acknowledge at the outset that the common distinction between "human" and "animal" research is a misnomer. Man (in a generic sense) is a biological phenomenon subject to the same basic metabolic and homeostatic principles as other organisms. From an evolutionary perspective man is as much an animal as any other animal; intellectually gifted perhaps, but nevertheless a byproduct of evolution. To be consistent, therefore, the dichotomy that has been erected between humans and animals would have to be extended to all other species (i.e., rat versus animal, chicken versus animal, frog versus animal).[1]

The classic paper by Hodos and Campbell (1969) makes it clear that the designation "subhuman" is also inaccurate and highly misleading. Evolution, in terms of historical relationships, is represented by a phyletic tree, not the proverbial phylogenetic scale. Although contemporary species share remote ancestors in common, it is doubtful that any contemporary species is ancestral to any other. Man, for example, did not evolve from the chimpanzee, although it is undoubtedly the case that we share an ancestor in common with chimpanzees and the other great apes. Extant species are represented by a horizontal dimension, not a vertical one. Technically speaking, a bumble bee is just as evolved as we are. Man's tendency to place himself on top is a reflection of preconceived notions of

[1]An alternative and much more accurate (and certainly less biased) distinction between humans and *other* animals is the designation *human* versus *nonhuman*.

human superiority, not evolution. The only thing that defines man as special, unique, or good is man. Indeed, it may be the case that the near universal tendency for humans (and even distinguishable subcategories of *Homo sapiens;* e.g., different races) to conceive of themselves as being fundamentally different represents what biologists call a reproductive isolating mechanism, which serves or has served to preclude hybridization.

But, before we examine the resistance that many people have to the use of nonhuman animals in psychological research, it is important to discuss briefly some of the principles of evolution.

EVOLUTION

Evolution is like sex. Most people think they understand it, but few really do. But not only is evolution like sex, in a very real sense evolution is sex. In terms of the way many people think about it, evolution is not based on "the survival of the fittest." In the first place, this is a tautology. By equating survival with fitness, the relationship between the two becomes circular. Consider the following scenario. "Why did these particular organisms survive? Because they were the most fit. But how do you know they were fit? Because they survived!" In other words, they survived because they survived and the concept of fitness becomes superfluous. Second, individual survival has nothing to do with evolution. Other things being equal, whether you live or die is inconsequential from an evolutionary point of view. Indeed, death is an inevitable consequence of life. The only thing that will ever happen to you with absolute certainty is death. Life is a terminal disease. Survival, therefore, is not the issue.

Evolution is represented by gradual changes in the composition of a gene pool over time. Thus, rather than being a question of whether you live or die, the issue is whether you reproduce. The composition of a gene pool at any particular point in time is a simple reflection of the differential reproductive success of individuals in prior generations. If everyone in a population has an equal chance of mating with everyone else and reproduction occurs at random, then the composition of the gene pool will remain static over time. Only if certain individuals have more or less of a reproductive advantage will changes occur. Therefore, evolution (which implies change) will occur only under conditions of nonrandom or differential reproduction. In this sense evolution is represented by the perpetuation and elaboration of the reproductively most viable configurations of genes. It is the continuation of genes that count, not individuals. As recounted by others many times before, a hen is an egg's way of producing another egg. In any other sense the hen is irrelevant. Individuals can be thought of as nothing more than a transient superstructure that provides temporary housing for and, through reproduction, perpetuation of genes (Dawkins, 1976).

Commonsense notions of fitness, in terms of strength, vitality, and resistance to disease, are not necessarily the basis for evolution. You could be the "fittest" person imaginable, but if you do not pass on your genes your contribution to evolution is zero. Evolution is based on differential reproduction. Sex, therefore, is the final common path for all evolutionary change. In order for anything to evolve it must directly or indirectly contribute to reproductive success. Rather than being represented by the survival of the fittest, natural selection can be thought of as a correlation between genotype and reproductive success. The rate of selection will be proportional to the size of the correlation, whereas the direction of selection would be determined by whether the correlation was positive or negative.

Evolution does not occur by design; it occurs only by selection, and the raw materials for selection are genetic accidents that occur in the form of mutations. Evolution is represented by the gradual accretion of self-perpetuating accidents. Differential reproduction is what brings order out of chemical chaos at the level of DNA.

In the long run, however, it is not so much a question of how many offspring you have, but rather how many offspring your offspring have. Moreover, because close relatives share a certain proportion of genes in common, your net reproductive success or inclusive fitness will also be affected by the long-term reproductive success of kin. The concept of kin selection has opened up whole new intellectual vistas and represents the conceptual basis for the new and growing field of "sociobiology" (Barash, 1977; Wilson, 1975).

The implications of this approach to evolution are far-reaching and profound. To the extent that individual differences in reproductive success are the key to evolution, anything that affects the likelihood of reproduction will have an impact on evolution. Birth control devices are an obvious case in point. With the advent of various means of contraception, the act of copulation can be selectively dissociated from reproduction. Take, for example, a would-be humanitarian attempt to distribute birth control devices widely in an underdeveloped country suffering from food shortages occasioned in part by a runaway population explosion. It seems reasonable to assume that the capacity to use a birth control device effectively requires, among other things, a minimal level of intellectual competence. Thus, other things being equal, individuals lacking the wherewithal to use such devices effectively will have a distinct reproductive advantage, and one might predict that the proportion of such genotypes in the population at large should increase over subsequent generations.

In industrialized countries where birth control devices are readily available, individuals now have a choice. Sex is no longer tied to reproduction. Birth control provides for recreational sex. One can participate in sexual activity without having to contend with the reproductive consequences. Therefore, coupled with easy access to abortion, the incidence of unwanted children should decrease. Selection in this instance would favor the development of maternal and paternal characteristics. That is, to the extent that increasingly the only people to have children are those who want children, there should be selection for the genetic underpinnings of the desire to have children. People who want children (for whatever reason) will have a reproductive advantage over those who are more neutral or do not want children.

Naturally, these examples presuppose the existence of a genetic influence on the desire to have children and on intellectual functioning and, therefore, many people might be tempted to object. But, in one sense it is impossible to imagine a trait that is not subject to genetic influences. In order to produce an effect on behavior, an environmental event typically must impinge on a receptor surface and impact physiological, neurological, and biochemical systems before it is translated into a change in behavior. Without an organism there would be no environmental influences. Genetic influences are what make environmental influences possible. There is no such thing as a pure environmental influence on behavior. Although genetic influences presuppose an environment, an environmental influence presupposes an organism. Although the relative contribution of genes and environment may vary from trait to trait, population to population, and species to species, it is inconceivable that either could ever equal zero. Any behavior that can be measured reliably can be shown through selective breeding to be subject to genetic influences (Hirsch, 1967). There has never been an instance in which selective breeding based on a behavioral trait has failed to produce an effect. Although it is true that there is no such thing as genetic determinism, by the same token there is no environmental determinism. All behavior is a reflection of both genetic and environmental components.

From this perspective it is clear that, whereas biologists often study the mechanics of evolution, psychologists have unwittingly been studying the dynamics of evolution. Whether organisms successfully reproduce and, therefore, gain genetic representation in subsequent generations, provide adequate care of offspring, and insure for the reproductive success of kin is in large measure predicated on how they behave. Likewise nest building, food gathering, competition for limited resources, detection and evasion of predators, and locating sources of water and shelter are highly dependent on psychological factors. To the extent that organisms have solved the basic problems associated with metabolism, respiration, digestion, and excretion, the direct impact of evolution is in many instances on behavior. When behavior becomes the focal point for selection, physiological and anatomical changes would be expected to occur only to the extent that they are required to support the behavioral changes necessary for eventual reproductive success. Under certain conditions one would expect continued selection for progressively more refined and sophisticated psychological capabilities. Evolution is as much the proper province of psychology as it is biology.

PROBLEM OF DUALISM

Although evolutionary theory is now widely accepted (although still poorly understood), there is a tendency to view evolutionary principles as being applicable only to man's anatomy and physiology. Do human behaviors and cognitive processes somehow transcend evolutionary influences? Witness the differential reactions of college students to courses in genetics as opposed to psychology. Practically everyone will admit that the principles of modern genetics are applicable to human inheritance. But when told that behavioral data gathered on white rats may be applicable to man, students are often sceptical and even insulted. Yet the entire science of genetics is based, in large part, on data obtained from fruit flies and garden peas. However you may feel about rats, it is clear that they are vertebrates, are mammals, and they are much more similar to us than fruit flies. In fact, the entire primate order derives its start from a group of primitive rodentlike insectivores.

When it comes to the use of models for human behavior, nothing is more compelling than the degree of similarity. In this regard one of us has argued that chimpanzees have the potential to create an identity crisis for *Homo sapiens* (Gallup, Boren, Gagliardi, & Wallnau, 1977). Recent evidence shows that in terms of immunological reactions, complex polypeptides, and even DNA itself, chimpanzees and humans are more than 99% identical. We are so much alike at the level of basic biochemistry that we are at least as similar to each other as members of sibling species (King & Wilson, 1975).

The difficulty in being able to accept nonhuman analogs to human psychological phenomena is based on a deeply engrained sense of species ethnocentrism and the age-old problem of the mind–body dichotomy or what has come to be known as dualism. Although it may be tempting to conceive of mental events as if they were somehow independent of biological phenomena, the evidence makes it abundantly clear that, in the last analysis, the human mind is nothing more and nothing less than an expression of complex neurochemical and electrophysiological events emanating from the nervous system in combination with sensory input. Just as centrally active drugs can produce dramatic changes in behavior, alterations of one's neuroanatomy can have a profound impact on psychological functioning. To talk about psychological processes as if they existed independent of a neurological substrate is to proclaim the existence of things that do not exist and is tantamount to mysticism.

The issue can also be thought of in terms of a clash between what Linden (1974) has called the Platonic and the Darwinian paradigms. The Platonic world view makes a sharp distinction between humans and (other) animals. From this perspective, *Homo sapiens* are held to be not only rational, but unique and somehow special. However, biologically speaking there is nothing terribly special or unusual about the human species. Indeed, among mammals we are unique more by virtue of our shortcomings than our advantages. Humans have lost the capacity to brachiate, we cannot run very fast, and we have no natural weapons (such as canine teeth, claws, spines, or horns). We are relatively hairless, our sense of smell is poor, human infants are more parent dependent over longer periods of time than those of any other species, and in comparison to other primates we are not very strong. Although we do have a fairly formidable cortex, cetaceans (e.g., porpoises and whales) have brains that, in terms of size and complexity, rival if not exceed our own.

The basic biological principles governing the metabolic, endocrinological, neurological, and biochemical activities in man are basically the same in many other organisms. Behavior, therefore, has become the last stronghold for the Platonic paradigm. For instance, Dobzhansky (cited by Desmond, 1979) argues that "if the zoological classification was based on psychological instead of morphological traits, man would have been considered a separate phylum or even kingdom." But the intellectual bankruptcy of dualism renders this kind of argument transparent. If we accept the proposition that, in the last analysis, behavior is nothing more than an expression of physiological processes, then to admit the biological but deny the psychological similarities between ourselves and other species seems logically inconsistent and indefensible. But logical problems or not, the "territorial imperative" is alive and well when it comes to the human psyche. Witness the controversy over the existence or nonexistence of the capacity for language in chimpanzees (Mounin, 1976). The debate that has arisen over the issue of language in nonhumans (principally chimpanzees) may improve our understanding of language, yet it is not clear that it will have the same effect in our thinking about man (Gallup, 1977).

SOURCES OF RESISTANCE

The history of science in part, and psychology in particular, has been one of bringing about gradual changes in man's conception of man, and man has not always felt too comfortable with what he has found. In a sense science has been rather degrading to humanity. The highlights of what has happened can be summarized as follows. First, contrary to the popular opinion of his day, Copernicus demonstrated that the earth was not the center of the universe. In fact, we later found ourselves inhabiting a small planet revolving around one of countless stars located at the edge of one of thousands of galaxies. Darwin was next, with arguments that man was a product of natural forces, with an evolutionary history, and a member of the primate order. Freud followed with the assertion that man was often irrational and many of our motives unconscious. Rather than logical, man is psychological. Now modern behaviorism seems to have reduced man to a machine, to be understood on the basis of little more than reinforcement contingencies. Indeed, according to the title of one book (Skinner, 1971) man is neither capable of "freedom" nor worthy of "dignity."

Understandably, this trend has met with considerable resistance. Perhaps the most apt attempt to identify and characterize this resistance is provided by Ardrey (1961, 1970). Ardrey contends that man's inability to see man for what he is is a consequence of the *illusion of central position* and the *romantic fallacy*. Copernicus notwithstanding, man still tends to conceive of himself as central, if not to the universe then certainly to biological phenomena. This is the doctrine of human uniqueness. But, under scrutiny, uniqueness is not enough. Each species is unique. For centuries man has sought to identify special and/or redeeming characteristcs that would distinguish people from other animals. Candidates for human uniqueness have included at one time or another such cherished notions as tool use, tool fabrication, culture, cross-modal perception, religion, language, reason, and self-awareness. It is important to reiterate, however, that the only thing that defines man as special, rational, or even good, is man. Indeed, recent findings (see Table 1.1) challenge many of these earlier views concerning the existence of distinctive human traits (Desmond, 1979; Gallup et al., 1977; Mason, 1976; Premack & Woodruff, 1978). To date, the differences appear to be a matter of degree rather than of kind. For example, Mason (1976) concludes, on the basis of a review of the literature, that the "essential terms of our uniqueness have yet to be defined" and that "time and time again we have imposed our own categories of experience on the (great) apes and have found solid evidence of correspondence [p. 292]."

Romanticism, on the other hand, boils down to the proposition that man is intrinsically good. But again, the same limitations apply. Would blue whales and Canadian seal pups agree? What about cattle waiting to be slaughtered or fish suffocating in the bottom of a boat? Or rats with a stomach full of rat poison? From an evolutionary point of view, good and bad are meaningless and ethics are species specific (Gallup & Suarez, 1980). At best, the concepts of good and bad are nothing more than states of mind, conditioned perhaps by selective pressure for progressively more elaborate and subtle forms of reciprocal altruism (Trivers, 1971). For better or worse, man is man.

For those who subscribe to the fallacy of romanticism, all the more obvious incidents of man's inhumanity to man, such as murder, rape, war, crime, and child abuse, are difficult to reconcile. This discrepancy between preconceived ideas about the way things ought to be and the way things are has ushered in what Ardrey (1970) calls the *age of the alibi*. In order to account for so many exceptions to romanticism one needs an excuse. A popular excuse used to be that certain people were under the influence of evil spirits. Nowadays the alibi is that man is corrupted by the environment. The environment has become the villain. Interestingly, psychologists, perhaps unwittingly, have provided the major vehicle for this excuse. One need not look very far in the literature to discover that the

TABLE 1.1
Selected Examples of Nonhuman Behavior That Challenge Preconceived Notions of Human "Uniqueness," or "Redeeming Characteristics" of Human Behavior

Source	*Phenomenon*	*Animal*	*Situation*	*Behavior*
Davenport & Rogers, 1970	Cross-modal perception	Chimpanzee	Laboratory	Subjects were able to match-to-sample across the visual and tactile modalities
Fisher, 1939	Tool use	Sea otter	Naturalistic observation	Hammer small mollusks against a rock resting on the chest
Gallup, 1970	Self-awareness	Chimpanzee	Laboratory	Subjects made explicit reference to their mirror image in terms of grooming, autostimulation, and responses to superimposed body marks with the aid of a mirror
Gallup, 1982	Consciousness	Chimpanzee	Laboratory	Capacity to become the object of one's own attention as evidenced by self-recognition
Gallup, 1982	Mind	Chimpanzee	Laboratory	Ability to monitor one's own mental states as evidenced by the attribution of mental states to other organisms
Goodall, Bandora, Bergmann, Busse, Matama, Mpongo, Pierce, & Riss, 1979	Murder	Chimpanzee	Naturalistic observation	Brutal group attacks on a lone conspecific that resulted in death; killing infants
Harlow & Suomi, 1971	Mental illness	Rhesus monkey	Laboratory	Depression, self-mutilation, and stereotypy induced by social isolation
Herrnstein, 1966	Religion	Pigeon	Laboratory	Stereotyped, highly ritualized responding that lacks an instrumental component and occurs under conditions in which the receipt of reward bears no relationship to responding
Itani & Nashimura, 1973	Culture	Japanese macaque	Naturalistic observation	Cross-generational transmission of food preparation behaviors (washing sweet potatoes to remove dirt; separating wheat from sand by immersion in water)
Köhler, 1927	Insight	Chimpanzee	Laboratory	Sudden solutions following periods of ineffectual problem solving (fitting sticks together to obtain food that was out of reach)
Kummer, 1968	Politics	Baboon	Naturalistic observations	Formation of coalitions during agonistic encounters

Lethmate & Dücker, 1973	Adornment behavior	Orangutan	Laboratory	Subjects adorned their bodies with objects (wearing a piece of lettuce like a hat) and examined the effect in a mirror
Menzel, 1972	Cooperation	Chimpanzee	Captivity	Cooperative ladder building and support
Menzel, 1979	Extrapolation	Chimpanzee	Captivity	Subjects not knowing the placement of hidden food could extrapolate its location on the basis of an informed chimpanzee's posture and initial movements
Morris, 1962	Art	Chimpanzee	Captivity	Painting and finger painting
Premack, 1971	Displacement and representation	Chimpanzee	Laboratory	Capacity to abstract and symbolize objects in their absence (using a blue triangle to represent an apple)
Premack, 1971	Quantification	Chimpanzee	Laboratory	Learned to use appropriately symbols for ''all,'' ''none,'' ''one,'' and several
Silk, 1979	Reciprocal altruism	Chimpanzee	Laboratory	Reciprocal food sharing
Teleki, 1973a	Reactions to death	Chimpanzee	Naturalistic observation	Agitated and peculiar responses following the accidental death of a group member
Teleki, 1973b	Hunting	Chimpanzee	Naturalistic observation	Systematic, male-oriented, cooperative hunting patterns that terminate in sharing
Thompson & Herman, 1977	Memory processes	Porpoise	Laboratory	Memory span and performance on probe recognition tasks that was highly similar to humans
van Lawick-Goodall, 1968	Attribution	Chimpanzee	Naturalistic observation	The mother of an infant injured while playing with another will attack the aggressor's mother
van Lawick-Goodall, 1968	Incest avoidance	Chimpanzee	Naturalistic observation	Avoidance of mother–son and sister–brother mating
van Lawick-Goodall, 1968	Politics	Chimpanzee	Naturalistic observation	Formation of stable dominance hierarchies based in part on psychological tactics
van Lawick-Goodall, 1968	Tool fabrication	Chimpanzee	Naturalistic observation	Preparation of twigs to fish for termites; use of crushed leaves as a sponge to obtain drinking water, as a means for extracting brains from prey, or as an aid to grooming
van Lawick-Goodall, 1971	Invention	Chimpanzee	Naturalistic observation	Banging kerosene cans together to augment an intimidation display by a low-ranking male who subsequently rose to alpha status
Woodruff & Premack, 1979	Deception	Chimpanzee	Laboratory	Intentionally withheld information, or provided misinformation to humans as to the location of a piece of favored food

supposed mechanism underlying such corruption is LEARNING. People, it is held, learn to be aggressive, they learn to commit crimes, they learn practically everything! By providing the alibi, however, it may turn out that psychologists will have done more to obscure our understanding of human behavior than to clarify it. Moreover, as an alibi, learning is a poor excuse. As detailed in the next section, learning is as much a consequence of biology and evolution as it is a means of corruption or intellectual liberation. Learning is a major vehicle to behavior change, but learning, itself, is not learned.

LEARNING IN EVOLUTIONARY PERSPECTIVE

The psychologist's continued preoccupation with learning as a means of affecting behavior change, and as a convenient scapegoat for social and psychological ills needs to be tempered by an evolutionary overview. Learning cannot emancipate man from his basic biological heritage for the quite simple, but at the same time compelling reason that learning is firmly embedded in that heritage. Moreover, as we intend to show, the distinction between learned and innate behavior breaks down under scrutiny. Learning is not an alternative to innate behavior, nor are the two orthogonal to one another. The entire range of reactions both realized and unrealized in all possible environments is influenced by genes. Even the specifics of what is learned may be constrained by information encoded in genetic material. Rather than a measure of intelligence, our capacity to learn is an index of genetic predispositions.

There are several points that should be kept in mind when thinking about the importance of learning in the determination of ongoing behavior. In the first place, *learning presupposes innate behavior*. Learned and unlearned components of behavior are not mutually exclusive. In fact, we contend that learning represents nothing more than the modification and elaboration of innate behavior. Without innate behavior, learning, as we know it, would be impossible. Conversely, if all human behavior were learned (as some people still believe) then learning would be impossible!

No doubt some readers will react quite negatively to these assertions but our justification is simple and straightforward. Most psychologists agree that many instances of more complex behavior are analyzable in terms of various combinations and permutations of classical and instrumental conditioning, which are thought to represent two of the most basic or elementary forms of learning. Yet it is easy to show that learning is predicated on the prior existence of unlearned, species-specific responses and stimulus–response connections. In classical conditioning, for example, the paradigm is one in which you must identify an already existent connection between a stimulus, referred to as the unconditioned stimulus (UCS), and a particular response called the unconditioned response (UCR). To take the proverbial Pavlovian example, food powder as the UCS when placed in a hungry dog's mouth elicits salivation as the UCR. To classically condition the salivary response, one identifies some other stimulus that the dog can detect, but that has no initial effect on salivation, such as a tone. This neutral stimulus, referred to as the conditioned stimulus (CS), is then systematically paired with food. That is, the dog is given repeated paired presentations of the tone followed by food. As a result the dog learns, in effect, to anticipate the receipt of food on the basis of having just heard the tone, and the tone or CS comes to elicit an anticipatory salivary reaction called the conditioned response (CR). Our point is simply that the development of this new stimulus–response connection requires an already existent unlearned or ''unconditioned'' connection between the UCS and UCR. Without at least a few unlearned (or unconditioned) stimulus–response connections to begin with, classical conditioning would be impossible.

Much the same holds true for instrumental or so-called operant conditioning. The application of reinforcement to increase the strength and/or probability of a response (or conversely punishment to decrease the likelihood of a response), presupposes the prior emission of the response. Reinforcement does not elicit or produce the response upon which it is made contingent, it merely affects the probability of its *reoccurrence*. You cannot manipulate the consequences of a response until the response occurs. Thus, in the absence of an already existent repertoire of responses, instrumental conditioning would also be impossible. Put simply, an organism without any unlearned behavior would be incapable of learning. A ''blank slate'' would forever remain blank. In fact, because learning represents the elaboration and modification of innate behavior, all learning could be viewed as being species specific and a unique adaptation to changing fitness needs that are partly niche specific.

Language acquisition in humans is a good illustration of the fact that learning presupposes innate behavior. Without the capacity to vocalize, verbalizing would be a mute point. Prior to the development of speech, human infants pass through a stage of spontaneous vocalizing often referred to as babbling. A phonetic analysis of this vocal output shows that contained in the babbling of the preverbal child are all the sounds needed to speak any human language (Bower, 1974). It is from this array of sounds that specific verbalizations later emerge. As evidence for the fact that the infant's vocal output is independent of specific auditory experiences or even the sound of human speech, congenitally deaf children pass through this same stage and babble indistinguishably from normal children (Lennenberg, 1969). It would appear, therefore, that not only are these proverbial preverbal sounds unlearned, but they are apparently independent of specific auditory experiences or even auditory feedback. The development of spoken language exploits an array of sounds that are maturationally determined and not the product of experience.

The second point is that *learning is a byproduct of evolution*. In its most rudimentary sense, learning is not learned. An individual's basic capacity to learn is not an acquired phenomenon. The ability to learn is as much a reflection of an organism's biology as its capacity for respiration, vision, metabolism, and hormone production. Learning evolved as a mechanism through which organisms could cope with transient, within-generation environmental changes, and as such represents a behavioral phenotypic trait subject to natural selection (Plotkin & Odling-Smee, 1979). It should, therefore, not be surprising to find that individual differences in learning ability are partly heritable, as evidenced by the result of selective breeding experiments employing different levels of learning as the criterion for reproduction (Bignami, 1965). By the same token, the rather substantial species differences in learning capacity are also genetically based. For instance, if a chimpanzee, rhesus monkey, cocker spaniel, guinea pig, chicken, and parakeet were all raised under identical conditions, there would still be substantial species differences in behavior and learning ability.

The ability to learn is a pervasive character of animal life. It is now reasonably well established that even single-cell paramecia are capable of rudimentary learning (Hennessey, Rucker, & McDiarmid, 1979). The omnipotent character of learning is a consequence of the fact that it has tremendous adaptive significance. Consider, again, classical and instrumental conditioning. Under natural conditions, classical conditioning enables organisms to anticipate recurring environmental events and take appropriate preparatory action. It would be highly advantageous for organisms to be able to predict and/or anticipate things before they happen. Similarly, although there are a few notable exceptions (e.g., saccharin, brain stimulation), most primary reinforcers serve to reduce basic biological drives and as such participate in the maintenance of homeostatic balances. Clearly, organisms that could learn to engage in the kinds of behaviors that minimize biological imbalances would have a distinct reproductive advantage. Conversely, most punishers carry the threat of tissue damage and/or deprivation from needed substances and as such are potentially harmful and disruptive of homeostatic states. Organisms that could learn to refrain from engaging in the kinds of behaviors that might not be in their best biological interests would also have an adaptive advantage. Thus, the ubiquitous nature of classical and instrumental conditioning should come as no surprise; there is every reason to believe that they are a byproduct of selective factors that have contributed to the reproductive success of organisms that could learn to cope with, adjust to, and anticipate changes in the environment. Habituation (as defined by a relatively permanent, stimulus-specific response decrement that occurs as a result of repeated elicitation) has also been suggested as an elementary form of learning. Habituation enables organisms to attend selectively to novel and potentially important stimuli. Without this capacity, the system would be rendered virtually ineffective by both stimulus overload and tendencies to monitor and respond continuously to redundant and insignificant cues.

To repeat, there is every reason to believe that learning is a consequence of evolution. As an evolved phenomenon, the ability and capacity for learning is genetically based. Although it is true that positive transfer, different learning strategies, and mnemonic devices can be employed to improve the rate of acquisition and retention, an individual's basic ability to learn can hardly be construed as an acquired trait. Learning is not learned.

Our third point is that *learning occurs only within relatively narrow genetically constrained corridors*. For instance, species-specific differences in effective unconditioned stimuli and the topography of unconditioned responses will clearly influence and set limits on the kinds of conditioned anticipatory reactions different organisms develop (Bolles, 1970; Seligman, 1970). Perhaps more importantly, what is reinforcing will also influence and establish boundary conditions on the specifics of what different species learn.

In the case of primary reinforcement, what constitutes a reinforcing stimulus is assumed to be independent of experience (i.e., the reinforcing properties of the stimulus in question are a consequence of innate mechanisms). Coupled with the fact that there are species differences in primary reinforcers, this imposes limits on learning (Hogan & Roper, 1978). Depending on what happens to be reinforcing, different behaviors will have different probabilities of being developed. For example, if bananas are reinforcing, certain behaviors and behavioral sequences will prove more effective than others in the eventual acquisition of bananas in the natural habitat as compared to gathering worms, termites, bamboo shoots, or herring. In other words, there are a number of ways to skin the proverbial cat, but the possibilities are finite and may not always transfer to other endeavors.

It is also important to note that as a stimulus event reinforcement is not an endpoint in most response episodes. Although it is frequently convenient to think of food as a reinforcer, it is not the food per se that has reinforcing properties, but rather the fact that food creates an opportunity to engage in a relatively hardwired consummatory sequence. It is the opportunity to eat the food that is reinforcing. The mere sight or smell of food would not be sufficient to sustain instrumental responding over extended periods of time. Reinforcing stimuli derive their reinforcing properties because they create the occasion for an innate response episode.

Other boundary conditions on learning are set by species-specific differences in genetically determined perceptual–motor capacities. Clearly, what any particular organism can or cannot detect (e.g., ultrasound in the case of bats or infrared radiation for pit vipers) will impose limits on what is learned. Likewise, because of species differences in response capacities and adjunctive anatomical specializations, learning to engage in certain behaviors may be more or less probable (flying, burrowing, swimming, brachiating, etc.).

There also appear to be *highly specific predispositions to learn* (see Rajecki's chapter, this volume). Although learning has tremendous adaptive significance, under certain circumstances it can produce maladaptive behavior. Judging from the development of certain behavioral and clinical anomalies in humans (e.g., autism, anorexia nervosa), learning can be as biologically harmful as it is advantageous. An organism whose survival is critically dependent on learning can ill afford to let the direction of learning be determined by chance factors. Indeed, it is now well established that what is learned is not arbitrary, but subject instead to genetic constraints and predispositions (Seligman & Hager, 1972).

Rats, for example, have an uncanny ability to learn to avoid poisoned bait. This is based on a highly specialized form of taste-aversion learning, which came as some surprise to many traditional learning theorists. In the first place, such learning seems to be modality specific. That is, rats can learn to associate taste but not audiovisual events with sickness. Conversely, rats can learn to associate audiovisual stimuli, but not taste, with shock (Garcia & Koelling, 1966). Rats not only seem capable of associating novel flavors with sickness in as few as a single trial (Garcia, McGowan, & Green, 1972), but they can form associations between tastes and illness when the onset of sickness follows the ingestion of novel food by over an hour (Garcia, Ervin, & Koelling, 1966). In the traditional classical conditioning paradigm the CS and UCS must occur and repeatedly reoccur in close temporal proximity to one another for conditioning to take place.

That the flavor and illness can be separated by a substantial period of time in the acquisition of taste aversions raises another paradox. Under natural conditions if the ingestion of tainted food is not followed by gastric distress for up to 30 minutes or more then chances are that the rat will have consumed familiar foodstuffs in the interim. Why then do they not develop taste aversions to familiar flavors, particularly as in this instance they would have been experienced in closer proximity to the onset of sickness? It appears that rats are predisposed to associate only new or novel tastes with illness (Revusky & Bedarf, 1967). As evidence for the generality of such specialized learning abilities, humans also seem capable of forming strong taste aversions to novel flavors in remote association to gastrointestinal distress (Seligman & Hager, 1972, p. 8). That the ability to form illness-induced aversions to foods is a pervasive phenomenon is further illustrated by the fact that in avian species where vision rather than taste is the primary sensory mode involved in food recognition, aversions are learned on the basis of visual characteristics of the food (Brower, Ryerson, Coppinger, & Glazier, 1968). The adaptive significance of such highly specialized associations should be obvious and provides a clear illustration of tbe impact of evolution on learning mechanisms (see Rozin & Kalat, 1971 for further discussion).

The acquisition of speech is another example of how genetic predispositions influence and direct the learning process. Young children are highly prepared to learn to speak. Language acquisition in human children does not require sets of carefully prearranged linguistic contingencies, rote learning, close supervision, prompting, or judicious training. Children often seem to learn language in spite of and not because of what they are formally taught. Rather than a measure of man's intelligence or superiority, language is a reflection of species-specific genetic predispositions. Although we are highly prepared to learn spoken language, we seem relatively unprepared to learn to read. Reading often takes considerable training and, as evidenced by the large number of "reading disabilities," represents the acquisition of a much more arbitrary skill.

Perhaps nowhere are the biological boundaries of learning better illustrated than in the development of human phobic reactions (Seligman, 1971). Contrast common human phobias for heights, snakes, insects, closed-in spaces, and the dark with such objects as electric outlets, stoves, bicycles, hammers, and matches. While the former stimuli were likely to have been critical to the reproductive success of our ancestors, the latter objects are evolutionarily irrelevant. But because they are a common form of aversive stimulation, the prevailing view of the development of phobias based on classical conditioning would lead one to expect a high incidence of phobic reactions to objects in the latter category. Yet, how many people do you know with a hammer phobia? That the object of many human phobias may be a reflection of our evolutionary heritage is nicely illustrated by the findings of Öhman and co-workers. They report that when using electrodermal responses as a measure of fear, greater resistance to extinction is found if pictures of common phobic objects such as spiders and snakes are used as conditioned stimuli, as compared with pictures of flowers and houses (Öhman, Eriksson, & Olofsson, 1975; Öhman, Erixon, & Lofberg, 1975). Acquisition also proceeds at a faster rate with common phobic stimuli (Hugdahl, Fredrikson, & Öhman, 1977). Likewise, analyses of human twin data suggest that both the strength and content of phobic fears may be subject to genetic influences (Torgersen, 1979).

From the standpoint of traditional learning theory it is difficult to explain why little children ever learn to walk. In the process of learning to adopt an upright posture and walk, children experience considerable aversive stimulation associated with tripping and falling, but few if any of us ever develop a walking or standing phobia. Moreover, it is difficult to argue that the reason children learn to walk is because of the enhanced mobility it provides, as during the acquisition of this skill walking is very awkward and crawling remains a much more effective, and certainly safer means of moving about. Walking, like language, is a species-specific predisposition.

To summarize, not only does learning represent the modification and elaboration of innate behavior, but learning itself is a byproduct of genetic instructions that have evolved because learning has such widespread adaptive significance. Not only does learning occur within relatively narrow genetically fixed corridors, but the specifics of what is learned appear subject to genetic constraints and biological predispositions. Learning is not a learned phenomenon. Although it may be appropriate to speak of the byproduct of learning as acquired, it is important to realize that what is learned is often far from arbitrary and no organism is completely malleable merely because it can learn. Organisms do not come into the world as blank slates nor is what is ultimately written on those slates a matter of chance factors. Learning can be a powerful vehicle for behavioral change and adaptation to altered environmental conditions, but learning per se can only be adequately understood in the context of evolution. Learning is not an alternative to innate behavior, it is an extension of innate behavior. There simply are no instances of learning that are free from collateral genetic influences.

A COMPARATIVE APPROACH TO LANGUAGE

The evidence to date strongly suggests that spoken language based on symbolic cues is sufficiently unique to man to qualify as a species-typical trait. However, language per se can hardly be construed as sufficient for placing man back on a Platonic pedestal. Language is a vehicle for the exchange of information, but it carries no intrinsic information in its own right. The human propensity to learn to speak, as we have seen, is a genetically based trait. Although speech can be legitimately used as one of man's defining characteristics, many of the precursors to speech and the cognitive and conceptual underpinnings of symbolic communication are not, as illustrated in Table 1, the exclusive domain of man (Desmond, 1979; Gallup, 1979; Premack & Woodruff, 1978).

On the other hand, it has recently become apparent that perhaps too much may have been made of the apparent capacity of both chimpanzees and gorillas to learn various forms of gestural and symbolic language. In spite of some grandiose claims (Gardner & Gardner, 1971; Patterson, 1978), much of the data remain fragmentary and anecdotal. Indeed, many results now appear to be a consequence of prompting, cuing, and imitation (Terrace, Petitto, Sanders, & Bever, 1979), and the status of symbolic language learning on the part of nonhuman primates will remain equivocal until more detailed and systematic observations are made. In fact, much of the performance of one chimpanzee named "Lana" (Rumbaugh, Gill, & von Glaserfeld, 1973) has recently been simulated on a computer, using simple principles of paired associate learning (Thompson & Church, 1980). However, the interest in and controversy about language learning in great apes has overshadowed a much more fundamental parallel between language and skilled movement sequences.

In its most fundamental sense, speech is merely a motor skill, albeit a complicated one. It is the intellectual capacities that underlie speech that are important, not the ability to utter articulate speech sounds. Why should verbal behavior be treated any differently from other kinds of behavior? Complex motor skills embody a *grammar* or deep structure that is strikingly similar to language (Hewes, 1973; Holloway, 1969). Just as the exact words and their sequences are pretty much irrelevant to the meaning of a message, the same is true of moving from point A to B in space. In each case there are sets of rules (a basic grammar) that govern the way in which units or elements can be combined to achieve an end. If motor movements were random, they would not be meaningful. Moreover, much like words in a sentence, the responses contained in a movement sequence exist as sets of sequential dependencies that effect the probability of subsequent responses (Fentress & Stilwell, 1973). Although frequently cited as a unique feature of language, *displacement,* as defined by the ability to represent objects and/or events in their absence, and *intention* are part and parcel of tool fabrication, a capacity that is well documented in chimpanzees (Beck, 1975; Teleki, 1974). Likewise, *reconstitution,* as an essential feature of language, is also an integral part of tool fabrication and even effective tool use. Chimpanzees, for example, show an uncanny ability to modify and/or pick appropriate tools to fit the occasion (rather than trying to find the occasion to fit the tool). Chimpanzees have been observed using sticks for such diverse purposes as termite fishing, leverage, olfactory probes, dental grooming, ladders, scratching, and hitting (Beck, 1975). The particular use of sticks, as well as the characteristics of the stick itself, is reconstituted by the chimpanzee to fit the unique requirements of the task at hand. Just as the verbal units contained in a message can be rearranged or reconstituted to form new messages, toolmaking, as well as tool use that is optional and flexible, involves a coherent rearrangement of parts of the environment as a means to an end. From this perspective a sentence is represented by a goal-directed motor sequence that follows certain rules, and is analogous to a flexible (as opposed to a stereotyped) movement sequence.

Menzel's (1973, 1975) classic work has shown that independent of any specific training, chimpanzees appear to have a natural language, based apparently on inferences they make about concurrent mental states among one another, using subtle gestures, postures, and eye movements. Menzel has demonstrated that chimpanzees can convey complex information to one another about the direction, probable location, and relative desirability (or undesirability) of different hidden objects that *only one* of the chimpanzees has seen. Moreover, after a brief experience with this kind of task, the uninformed chimpanzees begin to extrapolate the location of hidden food based on the leader's initial movements and often run ahead of him in an attempt to gain priority access. In instances of highly prized items, informed chimpanzees are subsequently observed actively attempting to "mislead" others as to the source of the object, and the followers react by abandoning an extrapolation strategy and proceed to keep the leader under continuous surveillance. Embodied in these activities are clear and unmistakable instances of displacement, intention, syntax, reconstitution, and possibly even deception (for more recent evidence of the chimpanzee's capacity for intentional deception see Woodruff & Premack, 1979).

As these examples illustrate, the near universal tendency to dichotomize behavior into the separate categories of "verbal" and "motor" behavior may be artificial, unnecessary, and even misleading in terms of the fundamental principles underlying such activities. This point is highlighted by the striking parallels between spoken and gestural language in humans.

Just as speech is lateralized in the human brain, recent evidence shows that vocalizations in primates, prompted by electrical stimulation of the cortex, can only be elicited by stimulating areas in the left cerebral hemisphere analogous to the human speech center (Petersen, Beecher, Zoloth, Moody, & Stebbins, 1978). This has prompted one writer to comment that "language was less a gift of the gods than an exploitation of the primate potential" (Desmond, 1979). It is also apparent that contrary to recent claims (Washburn & Benedict, 1979), some nonhuman primates show structural cerebral asymmetries similar to those found in the human brain (Yeni-Komshian & Benson, 1976).

In the last analysis it is not language per se that gives man such a tremendous intellectual advantage over other animals, but rather the accumulation and exchange of vast amounts of information that language makes possible. There is no evidence to suggest that the size or anatomy of the human brain has changed much in the past 10,000 years, yet our level of intellectual achievement has skyrocketed in that time. It is instructive to note that it took primitive man thousands of years to develop the arrowhead, and that today we can master the technology needed to send people to the moon or build atomic weapons capable of vast devastation. We are "intelligent" in large part because of the accumulating body of knowledge and the collective solutions of previous generations. The information pool underwrites our intelligence. Persons isolated from this pool of information (e.g., unwanted children sequestered away in attics) show no obvious signs of human intelligence.

The real breakthrough in human history was the invention of written language (which is another good example of the transparent dichotomy that has been erected to distinguish between verbal and nonverbal behavior). Writing provides for the relatively permanent storage of information, and once enscribed is not as subject to the inevitable distortion in meaning that characterizes spoken messages as they are passed on "by word of mouth." With the subsequent development of the printing press, man had achieved a means of information exchange no longer restricted by narrow dimensions of time and space, and the rapid dissemination of massive amounts of information became possible. The other obvious advantage man has over certain other big-brained mammals (e.g., cetaceans and elephants) are hands with opposable digits. Much of the evidence for human intelligence, such as our sophisticated technology, is a byproduct of our ability for precise manipulation. Hands are a vehicle for the implementation of ideas into reality.

MAN STUDYING MAN

In the sense of trying to maintain an objective and unbiased perspective, man is probably the least suited organism to study man. Psychologists, anthropologists, sociologists, and others simply cannot transcend their humanness for purposes of studying human behavior. As Jung (1958) has pointed out, to achieve an objective view of man would require the opportunity to see ourselves as we are seen by other species, which, in our view, is one of the major reasons for doing comparative psychological research. Although human behavior is human behavior, not chimpanzee behavior nor rat behavior, an understanding of the principles that underlie the evolution and behavior of both diverse and similar species can provide valuable information with important implications for man. Tinbergen's (1968) analysis of the role of submissive postures and appeasement gestures in inhibiting aggression, and the resulting implications for the development of long-range weapons, is an excellent case in point.

The real value of research on nonhuman species is not so much one of providing detailed and specific information that may be applicable to man, but rather to build an intellectual framework from which to conceptualize and interpret human behavior and human evolution. The only fundamental reason for doing research is to build a conceptual framework and to modify and/or check the way we view the world and ourselves in relationship to that world. All too often these days one gets the impression that certain questions (e.g., racial differences in IQ scores) are better left unanswered, and that psychologists and others ought to do only the kinds of research that are likely to yield answers they will feel comfortable with, let alone those that have "practical" implications. If science begins to subserve political and ideological interests, then it ceases to represent an enterprise of open inquiry and its scope becomes enslaved by preconceived ideas about the way the world should be, rather than the way it is. Science is predicated on the tenet that knowing is preferable to not knowing. Man's conception of man has been and, if we keep an open mind, will continue to be modified by man's conception of himself in relation to other organisms.

REFERENCES

Ardrey, R. *African genesis*. New York: Dell, 1961.

Ardrey, R. *The social contract*. New York: Dell, 1970.

Barash, D. P. *Sociobiology and behavior*. New York: Elsevier, 1977.

Beck, B. B. Primate tool behavior. In R. H. Tuttle (Ed.), *Socio-ecology and psychology of primates*. The Hague, Netherlands: Mouton, 1975.

Bignami, G. Selection for high rates and low rates of conditioning in the rat. *Animal Behaviour*, 1965, *13*, 221–227.

Bolles, R. C. Species-specific defense reactions and avoidance learning. *Psychological Review*, 1970, *77*, 32–48.

Bower, T. G. R. *Development in infancy*. San Francisco: Freeman, 1974.

Brower, L. P., Ryerson, W. N., Coppinger, L. L., & Glazier, S. C. Ecological chemistry and the palatability spectrum. *Science*, 1968, *161*, 1349–1350.

Davenport, R. K., & Rogers, C. M. Intermodal equivalence of stimuli in apes. *Science*, 1970, *168*, 279–280.

Dawkins, R. *The selfish gene*. New York: Oxford University Press, 1976.

Desmond, A. J. *The ape's reflexion*. New York: Dial Press, 1979.

Dobzhansky, T. Chance and creativity in evolution. (Cited in Desmond, A. J., *The ape's reflexion*. New York: Dial Press, 1979).

Fentress, J. C., & Stilwell, F. P. Grammar of a movement sequence in inbred mice. *Nature*, 1973, *244*, 52–53.

Fisher, E. Habits of the southern sea otter. *Journal of Mammalogy,* 1939, *20,* 21–36.
Gallup, G. G., Jr. Chimpanzees: Self-recognition. *Science,* 1970, *167,* 86–87.
Gallup, G. G., Jr. Review of Rumbaugh, D. M. (Ed.), Language learning by a chimpanzee. New York: Academic Press, 1977. *The Psychological Record,* 1977, *27,* 795–796.
Gallup, G. G., Jr. Self-awareness in primates. *American Scientist,* 1979, *67,* 417–421.
Gallup, G. G., Jr. Self-awareness and the emergence of mind in primates. *American Journal of Primatology,* 1982, *2,* 237–248.
Gallup, G. G., Jr., Boren, J. L., Gagliardi, G. J., & Wallnau, L. B. A mirror for the mind of man, or will the chimpanzee create an identity crisis for *Homo sapiens? Journal of Human Evolution,* 1977, *6,* 303–313.
Gallup, G. G., Jr., & Suarez, S. D. On the use of animals in psychological research. *The Psychological Record,* 1980, *30,* 211–218.
Garcia, J., Ervin, F. R., & Koelling, R. A. Learning with prolonged delay of reinforcement. *Psychonomic Science,* 1966, *5,* 121–122.
Garcia, J., & Koelling, R. A. The relation of cue to consequence in avoidance learning. *Psychonomic Science,* 1966, *4,* 123–124.
Garcia, J., McGowan, B. K., & Green, K. F. Biological constraints on conditioning. In A. H. Black & W. F. Prokasy (Eds.), *Classical conditioning II: Research and theory.* New York: Appleton–Century–Crofts, 1972.
Gardner, B. T., & Gardner, R. A. Two-way communication with an infant chimpanzee. In A. Schrier & F. Stollnitz (Eds.), Behavior of non-human primates. New York: Academic Press, Vol. 4, 1971.
Goodall, J., Bandora, A., Bergmann, E., Busse, C., Matama, H., Mpongo, E., Pierce, A., & Riss, D. Intercommunity interactions in the chimpanzee population of the Gombe National Park. In D. A. Hamburg & E. R. McCown (Eds.), *The great apes.* Menlo Park, Calif.: Benjamin/Cummings, 1979.
Harlow, H. F., & Suomi, S. J. Production of depressive behaviors in young monkeys. *Journal of Autism and Childhood Schizophrenia,* 1971, *1,* 246–255.
Hennessey, T. M., Rucker, W. B., & McDiarmid, C. G. Classical conditioning in paramecia. *Animal Learning & Behavior,* 1979, *7,* 417–423.
Herrnstein, R. J. Superstition: A corollary of the principles of operant conditioning. In W. K. Honig (Ed.), *Operant behavior.* New York: Appleton–Century–Crofts, 1966.
Hewes, G. W. Primate communication and the gestural origins of language. *Current Anthropology,* 1973, *14,* 5–24.
Hirsch, J. (Ed.). *Behavior–genetic analysis.* New York: McGraw–Hill, 1967.
Hodos, W., & Campbell, C. B. G. *Scala naturae:* Why there is no theory in comparative psychology. *Psychological Review,* 1969, *76,* 337–350.
Hogan, J. A., & Roper, T. J. A comparison of the properties of different reinforcers. In J. S. Rosenblatt, R. A. Hinde, C. Beer, & M. C. Busnel (Eds.), *Advances in the study of behavior* (Vol. 8). New York: Academic Press, 1978.
Holloway, R. L. Culture: A *human* domain. *Current Anthropology,* 1969, *10,* 395–412.
Hugdahl, K. Fredrikson, M., & Öhman, A. ''Preparedness'' and ''arousability'' as determinants of electrodermal conditioning. *Behavior Research and Therapy,* 1977, *15,* 345–353.
Itani, J., & Nishimura, A. The study of infrahuman culture in Japan: A review. In E. Menzel (Ed.), *Precultural primate behavior.* New York: Karger Basel, 1973.
Jung, C. G. *The undiscovered self.* New York: Harcourt Brace Jovanovich, 1958.
King, M. C., & Wilson, A. C. Evolution at two levels in humans and chimpanzees. *Science,* 1975, *188,* 107–116.
Köhler, W. *The mentality of apes.* London: Routledge & Kegan Paul, 1927.
Kummer, H. Social organization of hamadryas baboons, a field study. *Biblioteca Primatologica* (No. 6). New York: Karger Basel, 1968.

Lennenberg, E. H. On explaining language. *Science,* 1969, *164,* 635–643.

Lethmate, J., & Dücker, G. Untersuchungen zum selbsterkennen im spiegel bei orang-utans und einigen anderen affenarten. *Zeitschrift fuer Tierpsychologie,* 1973, *33,* 248–269.

Linden, E. *Apes, men, and language.* New York: Penguin, 1974.

Mason, W. A. Environmental models and mental modes: Representational processes in the great apes and man. *American Psychologist,* 1976, *31,* 284–294.

Menzel, E. W., Jr. Spontaneous invention of ladders in a group of young chimpanzees. *Folia Primatologica,* 1972, *17,* 87–106.

Menzel, E. W., Jr. Chimpanzee spatial memory organization. *Science,* 1973, *182,* 943–945.

Menzel, E. W., Jr. Natural language of young chimpanzees. *New Scientist,* January 16, 1975, 127–130.

Menzel, E. W., Jr. Communication of object locations in a group of young chimpanzees. In D. A. Hamburg & E. R. McCown (Eds.), *The great apes.* Menlo Park, Calif.: Benjamin/Cummings, 1979.

Morris, D. *The biology of art: A study of the picture-making behavior of the great apes and its relationship to human art.* London: Methuen, 1962.

Mounin, G. Language, communication, chimpanzees. *Current Anthropology,* 1976, *17,* 1–21.

Öhman, A., Eriksson, A., & Olofsson, C. One-trial learning and superior resistance to extinction of autonomic responses conditioned to potential phobic stimuli. *Journal of Comparative and Physiological Psychology,* 1975, *88,* 619–627.

Öhman, A., Erixon, G., & Lofberg, I. Phobias and preparedness: Phobic versus neutral pictures as conditioned stimuli for human autonomic responses. *Journal of Abnormal Psychology,* 1975, *84,* 41–45.

Patterson, F. G. The gestures of a gorilla: Language acquisition in another pongid. *Brain and Language,* 1978, *5,* 72–97.

Petersen, M. R., Beecher, M. D., Zoloth, S. R., Moody, D. B., & Stebbins, W. C. Neural lateralization of species-specific vocalizations by Japanese macaques (*Macaca fuscata*). *Science,* 1978, *202,* 324–327.

Plotkin, H. C., & Odling-Smee, F. J. Learning, change and evolution: An inquiry into the teleonomy of learning. In J. S. Rosenblatt, R. A. Hinde, C. Beer, & M. C. Busnel (Eds.), *Advances in the study of behavior* (Vol. 10). New York: Academic Press, 1979.

Premack, D. Language in a chimpanzee? *Science,* 1971, *172,* 808–822.

Premack, D., & Woodruff, G. Does the chimpanzee have a theory of mind? *The Behavioral and Brain Sciences,* 1978, *4,* 515–526.

Revusky, S. H., & Bedarf, E. W. Association of illness with prior ingestion of novel foods. *Science,* 1967, *155,* 219–220.

Rozin, P., & Kalat, J. W. Specific hungers and poison avoidance as adaptive specializations of learning. *Psychological Review,* 1971, *78,* 459–486.

Rumbaugh, D. M., Gill, T. V., & von Glaserfeld, E. C. Reading and sentence completion by a chimpanzee (Pan). *Science,* 1973, *182,* 731–733.

Seligman, M. E. P. On the generality of the laws of learning. *Psychological Review,* 1970, *77,* 406–418.

Seligman, M. E. P. Phobias and preparedness. *Behavior Therapy,* 1971, *2,* 307–320.

Seligman, M. E. P., & Hager, J. L. (Eds.). *Biological boundaries of learning.* New York: Appleton–Century–Crofts, 1972.

Silk, J. B. Feeding, foraging, and food-sharing behavior of immature chimpanzees. *Folia Primatologica,* 1979, *31,* 123–143.

Skinner, B. F. *Beyond freedom and dignity.* New York: Knopf, 1971.

Teleki, G. Group response to the accidental death of a chimpanzee in Gombe National Park, Tanzania. *Folia Primatologica,* 1973, *20,* 81–94. (a)

Teleki, G. *The predatory behavior of wild chimpanzees*. Lewisburg, Pa.: Bucknell University Press, 1973. (b)

Teleki, G. Chimpanzee subsistence technology: Materials and skills. *Journal of Human Evolution*, 1974, *3*, 574–594.

Terrace, H. S., Petitto, L. A., Sanders, R. J., & Bever, T. G. Can an ape create a sentence? *Science*, 1979, *206*, 891–902.

Thompson, C. R., & Church, R. M. An explanation of the language of a chimpanzee. *Science*, 1980, *208*, 313–314.

Thompson, R. K. R., & Herman, L. M. Memory for lists of sounds by the bottle-nosed dolphin: Convergence of memory processes with humans? *Science*, 1977, *195*, 501–503.

Tinbergen, N. On war and peace in animals and man. *Science*, 1968, *160*, 1411–1418.

Torgersen, S. The nature and origin of common phobic fears. *British Journal of Psychiatry*, 1979, *134*, 343–351.

Trivers, R. L. The evolution of reciprocal altruism. *The Quarterly Review of Biology*, 1971, *46*, 35–37.

van Lawick-Goodall, J. The behaviour of free-living chimpanzees in the Gombe Stream Reserve. *Animal Behaviour Monographs*, 1968, *1*, 161–311.

van Lawick-Goodall, J. *In the shadow of man*. Boston: Houghton Mifflin, 1971.

Washburn, S. L., & Benedict, B. Non-human primate culture. *Man*, 1979, *14*, 163–164.

Wilson, E. O. *Sociobiology: The new synthesis*. Cambridge: Harvard University Press, 1975.

Woodruff, G., & Premack, D. Intentional communication in the chimpanzee: The development of deception. *Cognition*, 1979, *7*, 333–362.

Yeni-Komshian, G. H., & Benson, D. A. Anatomical study of cerebral asymmetry in the temporal lobe of humans, chimpanzees, and rhesus monkeys. *Science*, 1976, *192*, 387–389.

2 Comparison of Human Behaviors

S. L. Washburn
P. C. Dolhinow
University of California at Berkeley

COMPARING HUMAN BEHAVIORS

In the study of evolution, including that of humans from ancestral primates, it has been traditional to begin with forms living in a remote period of the past and end with modern human beings (Valentine, 1978). This procedure has seemed logical, especially as the evolutionary sequences were arrangements determined by fossil evidence. In the early part of the 19th century it was the uncertainties about these orderings that formed major scientific barriers to the acceptance of theories of evolution (Ruse, 1979). It has also been customary to make comparisons among living and extinct forms according to a sequence suggested by the great chain of being, which, in the vertebrates, ran from fish to amphibians to reptiles to birds, and finally to mammals. Because the living representatives of these great groups do not constitute an evolutionary sequence, the problems that arise when one acts as though they did, are much greater than is usually realized (Hodos, 1970). Most students of evolution today enrich their understanding of fossils by studying living descendants, a method used by the late A. S. Romer and clearly illustrated in *The Vertebrate Story* (1959).

Although this paleontological/comparative method is, in a general way, useful, there can be major difficulties if investigation is limited to this alone. What if there are few or no fossils? How far can evolution be reconstructed from comparison alone? Some ethologists and sociobiologists apparently believe they are free to appeal to the processes of evolution while making a minimal effort to use either the fossil record or the evidence of comparative anatomy (Alexander, 1979; Barash, 1977; Chagnon & Irons 1979). The situation is further complicated when emphasis is placed on behaviors that can only be reconstructed in general ways.

Recently, the problems of interpreting the fossil record and of traditional methods of comparison have led to the use of immunology, electrophoresis, sequences of amino acids in proteins, and the comparisons of DNA (Dobzhansky, Ayala, Stebbins, & Valentine, 1977; Goodman & Tashian, 1976). In addition, there are now several methods of measuring absolute time (Press & Siever, 1974), and these, combined with molecular biology, are giving comparative and evolutionary biology a new basis. In 1859 Darwin (1859/1971) wrote, ''There is grandeur in this view of life.'' Surely the grandeur has been almost infinitely increased by the discovery of countless fossils, the measurement of time, and molecular biology.

If our goal is to understand human behavior, we must begin our study with human beings. This position is easily misunderstood and we stress that there are many other possible goals—the study of evolutionary processes themselves, of animal behavior, or of biology in general, to name just a few. The study of human health and disease depends heavily on the use of laboratory animals as models, yet it is often not possible to use human subjects. The remarks that follow are not meant to suggest that everyone should study human behavior or human beings, but rather to emphasize that *if* the goal is understanding humans, problem formulation and methods of investigation must be adjusted appropriately.

HUMAN BEHAVIORS

Human beings differ from other primates and from nonhumans in three very important general ways—locomotion, tool using, and language. Cognitive factors and social behaviors are also different, but they are more difficult to compare or reconstruct. Locomotion, tool using, and language may be put into an evolutionary order for the purposes of this discussion, reflecting the times in our evolutionary history at which each of the three developed.

Bipedal locomotion evolved close to 4 million years ago, as shown by the footprints found at Laetoli by Mary Leakey and her co-workers (Leakey & Hay 1979). Stone tools appeared in the archeological record by 2 million years (Isaac, 1979), and there is evidence of hunting at that time (Bunn, 1981). No satisfactory methods have been devised for determining the evolutionary origins of language, social behaviors, or cognitive factors. Although it is fascinating to speculate, even on a subject such as locomotion where there is a substantial fossil record, far more can be learned about locomotion by studying the behavior of people today than by looking at the fragments and bones that have survived from the distant past. The evolutionary sequence was: bipedal locomotion, stone tools, and, finally, large brains, precisely what Darwin anticipated in *The Descent of Man* (1871); his speculations have proven to be remarkably accurate. Today this order is supported by a considerable amount of information, ranging from tools and fossils to biochemical evidence and improved methods of dating, all of which enables us to see that the human brain is a product of the latter part of human evolution, long after our ancestors had assumed bipedal locomotion and become skilled tool users.

There are problems in comparing primate brains because the human brain is quite different from those of other primates. If ours were just larger than the brains of our relatives and ancestors, the interpretation of the fossil record would be straightforward. However, in addition to increased size, our brains have become internally reorganized in ways that are tremendously important for the very behaviors that must characterize our manner of life, a reorganization not reflected in the fossil record.

Because the body size of humans, chimpanzees, and orangutans is approximately the same, their brains may be directly compared without the complications that come from comparing animals with different body size. Brain/body size ratios and calculations based upon them show that human beings have by far the largest brains of the contemporary primates (Jerison, 1973, pp. 392–393).

The fossil record is so fragmentary that most of what is known about human behaviors must come from the present or the recent past. Compared to that of any other animal, the record of the developing complexity of human behavior is extraordinary. Proceeding at an ever accelerating rate from some 35 to 40 thousand years ago, human beings have made great technical progress, a record usefully summarized in Lenski and Lenski (1974) and Pfeiffer (1977). In a period of less than 12 thousand years, human knowledge and technology changed more than in the preceding 3 million years. Economically, foraging was replaced by agriculture and human power by machines. We are so much a part of this successful present that we must make an effort to be conscious of the fact that almost all technical advance is a product of human effort of the last 200 years.

Human evolution has been separate from that of other primates for at least 4 million years; our technical world has been present for only .00005% of that time. Even when the antecedents of agriculture at the end of the Upper Paleolithic Age are included in the duration of the modern world, it still occupies less than .005% of the time humans have been evolving. The means by which humans adapt are remarkably new and were developed long after humans had attained their present biological forms.

Two very different kinds of problems emerge from the study of human evolution—long-term evolutionary problems of the last 4 million years, and short-term historical problems of the last few thousand years. Understanding long-term problems involves the knowledge of the fossil record, comparative anatomy, and evolutionary theory. All the historical records concerned with changes in culture and learned behavior came after *Homo sapiens sapiens* had evolved to its present form.

These two classes of problems had at one time been mixed, and a common 19th-century belief was that biology was important in accounting for cultural and historical differences. This confusion has been revived by those who maintain that both genetic and environmental facts are always present, and that social scientists make a great mistake in trying to separate the two (Alexander, 1979; Barash, 1977; Wilson, 1977). We think that the separation of biological from historical problems is necessary because their entanglement leads to faulty biology and social science.

Language, if defined as cognitive factors plus speech, is literally a new world for humans, and appreciation of the nature of this experience is easily lost. Imagine that a species of mammal evolved a new kind of tooth or a new locomotor pattern so very successful that over only a few thousand years the behaviors of the species were revolutionized, that its numbers multiplied many times, its adaptations extended its range from the tropics to the Arctic, its locomotion extended its excursions from the bottom of the sea to other planets in space, and its technical powers were reflected in a change from the simplest sort of huts to huge metropolitan cities. These creatures could literally move mountains, and there would be no question but that their adaptations would be recognized as different and remarkable by scientists (including wandering ones from other planets) who would stress their importance to colleagues. When such changes are made possible by *learned* behaviors, however, some modern scholars have minimized their significance by the contention that human beings are unique only as any species is unique (Wilson, 1977).

To put our human ability in perspective, it should be noted that of the more than 200 species of primates today, only one speaks. In all the wonderful diversity of birds and reptiles and amphibians and fish, communication is extraordinarily limited, and only one kind of animal can speak of the past and plan for the future.

If a closer look is taken at our language, the word "language" itself is found to have several meanings. It may denote any system of communication, including everything from artificial codes to the so-called languages of birds, bees, porpoises, and chimpanzees (Sebeok & Umiker-Sebeok, 1980). The languages may be based on sounds, gestures, or combinations of the two, and analyzed by similarity in form, meaning, or underlying cognitive processes. A major complication in the study of these languages is the emotions of the scientists. It means a great deal to those studying ape communication to call it "language," although apes cannot speak. The most important behavior, speaking, is not there—even to a minor extent and not even after great efforts on the part of human beings. The emotional element is clearly shown in the recent controversies between Terrace (1980) and Rumbaugh (1980).

To make any progress in understanding language, the meaning of the word has to be clarified and, when necessary, limited in ways that are appropriate to different problems. One might define the problem as "speech" rather than language. This would then involve short sounds (phonemes), combined into words (morphemes), which are in turn combined into sentences (syntax). Such a sound system is unique to human beings.

It is because of the structure of our human brain that we learn to speak easily, and what can be communicated is of a new and entirely different order from that of any other animals. The human brain has no trouble in learning thousands of words, and the phonetic code allows new words to be coined as needed. Syntax, tenses, conditionality, and other linguistic features encourage the use of complex expressions that have absolutely no parallel in nonhuman communication. Because of the structure of their brains, chimpanzees cannot be taught to speak. Chimpanzees can make a wide variety of sounds, and there is no anatomical barrier in the larynx, pharynx, tongue, or face that prevents using these sounds in a form of speech. It should be remembered that it is the patterns of the individually highly variable sounds that are recognized in human speech.

The difficulty apes have in learning speech should be contrasted to their success in learning gestures. Gestural response to visual stimulus is a part of normal ape behavior, and using adequate systems of rewards, humans can greatly increase the number of an ape's gestures that carry meaning. Because apes do not naturally use any word (phonemes combined into morphemes), this ability cannot be instilled by training.

The ape–human difference may be explained by an evolution during which the adaptive success of some unknown primitive form of communication, based on sounds, led to reorganization of large parts of the brain (Geschwind, 1979). It is unfortunate that the date of this event cannot be determined from the fossil record, because the lack of such a date has produced a wide variety of opinions and endless heated debate. From what evidence we do have, however, it most probably had taken place before 30 thousand years ago.

Biology is important if the problem under consideration is the relation of the functioning brain to speech. If, however, the problem concerns differences among human languages, biology should be ignored. There is no evidence that basic biological differences were responsible for the rapid divergence of languages, the differences among contemporary languages, or the meanings of linguistic structures. The same general kind of neural mechanisms make it possible to learn very different languages.

When discussing the origin of language based on changes in the brain and articulatory mechanisms, we are concerned with biological evolution over a long period of time. Historical changes in languages as documented over the last few thousand years, or comparisons of contemporary languages, are another set of problems. Very different kinds of questions are being asked in these two instances, and biological evolution is relevant only to the first—the origin of language.

We agree with Alexander (1979) that organic evolution shaped the *capacity* for culture, yet we strongly disagree with his statement that organic evolution could have shaped the: ''actual expression of culture in different circumstances [p. 97]. Languages give perhaps the best example of a capacity to learn that is certainly biological and panhuman, but in which the actual form of the expression is the result of many different social histories. At issue here is that the rules of genetics and the methods for studying genetics are very different from the rules of learning and the methods of understanding learning. The "ignorance" of social scientists of which Alexander complains is simply their appropriate refusal to return to 19th-century confusions by mixing biological and cultural explanations.

Facial Expression and Language

The importance of initiating comparisons with what is known about human beings and human behaviors may be further illustrated by considering facial expressions. Humans learn to control their facial expressions, modifying even the first cries and smiles as time goes on, so that both are used in socially appropriate ways (Ekman, 1973). In our culture, smiling may be a habitual greeting, regardless almost of the smiler's inner feelings toward the recipient of the smile. Crying may be expected of mourners, even if some of those crying are not really grieving. This learning is possible because human facial expressions are controlled voluntarily by the cortex rather than involuntarily by the older limbic portions of the brain (Penfield & Rasmussen, 1952).

Damage to the human cortex in the relevant area results in facial paralysis. The importance of cortical control over facial expression, therefore, is demonstrated both by learning and by pathology. Monkeys have a small number of facial expressions and these are controlled by primitive limbic parts of the brain. Massive removal of monkey motor area cortex, which should control expression if the situation were the same as in humans, does not affect expressions of the face (Myers, 1978). Because the basic differences between human facial expressions and the expressions of the nonhuman primates are in the brain, studies that compare muscles and expressions without questioning the underlying neural control must inevitably be incomplete and may miss the point.

In children who are born blind, smiling develops normally at first, but then the expression declines rapidly (Eibl-Eibesfeldt, 1975, p. 450), showing not only that the initial development of this expression is biologically determined, but that the social environment is necessary for continued development and meaning. It should be remembered that apes develop far more rapidly than human beings. As a result, the impact of the social environment on the postnatally developing human begins far earlier than for other primates. The combination of relatively early birth and slow neural development means that the environment impacts the human organism in unique ways. If the smile is regarded as an adaptation to this early period of biosocial interaction, then one can see why the smile is unique to human beings both in its origin and subsequent elaboration. These points are missed if the problem of human smiling is approached by making comparisons based only on descriptions of nonhuman primate facial expressions.

Social Sciences

The social sciences are made possible and necessary by the products of the success of a new kind of biological adaptation. Comparisons of the social systems of humans with monkeys and apes show how limited social systems are without language and the cognitive abilities and speech that language allows. Human languages make human adaptation possible; human social behaviors, politics, economics, religions, all depend on the linguistic communication controlled by the human brain. The separation of human social behaviors from those of other animals, and of the sciences that study them, is necessary because of human biology, not because of some arbitrary action on the part of social scientists. Bipedalism, manual skills, and language form an evolutionary order in which both biology and behavior interact.

The first few million years of humanity may not have been very different from our nearest living nonhuman relatives. In terms of numbers, consider that 20 thousand years ago there may have been as many orangutans in southeast Asia as there were human beings in the entire world. Modern human demographic success is recent and is based on agricultural and technical progress, not on new biology.

The same logic we have used in considering speech may be used in contrasting the biological and social sciences. Just as a specialized human brain makes the learning of language easy, it allows social systems, which are learned, to change rapidly and to become remarkably different. Humans automatically learn the social system in which they grow up. There is no evidence of genetic influence on differences among social systems or on the rate of social change. The biology of the human actor is necessary for any human social system to develop at all, just as specific parts of the brain are necessary for language, and specialized legs for bipedal locomotion.

Wilson (1977, p. 131) has disparaged the importance of social sciences with the assertion that: "Human social behavior occupies only a small envelope in the space of realized social arrangements." He has envisioned that visitors from another planet, creatures who are far more intelligent and sensitive than we: "might find us uninteresting as a social species—just another cultural–linguistic variant on the basic mammalian theme—and instead turn to study the more theoretically challenging societies of ants and termites." The reply to this is simple: Termites can't talk. Speech is the biological adaptation that underlies knowledge, technology, social systems, and the acceleration of history. Speech makes it useful to separate the social from the biological.

We still see today the same kind of thinking when we read that we ought to look for the genes that may have caused differences among cultures, and that track cultural change. No evidence for these beliefs is presented. Actually, the direct evidence of history strongly suggests that social facts are most usefully understood as derived from previous social facts as modified by new social conditions (Bock, 1980).

Wilson (1977) states: "that sociology is truly the subject most remote from the fundamental principles of individual behavior. Advanced literate societies, the main concern of sociology, are the most removed in character from the kinds of social and economic systems in which the genetic basis of human social behavior evolved [p. 137]." This quotation shows that Wilson recognizes that the main interests of sociology are "emergent" and cannot be understood in genetic or biological terms. This quotation, in fact, strongly supports the positions of Durkheim, which Wilson dismissed in the preceding paragraph!

The difficulty arises from confusing the questions about the biological capacity for learning with the products of learning. Human beings have a capacity for learning social systems and this ability is the result of evolution, but, as noted earlier, the most extraordinary variety of systems may be learned and regarded as natural.

The social sciences are concerned with groups of people, institutions, and roles. They are primarily concerned with the present or recent history. For example, the role of teacher implies the existence of schools, which have changing social functions. A particular individual may be more or less successful, or more or less motivated, but the individual's personal characteristics do not create the role. The need for teachers comes from technical progress, not from the personal psychology of any individual teacher. Douglas (1979, pp. 60–61) has a particularly clear analysis of the importance of limiting the area being investigated and stresses that progress in the social sciences came only after biological and psychological explanations had been eliminated.

The actors in any human social system are real human individuals. The kinds of roles that can be played, however, are limited by the social system and are changed by history. For example, in the America of the 1800s and early 1900s, most people were farmers living in small communities. Today there are fewer farmers and many opportunities for people to do jobs that did not exist a century ago. The change lies in the roles defined by a social system, not in the altruism or selfishness of individuals. A government is made up of roles for president, senators, governors, and many others, and every generation learns the structure of its government as acted out by real people in history.

Alexander (1979, p. 95) makes the point that the human biological ability to adapt in a wide variety of cultural ways means that cultural adaptation *is* biology. The point is not that genes determine cultures, but that they determine the biology that makes possible the capacity for culture, and the capacity, variously expressed in cultures, is simply biological adaptation. Human social adaptation is adaptation of a new sort. The industrial revolution changed the human adaptation—in a few years there was more change than in the preceding hundreds of thousands of years. Among other goals, the social sciences attempt to understand technical and social change. To label this attempt biology does nothing to help in understanding the present situation. The progress of science and all the social changes this has made possible does not depend on human reproductive behaviors but on history (Bock, 1980). Calling technology adaptation (and therefore biology) does not help in understanding any technical problem.

To compare human behaviors one must start with the complications and richness of actual human behavior. At present this is described by social scientists, not by biologists. The methods social scientists use may need to be improved, but they are not biological methods.

CULTURAL EVOLUTION

The term *cultural evolution* is commonly used to describe a more rapid process thought to be parallel to biological evolution. The use of the word ''evolution'' to describe changes in learned behaviors and in gene frequencies, however, is misleading. The problem is the same as in the earlier discussed case of ''language.'' Evolution is a fashionable word and there are people who seem to think that applying it to history lends understanding and order that is intellectually powerful. This use also creates misunderstanding, as, for example, the frequent assumptions in the literature that change goes, in most cases, from simple to complex.

If we look at the Australian aborigines we see that they had one of the most complex social structures, religion, and totemic mapping of any known peoples, a far more complex system than those of contemporary Europe or America (Berndt, 1981, pp. 177–216). Eskimos would be at the opposite extreme in a scheme of simple to complex social structure comparisons. As usually used, the word ''culture'' includes technical and economic factors that have changed greatly in recent times, and in which purposive behaviors are easily demonstrated. Social beliefs in which function and purpose are harder to demonstrate may change much more slowly.

A strong case can be made that there has been change and great diversity in social behaviors, but no general trends of the sort that are demonstrable in biological evolution. Wilson (1977) describes cultural evolution as Lamarckian, but the word ''Lamarckian'' itself is misleading because Lamarck believed that use (effort), if continued over the generations, led to changed heredity. This is like asserting that the effort of learning to speak English would result in changing biology, so that over the generations it would become easier and easier to learn to speak English. ''Lamarckian'' is generally used to refer to the effects of use and disuse only, without regard to Lamarck's complex theory of biological evolution. The expression ''cultural evolution is Lamarckian'' confuses social structure, technical change, history, and vague evolutionary theory.

To take another example, consider human marriage, an institution that exists in a wide variety of forms, all of which involve social and economic obligations and learned customs. The latter may change in very short periods of time and may differ greatly from one culture to the next. For these reasons, attempts to arrange marriage customs in evolutionary orders are destined to fail. Although sexual behaviors are always important, the social, economic, or religious aspects of marriage cannot be predicted from human biology. Marriage customs are complex, diversified, and changing, but if we begin our investigation with animal behavior, treating marriage only as mating behavior, then its diversity is lost. Human marriage ends up being treated as a ''pair–bond.'' The bond is then compared to that of gibbons, marmosets, or even birds! The idea that it is useful to refer to human marriage as a ''pair–bond'' comes from the biases of our culture. We doubt that a Moslem would find it a useful expression.

It has been argued that the ringdove provides a better model than the rhesus monkey for studying human courtship because both humans and ringdoves form pair–bonds (Konner, 1977). What is ignored is the fact that birds mature in less than 1% of their life span and there is no period of nursing or close mother–infant relations. Hormone injections can stimulate or suppress parts of reproduction and rearing cycles (Lehrman, 1958). Using the behavior of ringdoves as models for human behavior is poor biology, and treating the diversity of human marriage customs as pair–bonding is meaningless.

This kind of comparison may be described as reductionist, but it may be more accurate to regard it as almost unbelievably superficial. Suppose our objective is to understand why gibbons form pair–bonds. We find that their social group is composed of one female, one male, and their young. To say they behave like ringdoves overlooks all the ways in which their biology and behavior is unlike that of ringdoves. Alternatively, gibbon behavior might be investigated on the assumption that understanding the pair is more likely to come from careful observation than from irresponsible analogy. Gibbon anatomy is adapted to climbing, hanging–feeding, and swinging. This is very effective judging by the numerical success of the species, but very ineffective as a way of moving long distances. Gibbons adapt to limited locomotor capacity by defending their food supply in a small territory, and because the territory cannot support many gibbons, this problem is solved by limiting the population to a pair and its young.

The extreme antagonism manifested by adult gibbons toward other gibbons is, most likely, a psychological mechanism for maximizing the likelihood that individuals of the species will keep spaced apart from one another. The morning call and large canine teeth in the females may be parts of the same adaptive behavioral complex. This is not the place to attempt to prove this view of gibbon biology and behavior, rather it is mentioned to stress two contrasting research strategies, one to compare very superficially, without regard to phylogeny and often on the basis of a term (such as pair-bond), and the other an attempt to understand in some depth the biological adaptations of both behavior and structure. Only when the latter pertains can comparisons be made on some substantial basis.

If nesting behavior is considered, it will be found that many birds and the great apes make nests in trees. The function of the bird nest is to provide a place for eggs and maturing young. The function of the ape nest is to provide a safe place for sleeping. Entirely different motor patterns are used by birds and apes for nest building. If we want to understand nests, the function of the whole pattern of use and building has to be analyzed; designating two very different kinds of adaptations as "nests" may be just as misleading as calling very different social adaptations "pair–bonds." Learning goes into an ape's ability to construct a nest—the infant ape watches specific adults, mainly the mother, while birds do not have the opportunity to watch—they are hatched in an already constructed nest. When the time comes for the bird to construct its own nest, it does so without observational learning.

Our eagerness to accept behaviors of nonhuman animals as homologous to human behaviors is based in part on our efforts to obtain animal models. When the importance of early experiences in human beings was recognized, it became obvious that an animal model was needed, because human infants could not be deliberately raised under conditions of deprivation. Macaques provided useful research subjects, and much has been gained from studies of their development and experimental manipulation of social experiences and bonds (Harlow, 1963; Hinde & McGinnis, 1977; Kaufman, 1973). These studies were useful, but it must be kept in mind that humans, born much less mature and being much slower to develop, are more affected by environmental deficits and abnormalities.

However, there are many kinds of monkeys. The rhesus macaque was for a long time "the" monkey of research; other macaques added to the macaque dimensions of development. The Indian langur monkey provided a very different research model from that of the macaques. The langur, for instance, has a very distinctive caregiving behavior. From the day of birth the langur infant has many important caregivers, and if the strong ties with its mother are broken experimentally when the infant is 6 months old, it seeks care from another adult in the group (Dolhinow, 1980). A surprising majority of such needy infants succeed not only in gaining substitute care for the missing mother, but after only 2 weeks' absence when the mother is returned to the group, most infants elect to remain with their adopted caregiver. This is a very different pattern from that of the macaques.

In contrast to the monkey situation, adoption of infants and children was common in early human societies (Sahlins, 1976, pp. 48–52). Adoption was important then because life was short and there were probably many orphaned youngsters because one or both parents died early, or at least well before an immature person could assume responsibility for its existence. It is necessary to be very cautious when comparing the monkey with the human condition with respect to adoptions, because human adoptions are regulated by social, economic, and most importantly, by language-based kinship relationships. Once again, the differences between the human and the nonhuman primates are critical in understanding the two ways of life. At a physiological level, a monkey model of distress upon loss of a major attachment object is valuable and appropriate (Kaufman, 1977; Kaufman & Rosenblum, 1967; Reite, Kaufman, Pauley, & Stynes, 1974; Reite, Seiler, & Short, 1978). Emotional responses are quickly translated into biological/physiological responses. The emotional and social responses, however, are too different to allow direct generalization from the nonhuman to the human, or vice versa, without very careful qualifications.

HUMAN NATURE

It is obvious that human beings have a biological nature that is, in significant part, different from that of other animals. Those chimpanzees that have been broght up with human children were given every opportunity and a tremendous amount of encouragement and teaching (Kellogg, 1968). But the apes have remained different from humans, regardless of investments in time, emotion, and effort by their human caretakers and teachers. Abilities for walking, talking, and thinking are clearly different. If, however, our interest is not in these very general human characteristics but rather in more specific traits, then the quality of human nature ceases to be so obvious. The most diverse customs may be explained or justified as existing because of human nature, and humans may be regarded as intrinsically good or evil. In fact, attitudes towards our "basic nature" have changed repeatedly in the course of history.

In *Human Nature and History,* Bock (1980) has reviewed the various ways human beings have regarded their own natures. He shows that the views of sociobiologists have close parallels with those of humanists and philosophers of long ago and that all views of human nature have been based on common sense—people thought about themselves and their fellow human beings and found human nature to be thus or so. But that is the source of the trouble. What is revealed is not agreement, but the most diverse of opinions. Humans can be selfish, altruistic, aggressive, cooperative, deceitful, honest, bad, and/or good; the terms used are very general. Rhinelander (1974) points to the importance of the philosophical understanding of humans, but states that there is no agreement on phrases such as "human nature," although, as he also notes, this term is frequently invoked.

The problems of describing human nature stem in part from the diffuse nature of the problem of definition, and in part from the way people have tended to look at humans. They have considered different kinds of analysis as appropriate to different levels of behavior. The cultural level may be peeled off, leaving the psychological, and then the psychological removed, leaving the biological. Geertz (1973) has pointed to the fallacies of this kind of thinking.

Human structure and function evolved together over millions of years. During development the human child will assume the behaviors of its group, their attitudes, and their language. In this sense the young human is "unfinished" (Geertz, 1973), and has to do a great deal of learning before becoming fully human. Thus the term "human nature" contains a large cultural element, and the precise meaning will differ from place to place and time to time. Sociobiologists appear to be convinced that there is a universal human nature and that the words they use have panhuman meanings. But commonsense words such as cooperation, aggression, or deceit have clear meanings only in particular cultures. The unfinished human does not become finished in some generalized *Homo sapiens* way, but in the habits of a particular way of life. There are no culture-free individuals, and the social expression of basic human biology may be modified to a remarkable extent.

We find that in some tribes men joyfully went to war, reveled in killing and torture, and sometimes cannibalism. These are as much part of human nature as are the opposite values of many contemporary peoples. What superficially may appear to be simple expressions of biological necessities, such as sexual behavior, turn out to be learned and patterned. The accepted normal way may include behaviors considered deviant or criminal in our culture, namely animal mating or rape. The ideal in one culture may be a crime in the next. The so-called common sense of human beings tells them that the customs they have learned are natural and that the customs of others may be wrong. It is the belief that human nature may be reduced to a noncultural biology that is fundamentally wrong. It is a profound misunderstanding of the nature of learned behavior and the biology of the animal that learns.

Sociobiologists write of the differences between males and females, and of the importance of the differences in determining behavior and fitness (i.e., reproductive success). One might think that culture would play a small role in influencing the production of young. For millions of years primate females have produced young, nursed, and carried them. A strong case could be made that such activities, obviously essential for the survival of the species, are a fundamental part of the nature of human females. The whole pattern changed, however, with control of conception, substitutes for mother's milk, and the availability of carriages and cradles. In the United States the average number of births declined from 7 per female in 1800 to 3.5 in 1900, declined in the depression years and rose dramatically after World War II, and has now fallen to an all-time low of less than 2. Despite the supposed instinct for inclusive fitness and of millions of years of mother–infant behavior, when given the option, many human females elect to have few or no children. The number is influenced far more by customs, wars, and depressions than by noncultural biological considerations. The critical biological importance of infants and millions of years of evolution did not produce a human nature in which biology rather than choice determined human reproductive behaviors.

In the matter of the IQ, the part of biological intelligence measured by IQ has long been considered to be largely determined by genes. However, recent studies have shown that test results are greatly influenced by the environment. A study done in Israel reveals that children of parents who came from Europe had an average IQ of 105, whereas the children of those who came from the Near East had an average IQ of 85. When the children of both groups were brought up in the kibbutz, the average climbed to 115. The test performance of both groups increased, with the lower group increasing 30 points (Bloom, 1969). This difference is approximately the same as that resulting from sustained intervention in children 1–6 years old (Heber, 1980), and a little less than the differences in average IQ between different occupations (Anastasi, 1962). It appears that relatively minor environmental differences may cause major changes in intelligence as measured by the IQ, and that the differences are established by 5 years of age. Although biology is certainly important in establishing a basis for human intelligence, it is not clear to what extent individual differences are the result of biology or environment. It seems certain that what appear to be major differences among groups are largely, if not entirely, environmentally determined.

We have used this illustration because ideas about human nature may be very important politically. The 1924 immigration laws were designed to keep out people from eastern and southern Europe because they were regarded as biologically inferior (a point of view supposedly proved by the IQ scores). The continuing misuse of biological beliefs for political purposes has been reviewed by Chorover (1979). The issue is not just that over the years views of our nature were used to justify slavery, torture, and social hierarchy, but that in the last 50 years in the United States, sterilization, brain surgery, and extreme use of drugs all have been justified by views of human nature.

Massive misuse of intelligence testing continues to support social hierarchies and to block equal education opportunity for all citizens (Chase, 1980). The view that one's own group's human nature is biologically superior was not the product of individuals regarded as either evil or ignorant. The use of intelligence tests to determine quotas for immigrants was supported by many scientists, even presidents of the American Psychological Association. H. F. Osborn, president of the American Museum of Natural History, was a leader of the eugenics movement (Chorover, 1979).

The problem is that human nature seems to be such a simple concept. We all have some intuition about our own nature and that of other people. It is hard to realize that, beyond the grossest conditions, human nature is a complex mixture of biology and culture that cannot be unraveled in any simple way. In a very real sense, all ways of life are in accord with human nature, and the differences between human cultures cannot be explained by appeals to human nature. In each culture human biology is modified, beginning with the health of the pregnant female and ending with the care of the aged.

SUMMARY

Many aspects of human biology evolved recently, long after the separation of ape and human lineages. The origin of the biological basis for language as we know it today may have been only 35 or 40 thousand years ago. From a comparative point of view this means that there are no living forms intermediate between apes and humans—all have been extinct for millions of years. The sharp distinction between human and nonhuman is the result of this evolutionary history.

If the study of human behavior is approached by evolutionary sequence or comparison of contemporary forms, the late events in human evolution are greatly underemphasized and reductionism seems almost inevitable. We believe that those who would compare human behaviors with those of other animals must start with a rich understanding of human behavior. Otherwise the significance of language, cognitive factors, and the great diversity of social behaviors are lost or minimized. Language is necessary for social behaviors, and it is the complexity of this new adaptive mechanism that makes the social sciences necessary from both practical and intellectual points of view.

REFERENCES

Alexander, R. D. *Darwinism and human affairs*. Seattle: University of Washington Press, 1979.

Anastasi, A. *Differential psychology: Individual and group differences in behavior* (3rd ed.). New York: Macmillan, 1962.

Barash, D. P. *Sociobiology and behavior*. New York: Elsevier, 1977.

Berndt, C. H. Interpretations and "facts" in aboriginal Australia. In F. Dahlberg (Ed.), *Woman the gatherer*. New Haven: Yale University Press, 1981.

Bloom, B. S. Letter to editor. *Harvard Educational Review*, 1969, *39*, 419–421.

Bock, K. *Human nature and history*. New York: Columbia University Press, 1980.

Bunn, H. T. Archaeological evidence for meat-eating by Plio-Pleistocene hominids from Koobi Fora and Olduvai Gorge. *Nature*, 1981, *291*, 574–577.

Chagnon, N. A., & Irons, W. (Eds.). *Evolutionary biology and human social behavior*. North Scituate, Mass.: Duxbury Press, 1979.

Chase, A. *The legacy of Malthus*. Urbana: University of Illinois Press, 1980.

Chorover, S. L. *From genesis to genocide*. Cambridge, Mass.: MIT Press, 1979.

Darwin, C. *Origin of species*. London: Murray, 1859. Reprinted in Modern Library Giants Series #G15. New York: Random House, 1971.

Darwin, C. *The descent of man and selection in relation to sex*. London: John Murray, 1871.

Dobzhansky, T., Ayala, F. J., Stebbins, G. L., & Valentine, J. W. *Evolution*. San Francisco: Freeman, 1977.

Dolhinow, P. C. An experimental study of mother loss in the Indian langur monkey (*Presbytis entellus*). *Folia Primatologica*, 1980, *33*, 77–128.

Douglas, M. *The world of goods*. New York: Basic Books, 1979.

Eibl-Eibesfeldt, I. *Ethology: The biology of behavior* (2nd ed.). New York: Holt, Rinehart, & Winston, 1975.

Ekman, P. Cross-cultural studies of facial expressions. In P. Ekman (Ed.), *Darwin and facial expression*. New York: Academic Press, 1973.

Geertz, C. *The interpretation of cultures*. New York: Basic Books, 1973.

Geschwind, N. Specializations of the human brain. *Scientific American*, 1979, *241*, 180–199.

Goodman, M., & Tashian, R. E. (Eds.). *Molecular anthropology*. New York: Plenum, 1976.

Harlow, H. F. The maternal affectional system. In B. M. Foss (Ed.), *Determinants of infant behavior*. New York: Wiley, 1963.

Heber, R. Discussion of Heber's work in *The legacy of Malthus* by A. Chase. Urbana: University of Illinois Press, 1980.

Hinde, R. A., & McGinnis, L. Some factors influencing the effects of temporary mother-infant separation: Some experiments with rhesus monkeys. *Psychological Medicine*, 1977, *7*, 197–212.
Hodos, W. Evolutionary interpretation of neural and behavioral studies of living vertebrates. In F. O. Schmitt (Ed.), *The neurosciences: Second study program*. New York: Rockefeller University Press, 1970.
Isaac, G. L. The food-sharing behavior of protohuman hominids. In G. L. Isaac & R. E. Leakey (Ed.), *Human ancestors*. San Francisco: Freeman, 1979.
Jerison, H. J. *Evolution of the brain and intelligence*. New York: Academic Press, 1973.
Kaufman, I. C. The role of ontogeny in the establishment of species-specific patterns. *Biological and Environmental Determinate of Early Development*, 1973, *51*, 381–395.
Kaufman, I. C. Developmental considerations of anxiety and depression: Psychobiological studies in monkeys. In T. Shapiro (Ed.), *Psychoanalysis and contemporary science*. New York: International Universities Press, 1977.
Kaufman, I. C., & Rosenblum, L. A. Depression in infant monkeys separated from their mothers. *Science*, 1967, *155*, 1030–1031.
Kellog, W. N. Communication and language in the home-raised chimpanzee. *Science*, 1968, *162*, 423–427.
Konner, M. Evolution of human behavior development. In P. H. Leiderman, S. Tulkin, & H. Rosenfeld (Eds), *Culture and infancy*. New York: Academic Press, 1977.
Leakey, M. D., & Hay, R. L. Pliocene footprints in the Laetolil beds at Laetolil, northern Tanzania. *Nature*, 1979, *278*, 317–323.
Lehrman, D. S. Induction of broodiness by participation in courtship and nest building in the ringdove (*Streptopelia risoria*). *Journal of Comparative and Physiological Psychology*, 1958, *51*, 32–36.
Lenski, G., & Lenski, J. *Human societies: An introduction to macrosociology*. New York: McGraw–Hill, 1974.
Myers, R. E. Comparative neurology of vocalization and speech: Proof of a dichotomy. In S. L. Washburn & E. R. McCown (Eds.), *Human evolution: Biosocial perspectives*. Menlo Park, Calif.: Benjamin/Cummings, 1978.
Penfield, W., & Rasmussen, T. *The cerebral cortex of man: A clinical study of localization of function*. New York: Macmillan, 1952.
Pfeiffer, J. E. *The emergence of society*. New York: McGraw–Hill, 1977.
Press, F., & Siever, R. *Earth*. San Francisco: W. H. Freeman, 1974.
Reite, M., Kaufman, I. C., Pauley, J. D., & Stynes, A. J. Depression in infant monkeys: Physiological correlates. *Psychosomatic Medicine*, 1974, *36*, 363–367.
Reite, M., Seiler, C., & Short, R. Loss of your mother is more than loss of a mother. *American Journal of Psychiatry*, 1978, *135*, 370–371.
Rhinelander, P. H. *Is man incomprehensible to man?* San Francisco: Freeman, 1974.
Romer, A. S. *The vertebrate story*. Chicago: University Press, 1959.
Rumbaugh, D. M. The language behavior of apes. In T. A. Sebeok & J. Umiker-Sebeok (Eds.), *Speaking of apes*. New York: Plenum, 1980.
Ruse, M. *The Darwinian revolution*. Chicago: University Press, 1979.
Sahlins, M. *The use and abuse of biology*. Ann Arbor: University of Michigan Press, 1976.
Sebeok, T. A., & Umiker-Sebeok, J. (Eds.), *Speaking of Apes*. New York: Plenum, 1980.
Terrace, H. S. Is problem solving language? In T. A. Sebeok & J. Umiker-Sebeok (Eds.), *Speaking of apes*. New York: Plenum, 1980.
Valentine, J. W. Evolution. In *Readings from Scientific American*, 1978.
Wilson, E. O. Biology and the social sciences. *Daedalus*, 1977, *106*, 127–140.

3 The Comparative Approach In Human Ethology

Irenäus Eibl-Eibesfeldt
Forschungsstelle für Humanethologie
Max-Planck-Institut für Verhaltensphysiologie
D-8131 Seewiesen, Federal Republic of Germany

THE STUDY OF ANALOGIES

When a flightless cormorant (*Nannopterum harrisi*) returns from fishing to relieve his mate from brooding, he approaches with a present: a beakful of seaweed, a small stick, or a piece of sea fan. Bowing and uttering calls, he approaches and passes the present on to his mate, who grabs it, often with a fierce movement, and adds the material to their nest. Only then will she allow the mate to stay and eventually take over the nest. Most observers of this event will intuitively be inclined to call this opening act of the relieving ritual a "greeting" and, indeed, functionally it is one; the ceremony of the cormorant serves to appease the partner. The tameness of these endemic Galápagos birds allows one to experiment; you can rob the oncomer of his present. He will act startled, but will continue after some hesitation on his way to the nest, evidently not realizing what has happened to him. But when, upon his arrival, his mate attacks him, he will retreat and search for a new present. When he finally returns with some little branch or some other object, he will be accepted by his mate (Eibl-Eibesfeldt, 1965).

It will not be difficult to recall human rituals similar in form and in function to that of the cormorant. Gift giving is common in rituals of friendly encounter and serves to bond and appease. It is, for instance, customary in central and western Europe for a guest to give flowers or other presents to the host upon arrival.

The greeting ritual of the flightless cormorant and the similar ritual of man are, of course, analogies (products of convergent evolution). They evolved in the cormorant independently as a phylogenetic adaptation and in man presumably as a result of cultural evolution. As independent developments, the similar greetings would be dismissed from further consideration by many. Again and again one reads that not much can be learned from the study of greylag geese or cichlid fishes that is relevant for understanding man, that such studies may have some aesthetic value of their own, but contribute little to the understanding of human conduct, and that if one really wants to get comparative data relevant to the understanding of man, one has to turn to nonhuman primates who are our close relatives. This view is based on the assumption that only homologies (similarities based on common ancestry) count, a view that is clearly a misconception. True, analogies tell us something different than homologies, but both certainly contain information of interest.

On the one hand, for those interested in the evolution of particular structures, homologies are the focal point of interest. The study of analogies, on the other hand, informs us about the selection pressures that shaped these structures and caused processes of behavior to develop along similar lines. In fact, the laws found by the study of analogies are of wider applicability, and thus more general and fundamental, than those discovered by the study of homologies. Should one, therefore, be interested in phenomena like ranking, territoriality, or mating systems, it is certainly advisable to study these in many different groups of animals, regardless of their phylogenetic relationship. The further apart they are, the better, because only then can we be sure that the regularities observed can be attributed to the more general laws of function and not derived from a genetic relationship (Wickler, 1967).

This approach is already widely used. Biotechnicians do not hesitate to compare structures of different origin, but serving the same function, in order to discover the principles underlying their construction. Indeed, for someone interested in the construction of wings, it is profitable to compare the flying organs of insects, birds, and bats, even though, in the first case, these organs are formed from a fold of the cuticula, whereas the wings of birds and bats are changed vertebrate extremities. Behavioral scientists too have not hesitated to study analogies in order to unravel the laws underlying the phenomena in question. The investigation of the incest taboo by Bischof (1972), the study of infantile attachment by Rajecki, Lamb, and Obmascher (1978), and the study of ritualization by Eibl-Eibesfeldt (1979b) may serve as examples of the fruitfulness of such endeavors. In addition, psychologists have widely applied the comparative approach for the study of learning processes.

Single movement patterns, whole strategies of social interactions, learning processes, perceptive and cognitive processes, as well as the results of complex behavioral interactions such as social structure, rank order, and the like, can all be investigated by the comparative approach, but one must always compare categories of the same type and the same level of complexity. The comparison of patterns of expressive behavior in humans and animals (expressive movements, rituals) reveals a number of general principles regarding the evolution of these behaviors. Signals and rituals as well as simple expressive movements that serve this function of communication must be conspicuous and unmistakable in order not to be confused with those of another species or those having another function. At the same time they must be concise and simple in order to allow the detectors of the recipient of the signal to easily adapt to them. Therefore, in the process of ritualization, movement patterns undergo analogous changes. Their amplitude is exaggerated and at the same time the movement is simplified. In addition, several movements may be fused into one pattern and be made conspicuous by rhythmic repetition. Typical intensity (Morris, 1957) makes the patterns more distinct and often additional organs develop to emphasize the movements. These and other changes occur in birds as well as insects or mammals whenever signals evolve. Analogous changes can be observed in the process of cultural ritualization. Because the selection pressures underlying the evolution of signals are in principle the same for humans and animals (i.e., the need for clear transmission of a message), culturally evolved rituals of man phenocopy many aspects of the phylogenetically evolved rituals of animals.

Thus, rituals of threat and aggression in many vertebrates allow animals to win a fight without running the risk of injury and without damaging the opponent. There are numerous examples of ritual fights (Eibl-Eibesfeldt, 1961); marine iguanas of the Galápagos Islands, for instance, engage in head-butting tournaments. After an introductory display, they try to push each other from the place by butting their heads against each other. After a while, one will finally give up by assuming a submissive posture, lying flat on his belly. The opponent then stops fighting and waits in a threat posture for his rival to clear off. It is clearly a tournamentlike way of fighting, and speculation has revolved around the question of how such behavior could evolve at all. A mutant who behaves ruthlessly by biting instead of head butting should have a clear advantage over a courteous fighter and thus his genes would soon be dominant in the population. This, however, is not found to be the case, due to a fairly simple regulatory principle that ensures that ritualized fighting occurs only as long as participants follow the rules of the game. Already in 1955, when I described the ritualized fighting of the marine iguana, I noted that nonritualized fighting (biting) does occur. Should a marine iguana fail to give the introductory display when entering another male's territory, the territory owner will rush toward him and bite him. It is also well known that cichlids will immediately turn to damaging ramming fights instead of ritualized mouth pulling when an opponent starts a damaging fight (Eibl-Eibesfeldt, 1955a). This strategy easily explains why a ruthless mutant cannot win. The mutant is confronted in turn with ruthless fighters and therefore has little chance (advantage). Maynard-Smith (1974) dealt with this phenomenon by developing different models including the model of the retaliator, which he found to be the best of the possible evolutionary stable strategies.

The question why ritualized fighting evolved is currently a focal point of sociobiological discussion. Ethologists originally assumed that by avoiding self-inflicted damage upon a species, population, or group member, groups with ritualized fighters would gain advantage over those where damaging fighting regularly occurs. This model of group selection was attacked with the development of alternative models that show that ritualized fighting could also evolve by individual selection. By giving a winner in a ritualized fight X points and the loser Y points, and X' and Y' points in a damaging fight, respectively, models that describe the most stable strategy can be developed. So far, however, these considerations are mainly speculative and the whole discussion of group versus kin or individual selection is lacking the sound basis of data.

What is of interest here, as far as the discussion of the comparative approach is concerned, is the fact that analogies in ritualized fighting are numerous among both animals and man. In man, analogous ritualizations can be observed in both individual and group aggression. Some of these occur as phylogenetic adaptations, and analogous ones as the result of cultural evolution. In all instances the tendency to avoid bloodshed is clear. A verbal dispute indeed saves a lot of trouble and can bring about the resolution of conflict without extensive harm to the parties involved. In primitive societies such as the Eskimos, Bushmen, or Yanomami, many of the conflicts are resolved verbally (Heinz, 1967; Hoebel, 1967), and one wonders whether verbalization of conflict or, more generally, the need for ritualization of social interactions was not one of the prime movers in the evolution of language.

With the development of armament, duelling was restructured to avoid serious damage, and with the development of new armaments, new conventions concerning their use had to be developed. Even in tribal cultures, strict rules governing duelling are common. The Yanomami practice several forms corresponding to different steps in the escalation of a fight. Chest pounding and side slapping are the more harmless forms after disputing. Clubbing comes next; here, the opponents take turns hitting each other on the head with hardwood clubs, and each in turn has to bravely brace himself to accept the blow. Deep gashing wounds and occasionally even fractures of the skull result, but they are rarely lethal. Spearing the opponent escalates the fight toward a more damaging one, but is still less dangerous than the final possibility—arrow shooting (Chagnon, 1968).

Group aggression in man has to be distinguished from individualized aggression, its typical human expression being war (see Eibl-Eibesfeldt, 1979a). War is a product of cultural evolution and at its beginning was originally destructive. It is fought without inhibition of killing, regardless of whether or not it is acting in defense or conquest. There is, however, a clear tendency to ritualize even warfare by conventions that serve to prevent utter destruction. There are conventions that regulate the use of arms, and conventions allowing surrender, armistice, and peacemaking. Again, these occur already in tribal societies. A well-known example is the stages of escalation in Tsembaga warfare (Rappaport, 1968; see further comparative data in Eibl-Eibesfeldt, 1979a). Even though we do not know the precise selective advantage of such ritualizations of warfare, their widespread occurrence suggests that ritualized fighting must have advantages over ruthless slaughter. And indeed recent history provides evidence that the ruthless, even though not lacking in bravery, have little chance to win, because the world will rise up against them. The strategy of the retaliator seems to work even on this level, a lesson to be learned.

The study of analogous development certainly allows us to develop hypotheses about the advantages of certain behavior, which then have to be tested in each special case. This holds true in general for the comparative approach. We never make any immediate inferences from the study of geese or fish as to rules governing human behavior, but we develop hypotheses that guide our attention, which in each case have to be tested by the study of man. A final example may illustrate the value of the study of analogies—one concerning ways in which individuals of various species create friendly ties or bonds with others.

Comparing behavior in various species (Eibl-Eibesfeldt, 1972), I was struck by the similarity of behavior patterns that serve to create bonds between adults, and patterns of interaction in the mother–child relationship. Apparently maternal behavior and infantile appeals have become ritualized into behaviors that serve to create bonding among adults. For example, dogs have two behavior patterns of submission, both of which are derived from infantile behavior. One consists of rolling on their backs and is derived from the pattern by which puppies invite their mothers to clean them. Indeed, the pattern functions in this very way; often the submissive dog urinates a little, thereby releasing sniffing and eventually cleaning by the dominant one. When cleaned, the submissive one may start to wag its tail, and what may have started as a fight, will end in play. The other pattern by which a submissive dog can invite amicable interaction is by pushing his snout against the corners of the mouth of the dominant animal. Young dogs also do this when they want to release feeding by regurgitation. Both these patterns are illustrated in Fig. 3.1.

Even birds, who developed parental behavior independent from mammals but in clearly analogous ways, use patterns derived from the repertory of parental behavior to bond to other adults. When a female herring gull or black-headed gull approaches a male, she uses food-begging movements to inhibit his aggression (Fig. 3.2). Similarly, both male and female sparrows beg like infants during courtship ceremonies. Adult birds groom each other, feed each other, and greet each other by presenting nesting material. All of these activities first evolved in connection with brood care.

These observations strongly indicate that the repertories of behavior that evolved in connection with the development of parnatal care constitute highly important preadaptations for the evolution of social interactions. When we turn to insects we discover again that only those species that developed maternal behavior gave rise to the evolution of higher social organization. Ants and bees

FIG. 3.1. Two infantile patterns used to express submission in the adult wolf: food begging (in the foreground) and infantile cleaning posture (from Eibl-Eibesfeldt, 1972).

FIG. 3.2. Ritualized food begging during the courtship of the female black-headed gull. It compares as a clear analogy (convergence) to the functionally related food begging in the wolf (after N. Tinbergen, from Eibl-Eibesfeldt, 1972).

indeed are bonded by patterns of mutual feeding derived from maternal feeding. Thus our thesis that higher social organization has its roots in maternal behavior is supported by the study of analogous developments.

In contrast, species that have not evolved brood care are incapable of interacting in a way that creates social bonds. Marine iguanas form aggregations and often one may encounter hundreds side by side basking on the rocks, but individuals interact little. They neither groom one another nor feed one another. Their social interactions consist mainly of aggressive displays, even during courtship.

In addition, comparative investigation reveals that neither sex, nor aggression, nor fear alone were sufficient preadaptations for the development of higher social organization. For example, fear in pelagic fish led to schooling, but to nothing more elaborate (Eibl-Eibesfeldt, 1955b, 1972). Yet this is not the whole story. On comparing social insects with mammals and birds a striking difference becomes apparent; the social insects seem to lack individualized bonds, in contrast to many birds and mammals. In these latter cases, individual recognition evolved as an adaptation primarily serving mother–child bonding and thus avoiding exchange of young by mistake. This is particularly necessary in animals where the young are of the altricial type (guinea pigs, many antelopes), carried around (primates), or where they are deposited outside a nest (sea lions, deer). This individualized (personal) bond gave rise to the development of the particular and intimate relationship between group members that we call love (Eibl-Eibesfeldt, 1972). Love by definition is characterized by a personal relationship, something that is lacking in the social insects. Personal bonds of graded intimacy, however, characterize the relationship of humans in the individualized group; that is, the group where every member knows his fellowman personally.

For most of history, human beings lived in such individualized societies and only recently have been faced with the problems of life in an anonymous community. Comparing analogous development in animals and humans is indeed a thoughtprovoking enterprise and fruitful as long as we remain aware that comparison is only the first step. We must then proceed to hypothesize about selection pressures that bring about these similarities and test these hypotheses as thoroughly as possible.

In cross-cultural studies, too, interesting analogies can be discovered. During my work with the Himba, a pastoralist Bantu group in the Kaokoland in Southwest Africa (Namibia), I observed every morning that, after milking, the members of the Kraal community went to the headman and offered him the container with the milk. He would take a sip or just dip his finger into the milk, and only after having done so were the others free to use the milk. This daily ritual has an important function. It reinforces obedience to the headman and allows the headman to subtly check the loyalty of the group members, both of which are important for the survival of the group. As cattle breeders, the Himba have to be on constant alert to defend their animals; records of raids by the neighboring

Hottentots are numerous. Furthermore, pastoralist people living in a fairly arid environment must be able to conquer new pastures in case of emergency. Because the Himba live scattered over a wide area, their ability to attack or retaliate depends on their skill to form military cadres on short notice, under the leadership of a hierarchy of headmen. The organization needed for such action cannot be built up instantaneously in a situation of emergency. It must be constantly maintained by daily rituals that enforce obedience (Gehorsamserweisung).

This hypothesis concerning the underlying function of the Himba milk ritual is backed by the discovery of analogous rituals in other cultures. In our own culture, for instance, rituals of obedience are practiced in our military, where efficiency of action also depends on obedience, which must be constantly maintained. The saluting of the flag in the morning and evening is one of several rituals that serve this purpose.

There are other analogies between the Himba and ourselves that are of interest and that provide additional support for our theses. During social gatherings the Himba engage in singing praises of the deeds of ancestral heroes, who either successfully defended the cattle or raided other groups. Here again we find an interesting parallel in European cultures in which heroic sagas are perpetuated in writing, poetry, and songs. Thus, heroic virtues are reinforced in cultural areas that are very different in many respects, but that both depend on a certain degree of military fitness for their survival. Similar rituals can be found in other cultures where warfaring plays a significant role, whereas nothing like this is found in cultures of a more pacifistic type lacking organized aggression. Bushmen, for example, have no rituals of obedience, nor do they engage in praise singing (Eibl-Eibesfeldt, 1976).

Concerning the question of causation, proximate and ultimate causes must be distinguished. Thus, the subjectively intended achievement, which often acts as a reward, needs to be distinguished from the ultimate function in the service of fitness. For example, the Trobriand Islanders perform elaborate rituals in connection with the harvest as described by Malinowski (1922) and recently reviewed by Keesing (1980). Among other things men compete with their harvest. The roots are piled up by those who cultivated them and the headman praises the most successful gardeners and distributes prizes accordingly. The harvest is then exhibited in special Yams houses, which are built in such a way, as to make the roots visible between the logs. Not all of the harvest must be consumed, but often quite an extensive amount is left to rot, and replaced on occasion of the next harvest.

Looking at proximate causes, achievement of status is certainly a strong incentive to work hard for a good harvest, but it is a means to another end. When I first witnessed the display, which so far has puzzled other observers, it struck me that the promotion of overproduction might function not only on the individual level of status achievement, but also on the community level. When harvests fail people have a reserve for themselves and for others. Indeed during another visit in 1982 we witnessed a patchy harvest on Kiriwina. Those who had planted early in the season had gotten a good harvest as they had caught the early rains. Others who planted later experience shortage which will not result in hardship because they can visit neighbors and they will be given yams from the harvest of those who have plenty. They will be able to judge whether their neighbors have enough to give, since it is all on exhibit. It is clear that such exchange is of enormous value not only in economic terms but also because it preserves the harmony of the island community.

From the point of view of individual motivation the yam display is one of the many forms of attaining status and accordingly can be compared with other conventions which serve the same end. On another level, however, this ritual can be seen as a means of social insurance through exchange and it can thus be compared with rituals of gift exchange like, for example, the Hxaro system amongst the !Kung Bushmen of the Kalahari (Wiessner, 1979, 1981).

THE STUDY OF HOMOLOGIES

When we speak of homologies, we generally attribute shared characteristics to common genetic heritage. However, as Wickler (1965) emphasized, we have to distinguish between phyletic homologies and homologies of tradition. Thus, homology indicates that only a common source of information was tapped underlying the patterns in question. However, whether this information was stored in the genome or as cultural knowledge is left open. Songs in bullfinches and language in man are examples of homologies of tradition; the silent bared teeth display of the Macaques and the smile of man provide examples of phyletic homologies.

In animal behavior research this question was no real matter of concern, because homologies of tradition play no significant role. But when dealing with human behavior we need to be more aware of this distinction, particularly if we engage in cross-cultural studies. Homologies found in behavior patterns of man and primates or even other vertebrates, however, can as a rule of fist be considered as phylogenetically acquired traits. Very early, Charles Darwin pointed out homologies in the behavior of man and the closely related chimpanzee. The facial expression of sulking is a striking example of a highly specific shared expression. If the hypothesis is correct, that sulking is a pattern of old biological heritage, then we should encounter this pattern under the same conditions in all human races. My extensive cross-cultural documentation indeed indicated this to be the case[1], as shown in Fig. 3.3.

---

[1]The documentation has been discussed in several publications (Eibl-Eibesfeldt, 1974, 1976, 1979a,b).

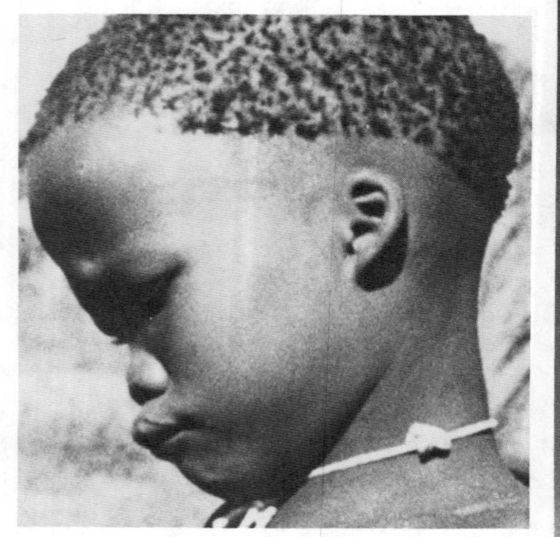

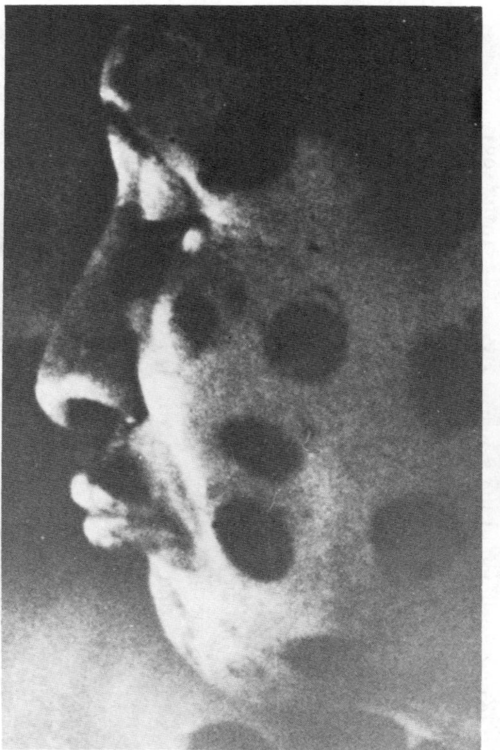

FIG. 3.3 Pouting (sulking) in (a) a bushman girl (!Ko Bushmen) and (b) an adult male Yanomami Indian (photographs by I. Eibl-Eibesfeldt); and sulking in (c) a chimpanzee (from Charles Darwin, 1872).

The criteria that Darwin applied for homologies were not defined, but it was clear that he, as well as his successors like O. Heinroth (1910) and Konrad Lorenz (1941), followed the criteria as used by morphologists. In other words, they treated behavior characteristics the same way a morphologist treats bodily structures when he or she uses them to investigate taxonomic relationships. Only recently were explicit criteria developed for defining behavioral homologies (Wickler, 1965). These three main criteria are in principle the same as those applied by morphologists:

1. *Special quality.* The special quality criterion refers to the fact that certain movement patterns are constant in form, meaning that the relative phase distance of the muscle contractions involved in an action remains constant. The motor pattern thus forms a transposable "gestalt." It is clear that such regularity does not imply rigidity. The sinuous wave passing through a fin can do so with different speed and amplitude, but the pattern nonetheless remains the same. This also holds true for learned movement patterns: Whether I write my signature fast or slow, or large or small—the gestalt of the signature remains the same.

But the special quality criterion alone often proves insufficient to trace homologies. Similar patterns often evolve independently and converge. These analogies can generally be ruled out, however, when the special form of the movement pattern is not dictated by function, as, for example, is the case in many expressive movements resulting from a phylogenetic (or cultural) convention between signal sender and receiver. Furthermore, movement patterns are more likely to be homologous if they are found in most members of a defined taxa (family, order), but less so if they are present in animals of very different taxa inhabiting the same ecological niche. Thus a number of bird species of very different families inhabiting arid habitats independently developed the capacity to drink by sucking up water, which is an unusual way for birds to drink. When this mode of drinking was first observed in both pigeons and sandgrouses, it was incorrectly assumed that it indicated a closer relationship between the two groups. Wickler (1965) corrected this view.

2. *Special position.* The special position of a pattern within a system often allows us to recognize structures as homologous, even if they do not appear to be very similar in form. Thus a morphologist will be able to classify bones that look very different in different vertebrate limbs by their position among the other bones. Movement patterns, in a similiar fashion, are often embedded in a very characteristic way in a sequence of movements. Figures 3.4 and 3.5 may serve to illustrate this principle.

3. *Intermediate forms.* The most important criterion for distinguishing homologies from analogies is the existence of intermediate forms linking homologous patterns by a series of intermediate stages. A well-known example is the paleontological evidence for the development of the hoof of the horse from the pentadactylid extremity of ancestral forms. When dealing with soft organs that leave no fossil traces, morphologists take advantage of the fact that living ani-

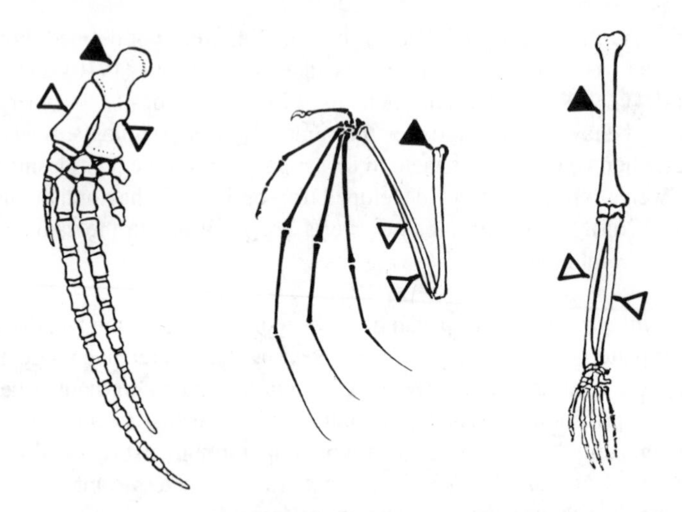

FIG. 3.4. By their position within a system even very dissimilar structures can be recognized as homologous, as is the case with the ulna, radius, and humerus as compared here in whale, bat, and man.

FIG. 3.5. The courtship displays of the mallard. The discrete movement patterns are the natural sequence in which these movements follow one another. Sequence and context, according to the criterion of special position (see text), would allow one to homologize movement patterns in related duck species, even if a single movement should deviate from the criterion of special form (from Lorenz & Van de Wall. 1960).

mals show graded similarities in their organs that can be ordered in a series reflecting different stages of evolutionary change (Fig. 3.6). Examples are both numerous and well documented. The development of the urogenital system from that in jawless fishes (*Petromyzon*) to that in mammals can serve as an example. In addition, because phylogeny in some cases is recapitulated in ontogeny, the study of ontogeny can be of additional help.

The very idea that evolution took place was in fact not discovered by the study of fossils, but by the comparison of graded similarities of living species. When Darwin studied the finches of the Galápagos Islands, he was struck by their resemblance and was no longer willing to accept them as independent creations. Behavioral scientists are restricted to the comparison of existing species. But behavior too shows graded differences from species to species that allow them to read the course of evolution. The study of ritualizations in particular has pro-

FIG. 3.6. The grunt whistle movement (#4 in Fig. 3.5) in four different species of ducks. From above to below: *Anas falcata, A. platyrhynchos, Lophonetta specularioides*, and *Chaulelasmus streperus*. The movement patterns again consist of a sequence that can be broken down into smaller action units that, however, are rigidly bound into one highly stereotyped pattern occurring in an all-or-none fashion with typical intensity. Variations in amplitude and duration are extremely small. The vertical line marks the moment the duck dips its beak in the water. In *Anas* the dipping phase overlaps (superimposes) with the raising and bending of the body and neck, and the two movements are clearly separate in *Lophonetta* and *Chaulelasmus*. Again, the different movement phases, at this level of integration, can be homologized by the criterion of special position (K. Lorenz & Van De Wall, 1960).

vided numerous examples of such gradations (Eibl-Eibesfeldt, 1979b), as shown in Fig. 3.7.

Structures can be seen as homologous if we look at them from a particular level of organization and as analogous if looked at from another level. Thus the flipper of a penguin and that of a dolphin are homologous if regarded as a vertebrate extremity. As flippers, however, they achieved their similar structure

FIG. 3.7. Linkage by intermediate forms reveals the origin of the male peacock courtship display from food enticing, which in turn derived from parental behavior. Upper left to right: Food enticing in the domestic cock (*Gallus gallus*). After scratching the ground, the cock steps back and, calling out, repeatedly lifts a morsel (or pebble as a substitute); hens are attracted by this display to search in front of the male, who then proceeds in courtship. The ring-necked pheasant (*Phasianus colchicus*) behaves similarly. In the monal pheasant (*Lophophorus impejanus*), the male pecks the ground and, upon the approach of the hen, he fans his wings and tail feathers, moving them slowly back and forth in the ''ecstatic display.'' The peacock pheasant (*Polyplectron bicalcaratum*) scratches the gound but does not offer food at this phase. Upon the approach of the female he relies on wing and tail displays instead. In rapid sequence the head moves back and forth. If the experimenter presents a morsel to this male, he will in turn offer it to the female, which indicates a persisting feeding motivation, not expressed except for head movements under normal circumstances. Finally, in the peacock, tail display has taken over to such an extent that without intermediate stages one would not recognize the origin of the display. The cock spreads his tail, shakes it, and steps a few paces back. Then he bends his ''fan'' forward and points with his beak downward. Thereupon the hen approaches and searches in front of him. In addition, young peacocks entice with scratching and pecking the ground (from Eibl-Eibesfeldt, 1972).

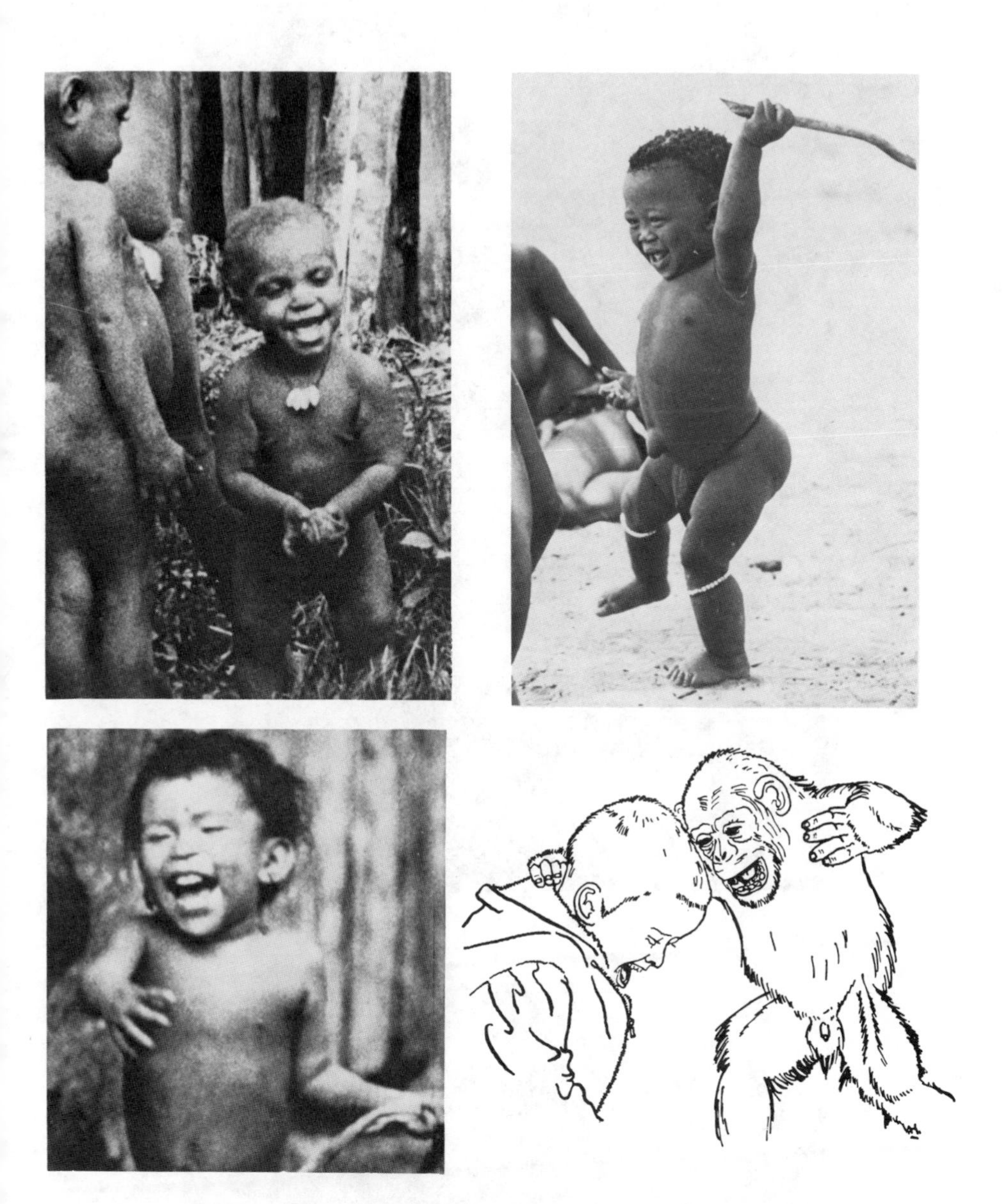

FIG. 3.8. Relaxed openmouthed display (''play face'') in (a) an Eipo boy (Westirian); (b) a !Ko Bushmen boy; (c) a Yanomami girl (photographs by Eibl-Eibesfeldt); and (d) boy and chimpanzee in playful interaction signaling with this homologous expression (from van Hooff, 1971).

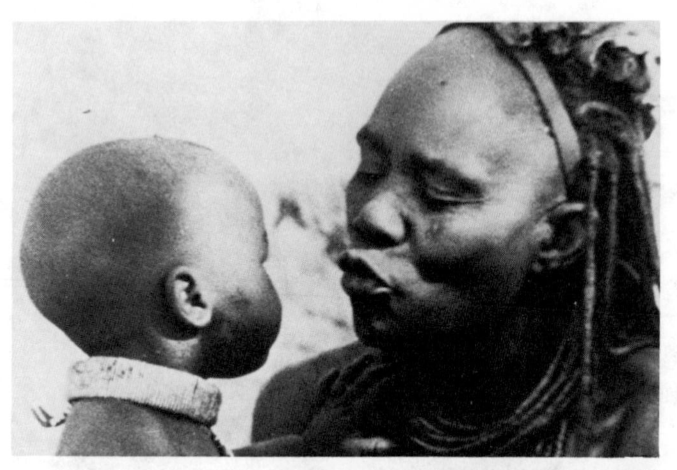

FIG. 3.9, a–e. Kiss feeding in the Himba (Kaokoland, Namibia, Southwest Africa). Grandmother and grandson are exchanging a morsel; the grandson passes the morsel over to her. From a 16-mm film taken at 50 frames per second by I. Eibl-Eibesfeldt. Frames #1, 60, 71, 109, and 146 of the sequence are shown.

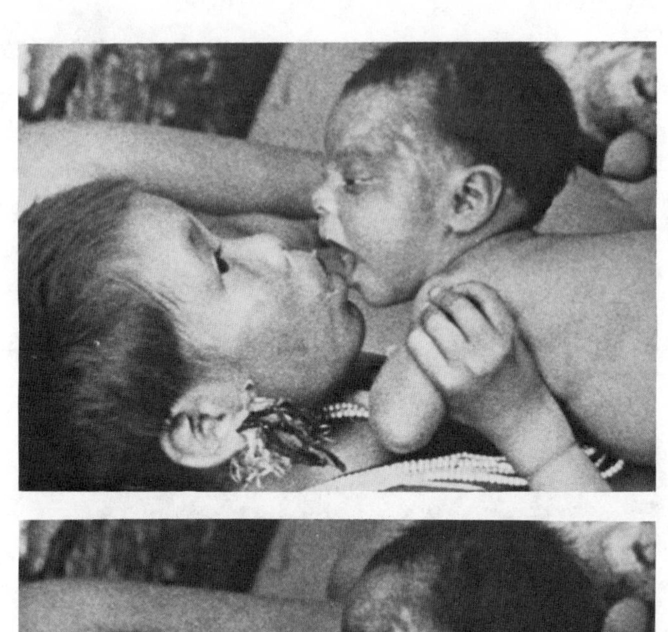

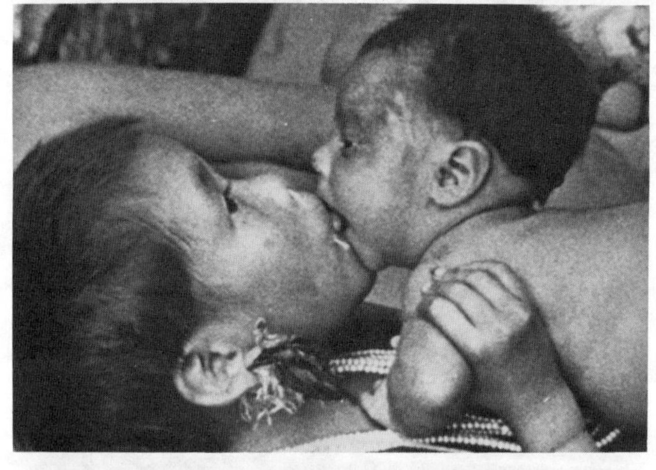

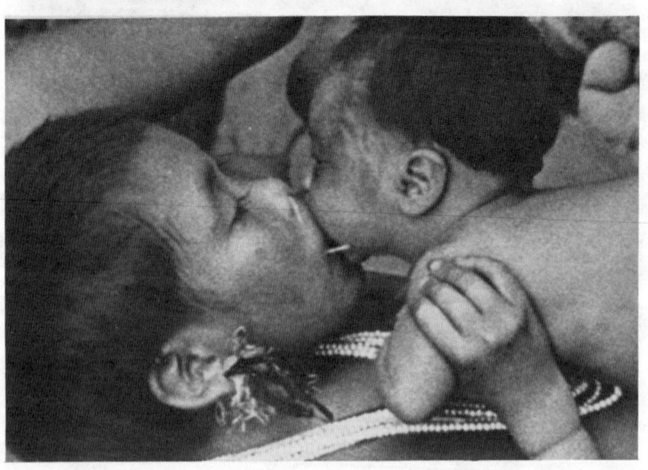

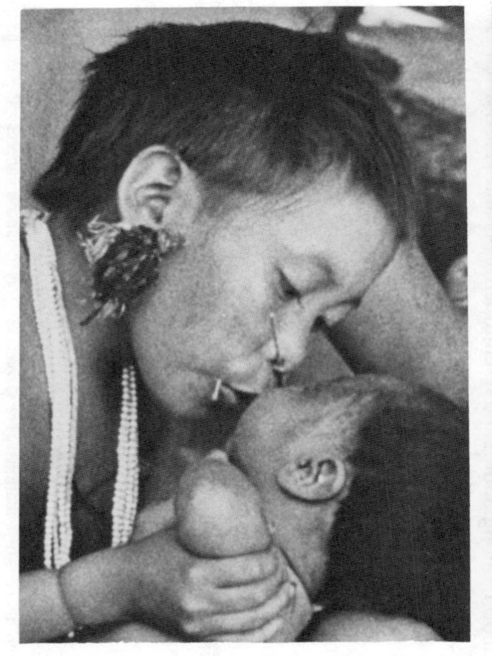

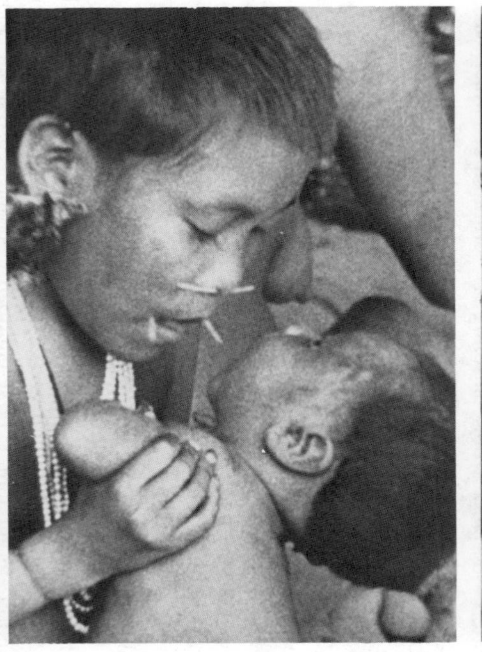

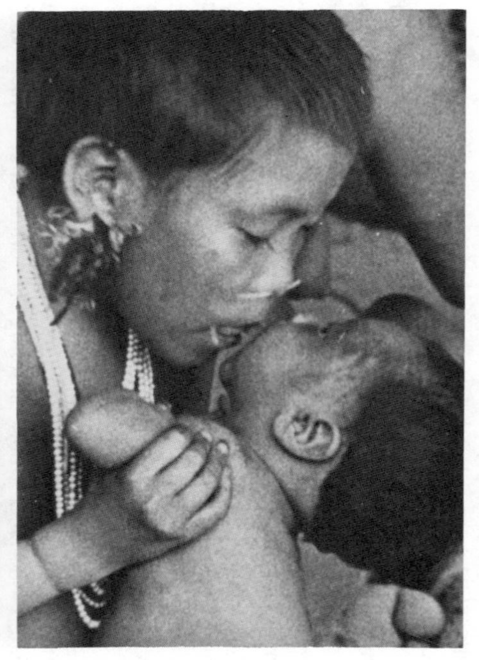

FIG. 3.10, a–f. Kiss feeding as an expression of affection. A Yanomami girl feeding her baby sister with saliva. From a 16-mm film taken at 25 frames per second by I. Eibl-Eibesfeldt. Frames #1, 54, 124, 127, 129, and 219 of the sequence are shown.

independently as an adaption for swimming in the water. Such types of analogies have been called *homoiologies* by morphologists. In addition, serial organs often show gradual changes within the same organism, such as the swimming, walking, and eating legs of crustacea, which change into one another in nicely graded stages. Such a serial homology has been called a *homonomy*.

With regard to behavior we often find that a functional behavior and the ritualized forms derived from it exist side by side. Thus the black woodpecker uses a movement derived from chiseling when calling his mate to relieve him from work. With slow pronounced movements he pecks against the rim of the nest entrance. This pecking to request relief is also used when taking turns in brooding; when the brooding mate hears the partner approaching, he or she pecks at the wall of the nest cave, thereby indicating readiness to take leave. Another movement pattern derived from chiseling is drumming, which in anthropomorphic translation indicates "a male is already at work here." In this manner he signals that the territory is occupied (Sielmann, 1958).

Ever since Darwin began to compare man's behavior with that of chimpanzees, primates have been a focal point of interest for behavioral scientists interested in the evolution of human behavior. In particular, the repertoire of expressive movement patterns in the two reveals homologies according to the criteria of homology just described. The relaxed openmouthed display of "play face," including the panting that accompanies excitation (Fig. 3.8) occurs in the same type of rough-and-tumble-play interaction in chimpanzees, toddlers, and small children. Van Hooff (1971) was even able to trace the evolution of this pattern by comparing a large number of primates including monkeys.

Kissing is another pattern of interest. For a long time it was thought to be a culturally evolved pattern of the western world. Rothmann and Teuber (1915), however, observed mouth-to-mouth contact and kiss feeding as a pattern of friendly interaction among adult captive chimpanzees, and van Lawick-Goodall (1968) noted similar behavior combined with embracing upon greeting in wild-living chimpanzees. Finally, we know that kiss feeding occurs in both orangutan and gorilla (Bilz, 1944). These observations strongly indicate that kissing in man may be related to the aforementioned patterns. A more careful cross-cultural examination in fact reveals that kiss feeding and kissing are universal patterns in mother–child interactions (Figs. 3.9 and 3.10). We have found these patterns in the Kalahari bushmen (!Ko, G/wi, !Kung), the Yanomami Indians, several Papuan groups (Eipo, Biamin, and others), Central Australian aborigines (Pintubi, Walbiri), Balinese, and others. It is difficult to say whether adults in a heterosexual context universally engage in kissing. But if not, it would not matter for our argument, as secondary cultural suppression of "Anlagen" (innate patterns) often occurs.[2] Kissing among the Japanese is said to be a recent introduction

---

[2]We often observe that movement patterns that exist as phylogenetic adaptations in man become culturally repressed in certain situations. The Japanese, for instance, suppress the exchange of eyebrow flashes among adults, whereas children are freely greeted in this way.

from the West, but there is a quote of an old Japanese reference in Krauss (1965), which says that lovers are warned not to insert their tongue in a woman's mouth during intercourse, as she might bite the tip off during orgasm!

If we look more closely at the movements involved in kiss feeding and kissing, we soon discover that the movement patterns are much alike in both events. The mother pushes the morsel with her tongue into the mouth of the child, who accepts it by opening his or her mouth and making sucking movements. Similarity of the movement pattern, the situational context, linkage by intermediate forms within the species, and intraspecies comparison all suggest that the pattern of kissing is derived from kiss feeding. In a wider context, this fits the more general rule that maternal behavior often becomes ritualized into patterns that are used to express affection or to create bonds.

When it comes into conflict with other cultural conventions, man often suppresses the expression of a particular behavior. In such cases he often translates the behavior into artifacts or language. An interesting example is provided by phallic displays. In a number of cultures, phallic displays occur during aggressive encounters. Male Eipo loosen the cord holding the tip of their penis gourd (Phallocrypt) and jump up and down, while calling out insults. During this display the gourd swings up and down. Phallic display also occurs as part of a startle response (Eibl-Eibesfeldt, 1976). Phallic displays have been observed in nonhuman primates (Ploog, Blitz, & Ploog, 1963; Wickler, 1967). When a group of vervet monkeys feeds, some males sit guard with their backs to the group. They expose their brilliantly colored red and blue genitalia, which become erect when strangers approach, posing a threat to members of another group. The males act as living border markers, so to speak. Their behavior can be interpreted as ritualized threat to mount, and indeed mounting occurs as an expression of dominance in monkeys and apes, in many other mammalian groups, and occasionally in man (Eibl-Eibesfeldt, 1980). More frequently, however, we find that man, instead of performing an action, conveys its message via artifacts. Phallic figurines are known in many cultures and serve to ward off evil spirits from a house and garden or, when worn as an amulet, to protect the body. Man also verbalizes phallic display: When Arabs curse an opponent they say: "The Phallus in your eye" (Eibl-Eibesfeldt, 1970; Eibl-Eibesfeldt & Wickler, 1968). Man's ability to translate actions into words thus effectively camouflages the phylogenetic roots of the pattern.

CONCLUSION

Studies of homologies as well as studies of analogies provide valuable insights for the understanding of human behavior, and both approaches can be combined as exemplified in the analyses of hunting adaptations and primate heritage in man by King (1980). In addition to primates, King turned to canid behavior in pursuit of his comparative approach. Man and the canid share many adaptations not to be found in all primates, such as the patterns of food distribution and all its consequences, deriving from the hunting of large prey.

The comparative approach can be usefully employed for all aspects of behavior. Movement patterns, cognitive capabilities, and learning structures have been compared across species. (See, for example, the recent contributions of J. Goodall, D. Fossey, J. Itani, T. Nishida, A. H. Harcourt, J. L. Popp, I. DeVore, R. S. Fouts, J. D. Bygott, W. C. McGrew, and others in D. A. Hamburg and E. R. McCown's book, *The Great Apes,* 1979.)

What one must bear in mind, though, is that the units of comparison and the level at which comparison takes place must never be confused. Motor patterns can only be compared with other motor patterns of the same level of integration, as exemplified by Lorenz' analyses of the courtship movements of ducks (Fig. 3.6). A comparative morphology of behavior can tell us about the common roots of behavior patterns by tracing homologies, whereas the study of analogies informs us about selection pressures that caused behavior in different species to develop along similar lines. Finally, the absence of analogies and homologies in behavior can provide us with useful information about how members of a species are unique.

REFERENCES

Bilz, R. Zur Grundlegung einer Paläpsychologie und ihre Anwendungen. I. Paläophysiologie. II. Paläopsychologie. *Schweizerische Zeitschrift fuer Psychologie,* 1944, *3,* 202–212; 272–280.

Bischof, N. The biological foundations of the incest taboo. *Social Science Information,* 1972, *11* (6), 7–36.

Chagnon, N. A. *Yanomamö. The fierce people.* New York: Holt, Rinehart, & Winston, 1968.

Darwin, C. *The expression of emotions in man and animals.* London: Murray, 1872.

Eibl-Eibesfeldt, I. Der Kommentkampf der Meerechse (*Amblyrhynchus cristatus* Bell) nebst einigen Notizen zur Biologie dieser Art. *Zeitschrift fuer Tierpsychologie,* 1955, *12,* 49–62. (a)

Eibl-Eibesfeldt, I. Über Symbiosen, Parasitismus und andere zwischenartliche Beziehungen bei tropischen Meeresfischen. *Zeitschrift fuer Tierpsychologie* 1955, *12,* 203–219. (b)

Eibl-Eibesfeldt, I. The fighting behavior of animals. *Scientific American,* 1961, *205* (6), 112–121.

Eibl-Eibesfeldt, I. Nannopterum harrisi (Phalacrocoracidae): Brutablösung. Encycl. cinem., E 596. Publikationen zu wissenschaftlichen Filmen, Institut fur den Wissenschaftlichen Film, 1965. (Film)

Eibl-Eibesfeldt, I. Männliche und weibliche Schutzamulette in modernen Japan. *Homo,* 1970, *21,* 175–188.

Eibl-Eibesfeldt, I. The cross-cultural documentation of social behaviour. *Colloques Internationaux du Centre National de la Recherche Scientifique,* 1972, *198,* 227–238.

Eibl-Eibesfeldt, I. *Love and hate. The natural history of behavior patterns.* New York: Holt, Rinehart, & Winston, 1974.

Eibl-Eibesfeldt, I. *Menschenforschung auf neuen Wegen.* München: Molden, 1976.

Eibl-Eibesfeldt, I. *The biology of peace and war.* New York: Viking Press, 1979. (a)

Eibl-Eibesfeldt, I. Functions of ritual: Ritual and ritualization from a biological perspective. In M. von Cranach, K. Foppa, W. Lepenies, & D. Ploog (Eds.), *Human ethology: Claims and limits of a new discipline.* Cambridge: Cambridge University Press, 1979. (b)

Eibl-Eibesfeldt, I. *Grundriß der vergleichenden Verhaltensforschung, 6th ed*. München: Piper, 1980.

Eibl-Eibesfeldt, I., & Wickler, W. Die ethologische Deutung einiger Wächterfiguren auf Bali. *Zeitschrift fuer Tierpsychologie,* 1968, *25,* 719–726.

Hamburg, D. A., & McCown, E. R. (Eds.), *The great apes*. Menlo Park, Cal.: Benjamin-Cummings, 1979.

Heinroth, O. Beiträge zur Biologie, insbesondere Psychologie und Ethologie der Anatiden. Verhandlungen des 5. Internationalen Ornithologenkongresses, Berlin, 1910.

Heinz, H. J. Conflicts, tensions and release of tensions in a Bushmen society. The Institute for the Study of Man in Africa, Isma papers No. 23, 1967.

Hoebel, E. H. Song duels among the Eskimo. In P. Bohannan (Ed.), *Law and warfare*. New York: Natural History Press, 1967.

Keesing, F. M. *Cultural anthropology: The science of custom, 6th ed*. New York: Holt, Rinehart, & Winston, 1980.

King, C. E. Alternative uses of primates and carnivores in the reconstruction of early hominid behavior. *Ethology and Sociobiology,* 1980, *1,* 99–109.

Krauss, F. *Das Geschlechtsleben des japanischen Volkes.,* 1965.

Lorenz, K. Vergleichende Bewegungsstudien an Anatinen. *Journal für Ornithologie,* 1941, *89,* 194–294.

Lorenz, K., & Van de Wall, W. Die Ausdrucksbewegungen der Sichelente, *Anas falcata* L. *Journal für Ornithologie,* 1960, *101,* 50–60.

Malinowski, B. *Argonauts of the Western Pacific*. London: Routledge & Kegan Paul, 1922.

Maynard-Smith, J. The theory of games and the evolution of animal conflicts. *Journal of Theoretical Biology,* 1974, *47,* 209–221.

Morris, D. "Typical intensity" and its relation to the problem of ritualization. *Behaviour,* 1957, *11,* 1–12.

Ploog, D., Blitz, J., & Ploog, F. Studies on social and sexual behavior of the squirrel monkey (*Saimiri sciureus*). *Folia Primatologica,* 1963, *1,* 29–66.

Rajecki, D. W., Lamb, M. E., & Obmascher, P. Toward a general theory of infantile attachment: A comparative review of aspects of the social bond. *Behavioral and Brain Sciences,* 1978, *1,* 417–464.

Rappaport, A. A. *Pigs for the ancestors*. New Haven: Yale University Press, 1968.

Rothmann, M., & Teuber, E. Einzelausgabe der Anthropoidenstation auf Teneriffa. I. Ziele und Aufgaben der Station sowie erste Beobachtungen an den auf ihr gerhaltenen Schimpansen. Abhandlungen der Preußischen Akademie der Wissenschaften, 1–20, Berlin, 1915.

Sielmann, H. *Das Jahr mit den Spechten*. Berlin: Ullstein, 1958.

van Hooff, J. A. R. A. M. *Aspecten van Het Sociale En De Communicatie Bij Humane En Hogere Niet-Humane Primaten*. (Aspects of the social behaviour and communication in human and higher non-human primates.) Rotterdam: Bronder-Offset, 1971.

van Lawick–Goodall, J. The behaviour of free-living chimpanzees in the Gombe Stream Reserve. *Animal Behaviour Monographs,* 1968, *1,* 161–311.

Wickler, W. Über den taxonomischen Wert homologer Verhaltensmerkmale. *Naturwissenschaften,* 1965, *52,* 441–444.

Wickler, W. Socio-sexual signals and their intraspecific imitation among primates. In D. Morris (Ed.), *Primate ethology*. London: Weidenfeld & Nicolson, 1967.

Wiessner, P. W. Hxaro: A regional system of reciprocity for reducing risk among the !Kung San. Ann Arbor, Mich.: University Microfilms, 1979.

Wiessner, P. W. Measuring the impact of social ties on nutritional status among the !Kung San. *Social Science Information,* 1981, *20,* 641–678.

4 Successful Comparative Psychology: Four Case Histories

D. W. Rajecki
Indiana University-Purdue University at Indianapolis

There has long been resistance to the idea that our understanding of human behavior can be enhanced by the study of nonhuman behavior and, as a social psychologist who has been drawn to experimental and scholarly research on animals, I am naturally sensitive to this resistance. Such negativism can be amply documented in academic essays (Montagu, 1968), certain of which manage to reach a rather strident pitch. For instance, Smith and Daniel (1975) assert that in looking back on the corpus of comparative animal research, they have the impression that it: ''represents one of the greatest wastelands of modern science or social science or pseudoscience,'' and that should a history of this activity be compiled, it would reveal a ''story of almost unparalleled stupidities and horrors which have, in toto, contributed little or nothing to man's knowledge of himself . . . , and very little to our knowledge of the animals that have been the objects of these experiments [p. 168].''

Further, the resistance to which I allude can be seen in other, more subtle forms. Satire is an effective form of criticism, especially if it can make its target squirm, and Woody Allen's recent (1980) contribution to this genre is well worth mention. Allen's brilliant piece tells, in dead pan, the story of a trio of crackpot scientists who are seeking a remedy for the embarrassing and dangerous human problem of choking on food in public restaurants. The research team's most heavy-handed member recommends that they start their project by conducting experiments on humans directly. As he would have it, convicts would be fed huge lumps of meat at short intervals (5 seconds) with accompanying instructions not to chew before swallowing!

However, the head scientist (who has kept a diary of all this) rejects his colleague's high risk approach, and instead embarks on a more conservative course by experimenting with animals. His fictive diary includes a progress report in which he excitedly proclaims that, by working around the clock, he was finally able to induce strangulation in a mouse:

> This was accomplished by coaxing the rodent to ingest healthy portions of Gouda cheese and then making it laugh. Predictably, the food went down the wrong pipe, and choking occurred. Grasping the mouse firmly by the tail, I snapped it like a small whip, and the morsel of cheese came loose. . . . If we can transfer the tail-snap procedure to humans, we may have something. Too early to tell [p. 37].

Bravo, Mr. Allen, a most exquisite sting! Yet, although I suppose that I should be relieved that the looney scientist described was credited with enough common sense to be cautious about applying this tail-snap procedure to the suffocating restaurant goers among us, can my colleagues (and I) be so innocently dense as to jump to the same caliber of vapid conclusion from our comparative work in wistfully hoping that we may have something? Is it true, as Smith and Daniel (1975) insist, that we have merely wreaked havoc in our comparative laboratories? Will it always be the case that we must hedge our claims concerning the relevance of animal data with the caveat that it may be too early to tell? I think not. Definitely not.

I make this claim in the face of several systematic criticisms of comparative psychology. An early one of these was Beach's (1950) contention that by then the field had become narrow to the point of disappearance as indicated by the small number of species selected, and the restricted range of phenomena studied. Later, Hodos and Campbell (1969) and Lockard (1971) successfully attacked a comparative psychology based on the assumption that there is a *scala naturae* (phylogenetic scale) in which various animal groups—for example, fish, reptiles, birds, mammals—represent levels of phylogenetic adaptedness, with man standing at the pinnacle of such a progression. Lockard went so far as to say that we were left with "the debris of what was once a traditional branch of psychology."

Still, work in comparative psychology goes on. This work involves several aims and methods as witnessed by the variety of contributions in this book from anthropology, ethology, genetics, philosophy, psychology, and zoology. But I do not presume to offer a comprehensive definition of what contemporary comparative psychology is or ought to be. Rather, I wish to ponder for a moment a pragmatic issue: the impact of comparative psychology on psychology in general. There is, I think, in most of us a tendency to evaluate scientific activity in terms of its relevance to the human condition. Although this is most certainly *not* the only justification for a comparative approach (as pointed out by Beach in 1950) it is a question that cannot always be ignored.

I would like to ascertain whether knowledge of animal behavior has made any real difference in the workaday world of the human laboratory or clinic, or in theories of human behavior. Any larger philosophical issues aside for the moment, have psychologists who deal with human problems and processes noticed the results of research on animals? Whether or not one is correct in speaking of animal and human behavior in the same breath, has there been any real impact of animal data on general advances in modern psychology? Actually, there are quite affirmative answers to my rhetorical questions. My thesis is that clear examples can be found of important influences of the knowledge of animal behavior on advances in a number of areas or subdisciplines in human psychology. Recall here that my aim is not to provide a broad definition of comparative psychology, but rather to seek illustrations of the general influence of nonhuman data. Of course, the pattern of influence, or the timing of influence vary from instance to instance. In some cases animal data stimulated new lines of research on humans, whereas in other cases findings from nonhumans clarified matters in areas of traditional interest. But whatever the nature of the influence, it *has* been felt, and it can be documented via a number of case histories.

Now as an historian, one has the responsibility for tracing cause and effect relationships in human affairs over time. To do so it is necessary to identify and emphasize certain events. Hopefully, the events the historian points to are the same ones that were involved in the historical progression, but that will not always be so. A person who was present and influential in an era may not always view himself or herself or the era as described by the historian. Nevertheless, this is the risk a writer of history must take, and I can only hope that my brief sketches have not excessively distorted events as they actually transpired in the advances I am describing.

In consulting my initial notes for this project, I find that I had jotted down nine widely different areas as candidates for case histories showing the positive influence of comparative studies. But because I wanted to go into some detail and depth on each, it became evident that I could not do justice to so many topics in the limited space available. As a compromise between breadth and depth, I settled on four prominent case histories. Two are from the domain of child psychology and development, and two are from an intersection of experimental and clinical psychology. Having two cases in each domain allows me to claim that comparative influences were not limited to single flukes. The two examples from child psychology are infantile attachment (Case 1), and dominance relations among peers (Case 2). Those from clinical/experimental psychology are learned taste aversions as therapy for alcohol addiction (Case 3), and learned helplessness and depression (Case 4). I take up these four cases in the order listed.

CASE 1: CHILD PSYCHOLOGY AND DEVELOPMENT—INFANTILE ATTACHMENT

Imagine that we observe a child of 1 year of age in the company of its mother. We see that the child voluntarily stays near the parent, exchanges smiles and vocalizations with her, and is upset if the mother leaves, but the child is forced to stay behind. All this is in marked contrast to the child's reaction to a stranger. The baby tries to avoid the newcomer, cries if the stranger persists in making contact, and is relieved at the stranger's departure. Based on these observations, one could conclude that the child is psychologically "attached" to the mother. Loosely defined, such an attachment can be thought of as an emotional bond, or an enduring relationship of the child to its mother. The question is, how might such a specific attachment come into being? In the first half or more of this century answers to this question were based on psychoanalytic (Freudian) and learning (Pavlovian, Hullian, Skinnerian) formulations. Interestingly, mainstream theorists in both the learning and analytic camps, although widely divergent on other points, had something important in common in their analysis of the ontogeny of filial ties in the child. That commonality was the assumption that there was a *physical* basis for the *psychological* state of attachment.

For the learning theorists, it was a straightforward matter to interpret social responsiveness on the part of the child as resulting from reinforcement contingencies present in social situations. There are several versions of such learning theories, but it will be sufficient to examine the general view expressed by Dollard and Miller (1950):

> In the first year of life the human infant has cues from its mother associated with the primary reward of feeding on more than 2,000 occasions. Meanwhile the mother and other people are ministering to many other needs. In general there is a correlation between the absence of people and the prolongation of suffering from hunger, cold, pain, and other drives; the appearance of a person is associated with a reinforcing reduction in the drive. Therefore the proper conditions are present for the infant to learn to attach strong reinforcement value to a variety of cues from the nearness of the mother and other adults [p. 91].

In other words, a consequence of the foregoing events is that the ministering parent or caretaker is in a position to acquire secondary reinforcing properties by virtue of his or her association with the delivery of primary reinforcers, be they either the provision of physically desirable things, or the removal of those that are physically undesirable. Ultimately, the child is attracted to the caretaker and engages in "attachment" behavior—e.g., seeking proximity, smiling, and vocalizing in response to parental overtures—in order to obtain the secondary reinforcers now inherent in the adult. (For further elaboration on learning theories of attachment see Rajecki, Lamb, & Obmascher, 1978).

Influential writers in the psychoanalytic tradition also posited that physical gratification was at the base of infant sociability. One such view was that of Anna Freud (1946) as seen in her formulation on the anaclitic (dependent need to be fed) origin of human attachment. According to Freud (1946), the child's love for its mother proceeds through stages, all of which hinge on the reduction in tension afforded by feeding. The first of these is the *narcissistic* stage in which the child's love is the love of the sheer pleasure of feeding. Next, the child enters a *transitional* stage when it recognizes that there is an external source of its pleasure, and his or her love shifts to the milk, breast, or bottle involved. Finally, when the child becomes aware that the caretaker is the ultimate source of tension reduction via feeding, it enters the *object relations* stage, and feels love for its mother. In summary, this sort of anaclitic attachment formulation has been labeled the by Bowlby (1969): "cupboard-love theory of object relations [p. 178]." Although the terminologies differ, the analytic analysis has clear affinities to the reinforcement model of the learning theorists, on these points at least.

Contributions from Animal Behavior

The assumptions of the learning and psychoanalytic theorists about the physical basis of infantile social bonds certainly sound plausible, but by now it seems these assumptions have become very much out of date and out of fashion. Indeed, the idea that physical care is at the root of human infant sociability has been currently listed, among others, as a "misconception about psychology among introductory psychology students" (Vaughn, 1977, p. 138). How can it be that the views of writers in such influential areas have ended up (somewhat ignominiously, one must think) as having no more status than as one of the misconceptions of undergraduates? The answer, in considerable part, is due to the relatively recent contribution of research on animal behavior to theory and research on this human behavior. Over the past two decades there have been rapid and dramatic advances in this comparative approach. To provide a framework within which to trace these advances, I organize my review around 3 salient years in that history: 1959, 1969, and 1979.

1959: A Beginning. By 1959, several radical (and now seminal) publications in the area of infantile attachment had appeared. Bowlby, in his 1958 review, attacked the analytic position that infantile social bonding was based exclusively on feeding or orality. Although largely theoretical, Bowlby's paper did review some data (or at least observations) with which he took the psychoanalytic theorists to task:

> A discrepancy between formulations springing direct from empirical observations and those made in the course of abstract discussion seems almost to be the rule in the case of analysts with first-hand experience of infancy—for example, Melanie Klein, Margaret Ribble, Therese Benedek, and Rene Spitz. In each case they have observed non-oral social interaction between mother and infant and, in describing it, have used terms suggesting a primary social bond. When they come to theorizing about it, however, each seems to feel a compulsion to give primacy to needs for food and warmth and to suppose that social interaction develops only secondarily and as a result of instrumental learning [Bowlby, 1958, reprinted in Bowlby, 1969, p. 367].

But most people, including theorists, are loathed to give up plausible-sounding interpretations, and one can well imagine that there might have been considerable reluctance to the acceptance of Bowlby's notions on the non-oral nature of infant–mother interactions. There is the real possibility that his views may have gone unheeded, and they may have become as obscure as earlier writing in this vein (see Bowlby's [1969] appraisal of the scientific fate of I. D. Suttie, who in 1935 had made an earlier attack on the oral basis of human sociality). But Bowlby's position did not sink into obscurity, and part of its buoyancy was due to contemporaneous reports of *animal* behavior written by Harlow (1958; Harlow & Zimmerman, 1959). In those papers Harlow provided the first really compelling experimental evidence that, in some primate species at least, the formation of social bonds is not necessarily dependent on feeding. His findings were that rhesus monkey infants formed attachments to cloth surrogate mothers regardless of whether they were fed on these surrogates, and did *not* form attachments to wire surrogates even if these were the dispensors of milk. In sum, whereas something like "contact comfort" might be involved, feeding and orality did not seem to be essential to the formation of infant monkey attachment. Thus, Bowlby's critique and Harlow's data on nonhumans represent the beginning of a reassessment of the alleged physical basis for psychological bonds.

1969: A Watershed. The years between 1959 and 1969 saw a continuation of research on attachment in young animals. For one thing, comprehensive reviews of the imprinting literature by Sluckin in 1965 and Bateson in 1966 indicated that precocial avian hatchlings, like their primate counterparts, did *not* require conventional reinforcers to develop strong bonds or attachments to any of a variety of artificial maternal surrogates. For another thing, studies of primates of various genera provided details of mother–infant interactions in experimental and naturalistic settings (see Bowlby's 1969 review). Taken together, these developments did much to bolster the growing conviction that something other than orality or conventional reinforcement schedules were the base of infant social ties. Indeed, this accumulation of information on animal attachment was an important component in two publications—one by Ainsworth; the other by Bowlby—that mark 1969 as something of a watershed in recent work on human infantile attachment.

Ainsworth's (1969) paper was an extended review of three major theories of infant attachment: the *psychoanalytic* and *learning* theories already noted, and what had become known as the *ethological* theory. Ainsworth's review is reminiscent of that of Bowlby in 1958 in that she found serious fault with theories of bonding based on orality, dependence, or reinforcement. However, what distinguishes her review from the earlier one is that Ainsworth had an additional decade's worth of avian, simian, and human data to bank on, and she put it to good use. Ainsworth's (1969) paper is also important because we find her making explicit comparisons between phyletic branches, and alluding to research on humans and animals in the same breath:

> Harlow's experimental work with rhesus monkeys (e.g. 1961; Harlow & Zimmerman, 1959) makes it clear that feeding gratification is not the primary basis for infant–mother attachment in this species, and comparable evidence is accumulating for other mammalian species, including the human. Thus, for example, Ainsworth (1963, 1967) and Schaffer and Emerson (1964) reported attachments formed by young infants to familiar persons who play no part in the infant's routine care, including feeding. That oral components of experience are conspicuous in the infant seems obvious, as does the fact that the feeding relation is an important aspect of mammalian mother–infant relationships . . . , but that these are crucial in the formation of the infant–mother tie is now clearly questionable [p. 980].

Further, Ainsworth (1969) notes that in general the psychoanalytic writers (and, by extension, the learning theorists also) treat the human infant as a passive recipient of environmental stimulation. She contrasts this assumption with her impression from the ethological (animal) literature that the young of other species are quite active in eliciting maternal behavior, and in seeking proximity on their own:

> It does not seem reasonable to assume that the human infant is passive while other infant mammals are active—and indeed direct observation of human infants convinced me (Ainsworth, 1963) that they are much more active and much less passively recipient than theoretical accounts have portrayed them [p. 981].

The third theory evaluated by Ainsworth (1969) was set forth in Bowlby's book of 1969. This book is the other important publication of 1969 to which I have alluded, and its importance lies in the fact that it contains Bowlby's formal and comprehensive presentation of a new theory of attachment. Basically, Bowlby argues that we can understand human infant attachment in terms of its function, not in modern technological or modern agricultural society, but rather in terms of that period of time in which man evolved *as Homo sapiens*. Most accounts seem to agree that early man emerged under conditions that supported a hunter–gatherer life-style, and it is under these particular environmental circumstances that other kinds of human "behavioral equipment" also evolved. In such primeval circumstances, mother–infant bonds would represent an advantage to the *species* (as well as the individual) because anything that kept the offspring close to the parent would reduce infant mortality, and thus more infants would survive to the point where they themselves could reproduce. Given that infant–mother bonds would contribute to the propagation of the species, there was, presumably, evolutionary pressure to select for mechanisms that would ensure such proximity of offspring to parent. One such mechanism, among others, would be psychological bonding on the child's part. Therefore, Bowlby's view relates infant attachment behavior simultaneously to man's evolutionary (genetic) history and to man's early environment; hence it is known as the ethological theory of attachment.

Bowlby's model does not rely on reinforcement or learning per se to account for attachment, but rather postulates an inborn "control" or feedback mechanism for attachment behavior. I have presented my understanding of Bowlby's word model as a flow diagram in Fig. 4.1. Information from the child's receptors as to the proximity of the attachment object (mom) is sent to an appraisal unit, where it is mingled with input from the child's psychological state. Depending on the appraisal of this combined information, various effector (response) systems can be activated, and a range of filial behavior can be anticipated. For

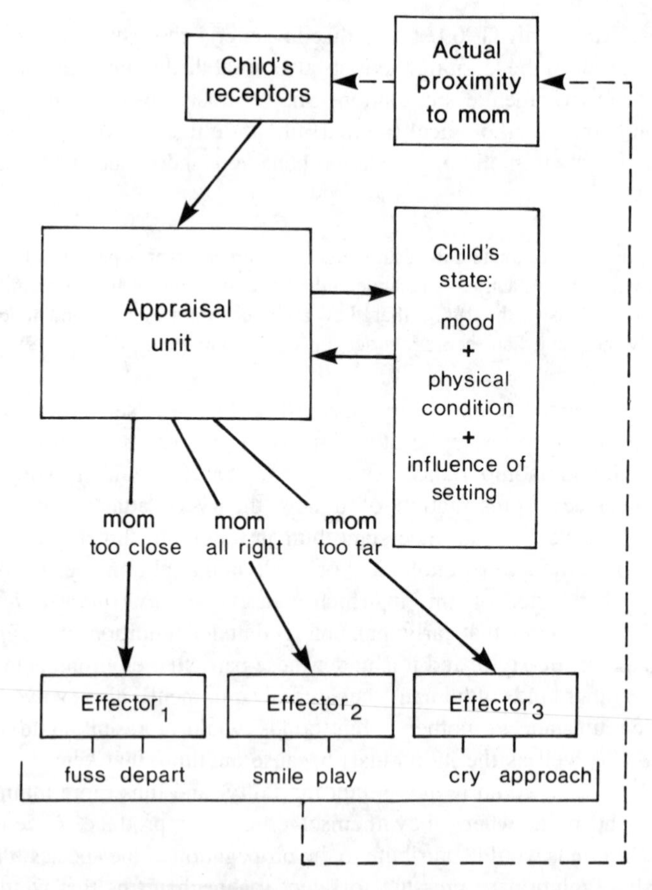

FIG. 4.1. Flow chart depicting Bowlby's theory of an inborn control or feedback mechanism for infantile attachment behavior. Note: The word "mom" in the figure refers to the child's mother or attachment object.

example, if the mother was 10 feet distant (receptor information) and the child was happy (state information), then the child might be content to interact at that distance by smiling or vocalizing. On the other hand, if the mother was 10 feet away (receptor information) and the child became frightened (state information) the child would strive to reduce the distance to the mother. A slightly frightened baby might be content to reach a point within an arm's length of the parent, whereas a badly frightened child might wish to be picked up and held.

However, a critical point to be made in the present context is that Bowlby legitimizes his ethological approach by an appeal to behavior seen among nonhumans. In an important section of his 1969 book (Chapter 11, "The child's tie to his mother: Attachment behaviour) he opens with an engaging pastoral:

> In the countryside in springtime there is no more familiar sight than mother animals with young. In fields, cows and calves, mares and foals, ewes and lambs; in the ponds and rivers, ducks and ducklings, swans and cygnets. So familiar are these sights and so much do we take it for granted that lamb and ewe will remain together and that a flotilla of ducklings will remain with mother duck that the questions are rarely asked: What causes these animals to remain in each other's company? What function is fulfilled by doing so [p. 180]?

This passage is followed by a review of findings on the mother–infant bonds in two monkey societies (rhesus and baboon) and two kinds of apes (chimpanzee and gorilla).

It is only after this extended allusion to the empirical literature on nonhuman infant bonding that Bowlby turns to human attachment behavior. As was the case with Ainsworth, he is firm on the relevance of the animal literature to the human literature:

> At first sight it might appear that there is a sharp break between attachment behavior in man and that seen in subhuman primates. In the latter, it might be emphasized, clinging by the infant to mother is found from birth on or very soon afterwards whereas in man the infant only very slowly becomes aware of his mother and only after he has become mobile does he seek her company. *Though the difference is real, I believe it is easy to exaggerate its importance* [Bowlby, 1969, p. 198, emphasis added].

Therefore, Bowlby's allusion to the nonhuman literature is not merely an attempt to show that findings from the human and animal domains might generalize to one another; it is an effort to *justify* the very application of the ethological model of infantile attachment to human beings.

1979: Do Old Theories Never Die? The decade from 1969 to 1979 witnessed a continued empirical and theoretical boom in the infant attachment business. On the empirical side, a great deal of new and exciting data were collected. For example, it was discovered that the human child's quality of attachment was related to his or her social competence as a toddler and preschooler, with securely attached infants being more competent than those who had been insecurely attached (Ainsworth, Blehar, Waters, & Wall, 1978; Easterbrooks & Lamb, 1979; Lieberman, 1977). This establishes that the continued study of attachment is warranted not only because it is an interesting phenomenon in its own right, but also because patterns of attachment are predictive of later childhood development.

As for the theoretical side, the reader is by now aware that learning theories of attachment have experienced sustained attacks (see also Rajecki, 1973, 1977). However, such theories, like any long-standing ideas, are sometimes difficult to dislodge, so it should not come as a complete surprise to learn that the 1970s saw the publication of both revised *and* new learning theories of early social bonding. For example, Gewirtz (1972) brought out an instrumental/operant learning model of infantile social dependence and attachment, and Hoffman and Ratner (1973) proposed a "reinforcement [classical conditioning] model of imprinting [with] implications for socialization in monkeys and men [their title, p. 527]." As made explicit in the title of the Hoffman–Ratner paper, these new versions of learning theory (as well as the ethological theory, of course) were not necessarily restricted to any given species and are open to tests with comparative data.

Given this undiminished interest in both learning and ethological theories of attachment (among others), some colleagues and I decided to take an up-to-date *comparative* look at these formulations so as to see how well the current theorizing fit the current available data. We organized our review (Rajecki, et al., 1978) around what we judged to be certain important facets or aspects of attachment phenomena: (1) the effects of *mal*treatment of the infant by the attachment object; (2) the ability of the infant to use the attachment object as a secure base from which to explore the broader environment; and (3) the effects of the involuntary separation of the infant from its object of attachment. Further, we tried to find data on these phenomena in a wide range of animal groups, including: (1) precocial hatchlings (e.g., chicks and ducklings); (2) puppies; (3) infant monkeys; and (4) human babies. That is, our outline of three phenomena in four animal groups generates a 12-cell (3 × 4) matrix, which we tried to fill with respectable data. Interestingly, we found a good deal of consistency over the animal groups within each of the three phenomena. Our general conclusions regarding the phenomena were as follows:

1. Maltreatment does *not* fundamentally influence attachment in the species for which data exist. We were mute on this point for human babies, due to the lack of experimental evidence.

2. Infants of all the species reviewed were able to utilize attachment objects as bases for exploration of and coping with the environment, but more data on puppies would be welcome.

3. Infants of all the species reviewed reacted adversely to involuntary separation from social objects, with intense behavior (distress vocalizations, agitation) occurring *very* soon after separation.

To return to our aim of evaluating theories, we felt that on the basis of our conclusions we could not endorse all current formulations with equal enthusiasm. Of course, the interested reader is invited to peruse Rajecki et al. (1978) for details, but our general evaluation was that whereas the Bowlby ethological theory made a pretty close fit with the data, none of the various kinds of learning theory were adequate. The problem with the learning theories was that they either did not have provisions for accounting for the phenomena that we studied, or the existing data flatly contradicted what such learning theories would have *predicted*. For instance, a learning theory would doubtless predict that an organism would learn to *avoid* an abusive agent rather than become attached to it. To the contrary, our scholarship uncovered reports (for better or worse, God help us!) on (1) chicks and ducklings that were frozen or given intense electrical shock in the presence of their object, or that were repeatedly bludgeoned by the attachment object itself; (2) puppies that were starved, or were beaten into submission by their human handlers; and (3) monkeys that were severely bitten and mauled by their biological mothers, or that were systematically tortured with noxious compressed air by their cloth surrogate mothers.[1] In all these cases attachment responses to the offending objects emerged or persisted! This grisly evidence, we submit, is compelling in questioning the validity of the learning position. Significantly, such evidence reveals the power of a comparative approach for the testing of general theories.

Attachment: Summary and Conclusion

Research on the attachment behavior of the young of various nonhuman animals has provided a rich source of facts and ideas over the last two decades. The comparative information available to the psychologists dealing with human social development has proved to be enormously useful and stimulating. Indeed, our very conception of the nature of human infantile attachment has changed as a consequence of contact with the animal literature. At mid-century the prominent

---

[1]All the illustrations just mentioned under points 1, 2, and 3 were planned scientific projects, duly reported in the literature. Full scholarly references to these can be found in Rajecki et al., 1978. The intentional maltreatment of laboratory animals for purposes of research is presumably constrained by the same general ethical considerations that guide research on humans. In short, before *any* research is begun, the investigator should weigh the so-called risk/benefit ratio. Do the potential benefits from the research outweigh the risks faced by the subjects? Although it is seldom easy to quantify all benefits and risks involved, it is to be presumed that a researcher who places his or her subjects at any sort of risk does so only because he or she feels—on some grounds—that the risk to the subject is outweighed by the potential benefit from the research.

view was that attachment behavior was the result of a mixture of operant and classical conditioning, no more, no less. However, by the fourth quarter of this century a widely accepted view (i.e., the ethological theory) is that current attachment patterns can be traced to the conditions surrounding man's emergence qua man. This means, ironically enough, that an integration of human and nonhuman data has revealed certain aspects of man's "human" nature more clearly than heretofore. What more could we ask of a comparative approach?

CASE 2: CHILD PSYCHOLOGY AND DEVELOPMENT—DOMINANCE RELATIONS AMONG PEERS

A child's first interpersonal relationships are with its parents, but by the time it reaches toddler and preschool age, peers become a significant element in social life. In fact, there has been a long-standing interest in peer relations in developmental psychology. Hartup's (1970) extensive review covered empirical literature on topics such as group formation, situational factors in group functioning, friendships, peer influences (e.g., on conformity), popularity, leadership, and status in the peer group. Regarding the last two entries in the preceding list, it is clear that developmentalists have been aware of power relationships between individuals, and one finds reference to the concepts of "ascendance– submission" and "dominance" in descriptions of children's social standing vis-à-vis one another (Gellert, 1962; Stott & Ball, 1957). Indeed, by 1970 Hartup could say that: "individual differences *inevitably* produce differentiation of status positions; children's peer groups *always* possess a hierarchical structure," and that "status differentiation is a *universal* attribute to group functioning" (p. 370, emphasis added).

Further, until the early 1960s developmentalists at least acknowledged that nonhumans also displayed such structured relationships, but apart from an occasional (and unfortunate) adoption of the misnomer "pecking order" (see Gellert, 1962, p. 169) to label this human pattern, cross-species comparisons were only *implicit*. Of course, Maslow's early writing (e.g., 1936, 1937) was available as a stimulus for comparative work on children's peer interaction, for he had published rather extensively on both human and nonhuman dominance relations. Still, his work in this domain does not appear to have been particularly influential, as indicated by its omission from later reviews on children's aggression (Feshbach, 1970) and peer relations (Hartup, 1970). Perhaps developmental psychologists shied away from Maslow's lead because of his emphasis on the relationship of dominance to postpubertal sex. In any event, although both monkeys and humans engage in sexual behavior, there doubtless remained the question of the relevance of simian sexuality to human sexuality. Therefore, even Maslow's work dealt with *implicit* connections between human and nonhuman dominance behavior.

However, by the latter part of the 1960s an *explicit* comparative approach to the study of children's dominance behavior began to surface, and the subsequent decade saw a proliferation of research that was stimulated and guided by advances in the study of animal behavior. The first of these advances was the emergence of systematic ethology (see Tinbergen, 1951), the subdiscipline of biology that places an emphasis on the study of the species-specific behavior of organisms in their natural settings (recall Bowlby's, 1969, ethological theory of infantile attachment in the preceding case). Insofar as possible, the ethological approach to the study of behavior capitalizes on naturalistic observation. In one of the earliest research reports on children stemming from an ethological perspective, Blurton Jones (1967) described his work thusly:

> It became obvious that one can study human behavior in the same way that Tinbergen . . . and Moynihan . . . and others have studied gulls, and van Hooff . . . , Andrew . . . , and others have studied nonhuman primates. This paper describes the crude preliminary observations which gave rise to this conclusion. It is a provisional, descriptive account, much like a report on the first season's ethological field work on any new species [p. 347].

Why regard children as a "new species?" Simply because in treating children as "human beings" in the past, human researchers and theoreticians had carried with them *assumptions* about human nature and human function that produced a confusing welter of concepts, incompatible operational definitions, and conflicting conclusions (see McGrew, 1972, pp. 1–11). The strategy of the ethologists, to the contrary, was not to begin with assumptions about behavior (e.g., its meaning or significance), but rather with a direct look at *the behavior* in question! As McGrew (1972) put it:

> Using such behavioral categories, ethologists directly record the behavior of normal individuals as it occurs. There is no need to resort to indirect measures: ratings, tests, questionnaires, projective techniques, interviews. Consequently, ethologists need not speculate about the applicability of [for example] doll play results or the validity of test scores to ongoing personal behavior. By recording behavior directly, ethologists see it functioning in real-life situations, and by using sophisticated and inconspicuous recording aids, they attempt to avoid disrupting ongoing behavior [p. 19].

The second development that contributed to the possibility of explicit comparisons between human and nonhuman dominance patterns occurred over roughly the same years that witnessed the emergence of ethology. From the late 1930s and onward primatologists—irrespective of their identities as anthropologists, psychologists, or zoologists—were learning more and more about *actual* nonhuman primate behavior in the field and in research colonies. One of the things that was abundantly clear from this research was that many primate taxa live in societies that are broadly based on some form of hierarchical structure. If Hartup could say of children that their peer groups always possess a hierarchical structure, in the same year Bernstein (1970) recognized that: ''strict status hierarchies are strikingly prevalent among macaques and baboons'' and that a number of writers have concluded that ''the status hierarchy is fundamental to primate societies [p. 72].'' Although he further recognized that the concept of dominance among primates had its critics, Bernstein's appraisal was that ''status relationships are, nevertheless, well structured in many primate taxa and are of great social significance [p. 75].'' In sum, whereas the ethologists had provided a justification and stimulus to study children's *behavior,* the primatologists provided a large catalog of nonhuman dominance behavior with which corresponding human behavior could be compared.

Behavior: Some Categories and Comparisons

Just what do primates *do* when they engage in dominance–subordinance behavior? The answer to this question is sometimes complicated, but if we examine, for example, Deag's (1977) description of the social life of a macaque troop we gain the impression that some animals consistently threaten, displace, and on occasion physically attack others (hence, the dominant monkeys) whereas different animals consistently avoid, escape from, or signal appeasement or submission to others (hence, the subordinate monkeys). Deag (1977) made an extensive study of a free-living troop of Barbary macaques in the mountains of Morocco and provided an overall summary of his observations. In general, consistent behavioral relationships were evident between troop members in pairwise comparisons. Animals avoided monkeys that had actively threatened them in the past; monkeys threatened others that had actively avoided them on other occasions (Deag, 1977). Furthermore:

> The frequency with which animals avoided others was more closely correlated with the avoided animal's rank than with their own, indicating that the subordinate was 'choosing' its [interactant] with greater care for rank . . . [and] . . . the frequency with which animals threatened others was more closely correlated with their own rank than that of the animal threatened, indicating that the dominant animal was not paying such close attention to the other interactant's rank [p. 468].

Thus, the initiation of threats and avoidances was not haphazard, with the key organizing factor being the rank of the dominant animal involved in the interaction.

To obtain a more detailed picture of the specific acts involved in dominance–subordinance behavior, we can turn to Table 4.1. The leftmost column of the table presents some relevant behavior categories culled from an ethogram

TABLE 4.1
Behavior Categories from Ethological Research on Monkey, Ape, and Human (Children) Social Conflict and Dominance Patterns[a]

Japanese monkeys: (after Fedigan, 1976)	*Chimpanzees: (after van Lawick-Goodall, 1972)*	*Preschool children: (after Abramovitch & Strayer, 1978)*
	Dominance Behavior	
Visual and vocal threat: consists of the following agonistic signals: stare, gape, and growl Lunge or bluff charge: a plunge forward towards an opponent in an agonistic encounter Displace: one monkey moves toward another who immediately oves out of the former's way Take away: to remove a desirable object, such as a food item, from another monkey Chase: to pursue another animal with accompanying agonistic signals Pull or push: to attempt to move another monkey by applying pressure Cuff: the first monkey hits the second with the flat of the hand Pinch or grab: to take hold of another's body by the hand and squeeze to the point of causing pain Bite: to seize another with the teeth	Glare: lips compressed, animal stares fixedly at another individual Arm raising: the forearm or the entire arm is raised with a rapid movement, the palm of the animal is normally oriented towards the threatened individual Hitting away: a hitting movement with the back of the hand directed towards the threatened animal Frustrate[b]: fear of another individual prevents (a) chimpanzee from obtaining a desired objective (as when) a dominant individual takes food in the possession of a subordinate Hair-pulling: attacked animals lose fairly large handfuls of hair during a fight Slapping: downward movement of the arm in which the palm of the hand slaps the body of the other Stamping on back: attacker seizes the victim by the hair and endeavors to leap onto its back. If this is achieved, attacker then stamps on the victim with both feet Lifting, slamming, and dragging: the victim may be lifted bodily from the ground and slammed down again Biting: during an attack the aggressor often puts his mouth to the body of the victim as though biting	Threat gestures Face and body posture Intention hit Intention kick Intention bite Object/position struggles Displace without contact Displace with contact Physical attack Chase Push–pull Hit Kick Wrestle Bite

(*continued*)

TABLE 4.1—*Continued*

Japanese monkeys: (after Fedigan, 1976)	*Chimpanzees: (after van Lawick-Goodall, 1972)*	*Preschool children: (after Abramovitch & Strayer, 1978)*
	Subordinance Behavior	
Fear grimace: a submissive signal in which the lips are retracted from the teeth, with the teeth clenched Avoid: a monkey notices another in its path or coming in its direction and changes its movement pattern in order to avoid encountering the latter monkey Seek aid: an animal in a dispute screams and looks repeatedly toward an uninvolved monkey for support	Grinning: lips are parted, the corners drawn back, and an oblong expanse of closed teeth are shown Presenting: in a nonsexual context occurs when a subordinate individual turns its rump toward a higher ranking one Bowing, bobbing, and crouching: involve various degrees of limb flexion . . . so that (the) body is close to the ground Screaming: some screams have a rasping quality, some are long and drawn out Submissive behavior (help seeking?): the subordinate, after being attacked or threatened, approaches a third individual of higher social status than the aggressor and makes gestures similar to appeasement gestures	Submission Cry–scream Cringe Hand-cover Flinch Withdrawal Requests cessation Help seeking Seeks adult help Seeks child help

(Reproduced from Rajecki & Flanery, 1981.)

[a] The entries are quotes from the original sources, rearranged here under dominance and subordinance categories.

[b] Strictly speaking, the van Lawick-Goodall (1972) paper was about expressions in chimps, and not about social conflict and dominance per se. Therefore, van Lawick-Goodall does not name any response pattern as ''frustrate,'' but her descriptions clearly indicate that object/position struggles do occur in chimpanzee groups.

(description of the behavioral repertoire) of members of a troop of Japanese macaques living in seminaturalistic circumstances (Fedigan, 1976). These categories are typical of those widely employed in the monkey literature (cf. Richards, 1974; Rowell, 1974), In the next column of Table 4.1 are comparable behavioral categories from a study of the social expressions of free-living chimpanzees (van Lawick-Goodall, 1972). The final column shows behavior categories used in studies of dominance in preschool children (Abramovitch & Strayer, 1978). Interestingly enough, there are several one-to-one correspondences in the entries in the three lists. Even *more* interesting is the fact that these researchers did not merely borrow categories from one another, they rather arrived at their classifications independently. For example, according to Fedigan (1976): ''the development of this ethogram was based on 6 weeks of preliminary observation of this troop of Japanese macaques.'' Similarly, Abramovitch and Strayer (1978) report that their ''social agonism inventory was developed from repeated observation of videotape episodes of social conflict collected during the initial 6 weeks of observation.'' Given the similarities in the classification schemes in Table 4.1, the time certainly seems ripe for explicit comparisons of nonhuman and human dominance patterns and their implications for peer relations.

Dominance Patterns Compared

A great part of the interest in rank, dominance, status, and other such concepts lies in the assumption that knowledge of such relationships would predict other social relationships or serve as a single, unifying principle for understanding social structure. Although this assumption does not always hold (see, for example, Bernstein, 1970), dominance relations *are* important in primate groups, and at least some aspects of social life are bound up with the status hierarchy. In the next section I outline several of these important aspects and cite evidence from monkey, ape, and human groups to indicate the role or influence of dominance in each. A fuller treatment of these (and other) aspects is available in Rajecki and Flanery (1981).

Dominance and Resource Allocation. ''Might is right,'' or is it? In simian societies it often is. The research of Richards (1974) with captive rhesus monkeys clearly indicates that dominant animals (as identified by threats) had priority of access over subordinates (as identified by avoidances) to dietary items such as liquids and solid food. Similarly, Wrangham (1974) found the same relationship of rank to food access in apes (chimpanzees) at artificial feeding stations located in the animals' home range. Rank is also related to male sexual access to females in monkeys. Kaufman (1967) identified the male dominance hierarchy in a troop of free-ranging rhesus in terms of exchanges of aggressive and submissive acts and subsequently noted that the highest ranking males were in close proximity to the troop's females about 90% of the time, whereas the lowest ranking males were seen in this area only about 14% of the time.

Children also compete with one another over any number of things (as many haggard parents can tell you), so much so that Smith and Green (1975) estimated that over 70% of all aggression between peers was over some property. Of special interest to us, Strayer and Strayer (1976) identified a dominance hierarchy in preschoolers in terms of attacks and threats (see Table 4.1), and then found that the outcomes of 76% of all property fights among these kids were predicted by the attack–threat ranking. As was the case with monkeys, the huge majority of wins went to the previously identified dominant child in the conflict. Spatial rights are also related to rank in children's groups. A number of writers (Deutsch, Esser, & Sossin, 1978; Savin-Williams, 1976, 1977; Sundstrom &

Altman, 1974) all reported that the most desirable areas (by one or another standard) in the children's environments were generally occupied or controlled by those high in their respective dominance hierarchies.

Dominance in Relation to Situational Demands. When the environment of nonhuman primates *deteriorates,* their dominance relations can be intensified. Captivity can be viewed as one form of environmental deterioration, and in general confinement results in an increase in the number of agonistic exchanges between monkeys and between apes. In turn, there may be an intensification of the dominance hierarchy (see the reviews of Gartlan, 1970; Hamburg, 1971). Ironically, apparent *improvements* in the environment can also intensify dominance relations. One kind of improvement is the artificial provisioning (feeding) of monkeys and apes. Southwick, Siddiqi, Farooqui, and Pal (1976) noted a two-to-sixfold increase in rhesus macaque aggression when they were fed small food items (rice grains, peanuts, beans), and especially when fed large food items (carrots or bananas) by the local human population. This sort of provisioning effect has also been seen in chimps who showed 10 times as much aggression on provisioned days as on provision-free days (Wrangham, 1974).

The amount of resources available similarly influences the amount of agonistic interaction in children. Smith (1974) varied the number of duplicate sets of toys (1, 2, or 3 sets) available to groups of children of constant group size, and found a clear result: the fewer the toys the more the conflicts. And the environment children play in also affects social conflicts. Kids observed in an outdoor setting in the summer exhibited only half the rate of object–position struggles (and half the rate of counterattacks) as did a similar group of kids observed indoors in the winter (Strayer, Chapeskie, & Strayer, 1978; Strayer & Strayer, 1978). This difference was attributed to the spatial constraints of the indoor setting, and the many toy-related activities held there.

Dominance and Social Signals. Fedigan's (1976) list of behavior categories in Table 4.1 includes postures and expressions that convey threat and submission, and a catalog of related expressions is available for chimpanzees (van Lawick-Goodall, 1968, Plates 10 & 11). We have already noted that such threat and submission signals are related to such outcomes as preferential access to food items (Richards, 1974) and sexual opportunities (see Kaufmann, 1967; Kolata, 1976). No further documentation is necessary except to refer the reader to Wickler's interesting set of photos that illustrate that an exchange of such signals can substitute for more violent forms of conflict in object–position struggles in crab-eating macaques (Wickler, 1967, Plate IV).

The facial expressions of children also predict success in object–position struggles. Camras (1977) induced kindergarten-aged children to compete for access to an attractive plaything in a sort of tug-of-war contest. She found that aggressive facial expressions (lips pursued and thrust forward, lowered brows, staring) on the part of a current holder of a plaything were positively related to persistent resistance to the surrender of the object. Furthermore, the children who were the recipients of such expressions (the competitors for the plaything) were relatively hesitant about making renewed efforts to wrest the object away from the current holder. In related research, Zivin (1977) found that children's so-called "plus-face" expressions (raised brow and chin, eyes wide open) were related to wins in preschoolers' spontaneous conflicts, whereas the so-called "minus-face" (furrowed brows, lowered eyes and chin) predicted losses.

Dominance and Attention Structure in Groups. Deag's (1977) data on Barbary macaques cited previously indicated that it was the dominant animal's rank that is the key to the course of events in that species' agonistic exchanges. In a real sense, this means that dominant animals bear more watching than do subordinates, a pattern of monitoring that has been termed "attention structure" by Chance (1976). Such primate attention structures are fairly well documented, with an interesting example offered by Keverne, Leonard, Scruton, & Young (1978). These researchers noted that, in talapoin monkeys, low-ranking individuals scanned others in the colony more often than did members of higher rank. Most interestingly, when there was a shift within the dominance hierarchy (i.e., an individual animal gained or lost in social rank) there was a corresponding shift in the attention pattern!

For children, perhaps the clearest finding on attention structure in relation to rank was provided by Abramovitch (1976). Initially, she identified a dominance hierarchy among a group of preschoolers by recording wins and losses in property fights. Following this, attention structure was checked by recording the direction of social glances (who looked at whom) exhibited by the children. Children's social rank and number of glances received were then compared, and a result similar to that seen in monkeys was obtained: High-ranking children received a disproportionately large amount of attention, whereas subordinate kids got a disproportionately small amount of attention.

Dominance Relations: Summary and Conclusions

In the foregoing section, we note that there is an impressive similarity between simian dominance patterns and those seen in groups of children. Dominance-related phenomena are implicated in important features of the child's social life, including: (1) the gain (or loss) of scarce resources; (2) societal consequences of the adequacy of, or opportunities in the environment; (3) the role of social signals in social conflict; and (4) the relation of rank to social monitoring (attention structure) in the group. It is also worth pointing out that although the foregoing review focused on children (ages ranged from preschool level to early adolescence), dominance relations in humans are by no means limited to the young. In an effort to apply ethological methods to the study of an adult, captive human group (inmates in a single large room in a state prison), Austin and Bates (1974, p. 449) used the following ethologically based behavior categories to record the social activities of male prisoners:

Aggressive contact: Hit, slap, push, kick, etc.
Stare: A fixed gaze directed by one inmate toward another.
Take from: The act of one inmate taking something from another.
Avoid: The act of one inmate withdrawing upon the approach of another; also the act of one inmate circling around, rather than moving directly in front of another.
Avert gaze: The act of one inmate avoiding eye contact with another.
Grimace: The act of one inmate presenting a nervous smile toward another.

Obviously, these categories have affinities to the lists in Table 4.1, and based on this scheme Austin and Bates were able to identify a clear social hierarchy in the prisoners' group. Moreover, this hierarchy was associated with two forms of spatial privilege in the prison setting. First, dominance was related to physical mobility (r = .77), with rank thus predicting the freedom of movement of a prisoner within the area of confinement. Second, rank was related to preferred sleeping facilities, assuming that a single bunk is a more desirable bed than is a double bunk (bunk bed). Prisoners in the top half of the hierarchy occupied a disproportionate number of single bunks, whereas those in the bottom half of the hierarchy were found to occupy a disproportionate number of double bunks (Austin & Bates, 1974).

Therefore, it can be seen that the application of ethological methods—derived from the study of naturalistic animal behavior—to the study of this form of human social behavior reveals that, like several animal species, people existing in relatively small, stable groups sort one another out in a hierarchical fashion, and that the resulting hierarchy has clear implications for activities within the group. Of course, such findings take nothing away from the importance of a sociological analysis of human status and rank based on role, occupation, income, or heritage. Still, the dominance hierarchy seems to be a pervasive phenomenon in the primate world, and the nature and consequences of this phenomenon demand further comparative study.

CASE 3: EXPERIMENTAL/CLINICAL PSYCHOLOGY—LEARNED AVERSIONS

Alcoholism is a serious and perennial form of drug abuse, and over the years any number of remedies for this problem have been sought. One of the hoariest forms of attempted solution is based on the possibility of the alcoholic forming an aversion to the drink. One need not be versed in the principles of Pavlovian (classical) conditioning to follow the logic of this idea, and, indeed, applications of this method antedated Pavlov by centuries. Essentially, the idea behind the therapeutic effect of learning an aversion is as follows: Assuming that alcoholics drink because they have a pleasurable reaction to ingesting alcohol, if we make this ingestion an unpleasant experience at least some of the time, then the drinker's desire for the substance may be reduced.

To put it another way, this time using Pavlovian terminology, if we can associate any strong unconditioned stimulus (UCS) that automatically induces a markedly unpleasant reaction (UCR) with a to-be-conditioned stimulus (CS)—such as the alcoholic's favorite beverage—sooner or later the properties of the drink may come to elicit a conditioned response (CR) of discomfort. To accomplish such conditioning we can present the CS and immediately follow it with the UCS. Then, just as Pavlov's proverbial dog could be trained to salivate to the sound of a bell alone after it had been previously paired with the presentation of food, the taste of alcohol alone may elicit discomfort in the drinker after it has been previously paired with the presentation of something terribly aversive. Hence, if alcohol makes the person uncomfortable, he or she may drink less of it.

Finding "the" Aversive Unconditioned Stimulus (UCS) for Aversion Therapy

As noted, the logic behind conditioned aversion as a therapy for alcoholism is straightforward. The only trick is finding an appropriately aversive UCS, and associating it in time with the CS of alcohol. Franks (1966) recounts that in medieval times dead spiders in the drinker's glass, beatings, and purgings were employed, but he does not report the clinical success of these choices. In this century, dead spiders seem to have gone out of fashion as the UCS in aversion therapy, to be replaced by one or another of several other forms of aversive stimuli. But even modern therapists have had a hard time settling on the preferred UCS, and a number of presumably obvious choices have met with very uneven success. In fact, as late as 1975 it was quite unclear if *any* of the traditional preparations were suitable as effective unconditioned stimuli! However, in 1966 there appeared two short papers by John Garcia and his coworkers that by now have taken psychologists a long way toward resolving the dilemma of the choice of materials for human aversion therapy.

Garcia's papers were on learned aversions, all right, but the interesting thing in this context is that they did not report on human behavior, but rather the behavior of the laboratory rat. In the sections that follow, I sketch the difficulties of the aversion therapists in finding a workable UCS, and show how Garcia's rat studies entered into a solution to the important problem of treating alcoholism.

Emetics as the UCS. Drugs that induce nausea and vomiting—emetics—were the first widely used aversive stimulus in modern aversion therapy. At some point after the emetic had been administered, the patient was made to smell and swallow some sort of alcoholic beverage, and in this fashion the UCS of the nauseating emetic was paired with the CS of alcohol. That is, the patient experienced more or less violent illness at about the same time he or she experienced ingestion of alcohol, which is the essence of the learned aversion approach. A series of such conditioning sessions were commonly employed in therapy. The question is, how effective was all this?

Unfortunately, critical evaluation of the early emetic-based aversion therapy programs was not especially positive. No one said that the patients of the aversion therapists did not improve. What critics doubted was: (1) how much they improved; and (2) why. The first problem arose because the therapists involved were probably better clinicians than they were experimental psychologists, and therefore their attempts to evaluate their own programs appear to have been seriously flawed in one way or another (Elkins, 1975; Franks, 1966). Many of the claims for patient improvement were suspect because of the absence of appropriate experimental designs, control groups, and follow-ups. Critics pointed out that improvement could have been due to clinical experiences other than the conditioning, spontaneous remission, or unknown factors. And a second problem was no less serious. Franks (1966), for one, came to doubt that emetics were even *appropriate* in the attempt to apply principles of classical conditioning to the control of alcoholism. He pointed out, among other things, that it is difficult to specify the precise onset of the emetic-based UCS, so precise pairing with the CS of alcohol is difficult. Therefore, therapists sought other forms of UCS for their treatment programs.

Apnea as the UCS. Apnea is a drug-induced, temporary muscular paralysis that includes the inability to breathe. During the 60- to 90-second period of complete paralysis, the patient remains fully conscious, with intellectual and emotional functioning unimpaired, but is unable to communicate anything, including his distress. According to Franks (1966), most patients find this experience to be terrifying, especially if the person has been given *no* information or forewarning about the nature of apnea! Hence the aversive state of apnea might serve as a potent aversive UCS in aversion conditioning, and, indeed, Sanderson, Campbell, and Laverty (1963) put alcoholic patients through apneic conditioning. Just before the paralysis was induced, the patient was asked to sniff and taste an alcoholic drink, and during apnea the therapist placed drops of the liquid in the patients mouth. Following apnea the patients were assessed for immediate reactions to alcohol, and their long-range posttreatment history of drinking was also recorded.

Unfortunately, the effects of apneic conditioning were mixed, at best. Some patients were even more avid for a drink immediately following the paralysis than before (Elkins, 1975) and apnea alone—without alcohol accompaniment—was about as effective in reducing drinking as apnea induced in association with alcohol (Franks, 1966). Of course, another brake on the use of this method is the ethical concern surrounding the hazards of the procedure itself. Again, therapists sought other forms of stimuli to serve as the UCS in their conditioning programs.

Noxious Faradic Stimulation as the UCS. Franks (1966) used his gloomy assessment of emetics and apnea in aversion therapy as a springboard to advocate noxious electrical shock. In Franks' view, the pairing of the taste of alcohol with unpleasant electrical stimulation as the UCS would have several advantages, including precise control with regard to onset, intensity, and duration, thus assuring a pairing of the UCS with the CS. This sort of recommendation (see also Rachman, 1965; Rachman & Teasdale, 1969) was well received, and a number of applications of the shock procedure were undertaken in the context of aversion therapy for alcoholism.

Alas, electrical shock as the UCS in aversion therapy failed to live up to its billing. According to Elkins' (1975) penetrating review, clinical applications of this procedure were characterized by many of the same deficiencies encountered in the era of emetic use. In several cases where shock was employed, there were inadequate research designs or control groups so that an accurate assessment of the result of conditioning per se was impossible. In these cases the influence of electrical shock was confounded with the influence of other theraputic practices such as counseling and posttreatment social pressures. Worse, where effects could be assessed, they were found to be absent altogether, or unimpressively weak. For example, Vogler, Lunde, Johnson, and Martin (1970) recorded time to relapse of alcoholics who received conditioning compared with those in various control groups. On average, the control patients relapsed in 9 days, whereas conditioned patients succumbed in 19.5 days. As Elkins (1975) points out, this difference, although statistically significant, is largely unimpressive from a practical standpoint.

What Next as a UCS? If emetics, apnea, and electrical shock do not work in aversion therapy, what on earth will? Does this history of only marginal success mean that the once promising therapeutic technique of aversion therapy is to be abandoned? Happily, there is an optimistic answer to this question, and that answer is largely based on research on nonhuman animals.

The Nature of Learned Aversions in Rats

In 1966, Garcia and his coauthors (Garcia, Ervin, & Koelling, 1966; Garcia & Koelling, 1966) published findings from studies on rats that ultimately were to stimulate research and theorizing that would open anew the possibility of aversion therapy for alcoholism. Recall that the aversion therapists had generally been considered to be working within a Pavlovian framework. To the contrary, the Garcia–Ervin–Koelling papers provided a challenge to two of the basic tenets of Pavlovian conditioning. These tenets were that of the principle of equipotentiality of conditioned stimuli, and the contiguity rule for the association of the CS with the UCS.

Equipotentiality. The equipotentiality principle states that *any* heretofore ineffective stimulus (CS) can serve as a substitute or signal for rewarding or punishing events in the environment (UCSs). For instance, Pavlov's dog learned to salivate to the sound of a bell after that sound had been associated with food, and thus the sound of the bell can be thought of as a substitute for the taste of the food. According to the equipotentiality principle, any *other* perceptible stimulus could also serve as this sort of substitute. Presumably, dogs can be trained to drool at the *sight* of lights, the *feel* of tactile pressure, the *smell* of perfume, or any other stimulus that the researcher can provide.

Contiguity. The rule of contiguity states that in order for classical conditioning to occur, the conditioned stimulus (the substitute signal) must be temporally associated (paired) in the proper fashion with the unconditioned stimulus. Research had indicated that an effective method of pairing the two is to present the CS first, and quickly follow it with the UCS, perhaps within a half-second. This method means that over a series of pairings the CS "predicts" the UCS, or, in other words, serves as a signal for the UCS. Sooner or later one reacts to the CS as if it had acquired some property of the UCS. Should one unfortunately present the UCS first, or delay the presentation of the UCS after the CS has occurred, classical conditioning will be impaired. Surely, delay of the presentation of the UCS for more than a few seconds would not produce much learning.

Garcia's Findings, to the Contrary. The news in Garcia's 1966 papers was that the Pavlovian tenets of equipotentiality and contiguity might not apply to learned taste aversions. Firstly, in terms of the contiguity rule, Garcia was able to demonstrate that the presentation of saccharine-flavored water (CS) followed by apomorphine-induced nausea (UCS) produced marked future avoidance of the saccharine flavor in rats, even when the CS–UCS interval was as much as 30 minutes or longer! Thus, in contrast to the rule of contiguity in Pavlovian conditioning, learned taste aversions seemed to follow a time line that indicated that they may be mediated by some other mechanism of learning.

Secondly, Garcia showed that the idea of equipotentiality of conditioned stimuli did not necessarily hold for learned taste aversions in rats. In another experiment he paired cues (CSs) from an external source (a click sound) and an internal source (sweet flavor) with the consequences (UCSs) of either an external sort (exteroceptive pain) or an internal sort (properioceptive nausea). This experimental setup is shown in Fig. 4.2 (after Garcia, Clarke, & Hankins, 1973). As the rat *tasted* sweet liquid and simultaneously *heard* a click sound it felt either *pain* from an electrical shock, or *nausea* from an internal source. In other words, there were two CSs in the situation that were paired with one or the other of two UCSs (i.e., click & sweet + pain in one instance; and click & sweet + nausea in the other). The question is, what kind of learning would take place in terms of

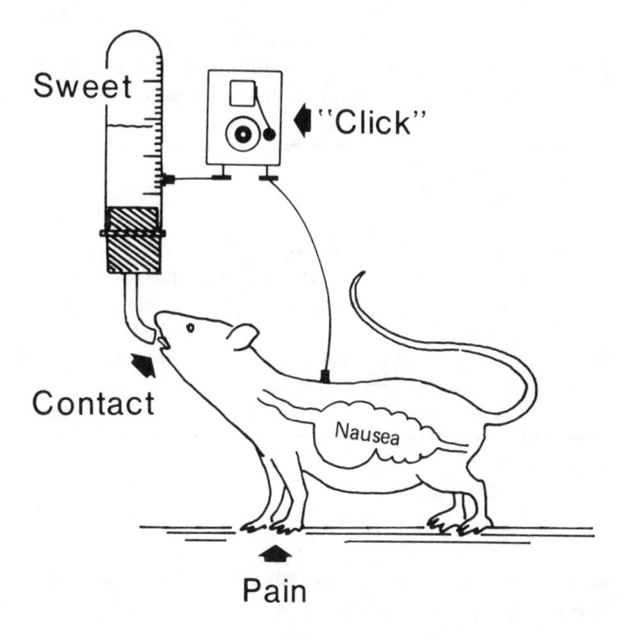

FIG. 4.2. Situation for testing the equipotentiality of conditioned stimuli. The click and the sweet are types of CS; the pain and the nausea are types of UCS (see text). Reprinted from Garcia et al. (1973).

which CSs might come to signal (substitute for) which UCS? This theoretical question can be put more clearly as a series of queries:

1. Would the *taste* (sweetness) cue allow the rat to avoid *nausea* in the future? If it tasted sweetness would it stop drinking and thus avoid nausea?
2. Would the *sound* (click) cue allow the rat to avoid *nausea* in the future? If it heard a click would it stop drinking and thus avoid nausea?
3. Would the *taste* (sweetness) cue allow the rat to avoid *pain* in the future? If it tasted sweetness would it stop drinking and thus avoid pain?
4. Would the *sound* (click) cue allow the rat to avoid *pain* in the future? If it heard a click would it stop drinking and thus avoid pain?

The experimental answer to these queries are found in Table 4.2. That is, one can say "yes" in reply to #1 and #4, but not to the remaining queries. Apparently, internal cues like taste are not good substitutes for external consequences such as the pain of electrical shock, and neither are external cues like sound good substitutes for internal consequences like nausea. It therefore seems that rats are best able to use external cues to cope with external events, and internal cues to

TABLE 4.2
Results of Research on Rats Concerning the Equipotentiality (See Text) of Internal and External Cues for Internal and External Consequences[a]

	Consequence	
Cue	*Internally Mediated Nausea*	*Externally Mediated Pain*
Internal (taste)	In future will not drink with flavor cue present, thus avoids internal nausea	In future will drink with flavor cue present, thus suffers external pain
External (sound)	In future will drink with click cue present, thus suffers internal nausea	In future will not drink with click cue present, thus avoids external pain

[a] (After Garcia et al., 1974.)

cope with internal events. In short, this demonstration by Garcia indicates that conditioned stimuli are not equipotential where learned taste aversions, or the avoidance of electrical shock are concerned.

Summary: The Implications of Garcia's Rat Studies. Garcia, Hankins, and Rusiniak (1974) draw a contrast between avoidance learning in the case of externally referred problems, and that of taste aversions that are internally referred. In the case of avoidance learning, the organism seems to acquire an if/then association of the sort, "if click, then shock," in a specific time and space orientation. To the contrary, aversion learning seems not to be so much of an if/then association as it is a change in motivation or incentive. The rat who has ingested poison along with saccharine later reacts as if saccharine itself were unpleasant, and the rat will refuse saccharine in places where it never made the animal ill. Thus: "time–space contextual information, so important in the external sphere, is dispensed with as unnecessary for adjusting incentives in the internal sphere" (Garcia et al., 1974, p. 828). These and other distinctions led Garcia et al. (1974) to a consideration of fundamental differences in coping with the *milieu interne* and the *milieu externe*.

In the milieu externe the organism must confront difficulties and seek its own resources based on time and space discriminations measured in milliseconds and millimeters. On the other hand, for reactions to slow-acting nutrients and toxins in the milieu interne, discriminations measured in milliseconds are of little value. Therefore, Garcia is more in favor of a dual internal–external specialization of adjustment mechanisms than a global process such as classical conditioning. What works on the outside might not work on the inside, and vice versa.

The Milieu Interne and Aversion Therapy for Alcoholism

Garcia's rat studies (Garcia et al., 1966; Garcia & Koelling, 1966) are seen as a landmark in the field of learning mechanisms in food selection (Rozin, 1977). But of what use are they to us in our quest for a workable aversion therapy for the control of alcohol consumption? For one thing, knowledge of the features of learned taste aversions in nonhumans prompts a reconsideration of procedures for humans. Recall that the early aversion therapists were thought to be working within a Pavlovian paradigm in which the to-be-conditioned stimulus (alcohol taste) was placed in association with any of a host of states including nausea, vomiting, apnea, electrical shock, or whatever. However, the rat research instructs us that a cue like taste is probably more easily substituted for an internal than an external consequence (see Table 4.2). Therefore, if we were to select a workable consequence from the preceding list for the sake of renewed attempts to produce learned aversions to taste in humans, internal consequences would be better choices than the others. Holding off on shock and apnea for the moment, the rat work indicates that the use of emetics may be the preferred aversive stimulus after all! In fact, following publication and recognition of Garcia's work, certain clinicians began to appreciate the need to reevaluate the emetic-based approach, and the possibility of putting the emetics back in action.

Modern Emetic Therapy. One of the first good experimental studies of learned aversions was that of Bernstein (1978), who was able to show that children indeed formed a dislike for an otherwise attractive ice-cream flavor after ingestion had been followed by aversive internal states. The children were victims of cancer, and the aversive internal states were induced by their routine chemotherapy. In fact, Bernstein found evidence that such aversions to a taste lasted for several months after a single pairing with nausea and vomiting. Children in control groups who received ice cream alone, or chemotherapy alone, did not form such aversions (and see also Lamon, Wilson, & Leaf, 1977, for further evidence concerning learned aversions to nonalcoholic substances). At about the same time Boland, Mellor, & Revusky (1978) published compelling evidence that aversion training was clearly effective in reducing alcohol consumption, and their report is recognized as the groundbreaker in this modern era. However, for purposes of explication I have chosen to review the therapy and program evaluations of Baker and Cannon (1979) and Cannon and Baker (1981), as their reports are especially interesting.

Patient #1. Baker and Cannon (1979) offer a detailed report of two case histories in their clinical setting that provide a good description of current aversive therapy. The first patient in this report was a 29-year-old male with a history of drinking a fifth of bourbon or vodka per day. He suffered from various physical and psychological complaints associated with alcoholism and had been arrested a number of times for driving while intoxicated. In a given aversion session the patient was given syrup of ipecac and emetine as emetics. Over the next 20 minutes he was asked to drink some 80 fluid ounces (2400 ml) of alcoholic beverages, along with some 60 ounces (1800 ml) of water. The patient was able to ingest this volume of liquid because he was regurgitating from 17 to 29 times in each five such sessions. Following a session the patient was given a drink of potassium antimony tartrate in beer to prolong nausea and was restricted to bed for a period of 4 hours. A cloth soaked in various spirits was placed next to the patient's head as long as he remained sick to his stomach.

This procedure was evaluated by comparing the patient's pretreatment consumption with that following therapy in terms of inpatient taste and preference tests. In these the patient was free to imbibe as much of a given drink as he pleased, and the volume of liquid consumed was covertly recorded. This patient's pretreatment consumption was about 70 ml per beverage type, but this dropped to below 30 to 40 ml following treatment. The patients postdischarge drinking was also recorded for a 9-month period and showed a similar decline compared with the pretreatment period. During the follow-up interval, he was abstinent for 238 days, drank moderately. for 29 days, and was drunk a total of only 7 days (Baker & Cannon, 1979, p. 234).

Patient #2. The second patient was also a male in his late twenties with a diagnosis of alcohol addiction, chronic. One of the man's favorite drinks was bourbon, so Baker and Cannon (1979) used bourbon in a series of five conditioning sessions as previously described. However, this patient was given taste tests throughout the sequence of aversive associations. Prior to therapy this patient voluntarily consumed an average of 80.5 ml of a bourbon drink per baseline test. After the first session his voluntary consumption dropped to 50 ml, and following the third session consumption was down to about 30 ml. By the end of the fifth conditioning session he was voluntarily ingesting less than 10 ml of bourbon beverage. Clearly, this patient's inclination to drink was negatively related to the amount of aversion therapy he had received.

Emesis Versus Shock in Aversion Therapy. It follows from Garcia's general rat findings (see Table 4.2) that emesis (an internal consequence) should be more effective than electric shock (an external consequence) in human learned taste aversion therapy, as taste is an internal cue. The findings of Baker and Cannon (1979) support the claim that clinically induced emesis *can* result in learned aversions with a therapeutic result, but their data tell us nothing about the potential *advantage* of emesis over shock, if any. Therefore, these researchers set out to fill this gap in the literature with a well-designed study.

Cannon and Baker (1981) randomly assigned a half dozen or so alcoholic patients to each of three experimental groups. Patients in one group were given five sessions of emetic-based aversion therapy as already described. Patients in a second group were given 10 sessions of aversion conditioning using very painful electrical shock as the unpleasant stimulus. The final group of alcoholics received neither emetic nor shock therapy and served as the control for any effects in the other two groups. As was the case in their earlier work, Cannon and Baker (1981) measured affinity for drinking via inpatient taste tests before and after therapy for patients receiving either form of conditioning, and at corresponding points in time for persons in the control group.

The results of these procedures were quite clear. There was no reason to predict that the consumption of the control patients would change in the absence of therapy, and it did not. To the contrary, the emesis group showed the typical sharp decline for this treatment, with their posttreatment consumption reduced by 40 ml. Indeed, the advantage of emesis over shock was also clear in that the posttreatment measure of the shocked patients showed only a 10-ml average drop when compared with their pretreatment consumption. Apparently, rats and human alcoholics are similar where learning cues to internal and external consequences are concerned.

Learned Aversions: Summary and Conclusion

It seems that the original proponents of emesis as the basis for an aversion therapy were right after all. Unfortunately, the therapeutic use of emetics went out of vogue (for some) because these early workers did not fully grasp the distinction between *milieu interne* and *milieu externe* in the application of their methods, and did not do a good job of convincing critics of the efficacy of their procedures. The subsequent shift from emetics to shock proved to offer little improvement in aversion therapy, and a viable replacement for shock does not seem to have been at hand. Fortunately, Garcia's papers had appeared by this time, and his research with animals convincingly reopened the possibility that emesis or nausea—as internal consequences—might yet be the preferred choice of aversive stimulation for the conditioning of internal cues like taste. By now, careful research with human populations has revealed that emetics are quite useful in producing learned taste aversions.

Still, emetic-based aversion therapy is hardly a panacea for the major problem of alcoholism in our society. Its cost and difficulty of application seem to be prohibitive in all but the most severe and intransigent cases, and even where affordable, it is contraindicated by physical ailments that are not uncommon among chronic alcoholics. Nevertheless, such therapy does work, and one basis for this success stems from our knowledge of learned taste aversions in nonhuman animals.

CASE 4: EXPERIMENTAL/CLINICAL PSYCHOLOGY—LEARNED HELPLESSNESS

Psychological depression is a widespread and costly disorder. Almost everyone becomes mildly depressed occassionally, some 12% of the adult American population seeks professional attention for more severe depressive episodes, and one of every 100 depressive illnesses ends in suicide (Williams, Friedman, & Secunda, 1970). Here is a problem that warrants attack on all fronts, including a comparative approach. Indeed, researchers have long sought animal models for human depressive disorders, for if depression can be induced, studied, and relieved in animals, such techniques may provide insights into means for dealing with human depression.

Criteria for Animal Models of Depression

But how would one go about selecting behavior to examine in the animal domain? Do, in fact, animals even get depressed? Guidelines to assist in this selection were set out by McKinney and Bunney (1969). They argued that an animal model for a human disorder should have the following characteristics:

1. Agents that predispose or cause the problem in humans should be effective in inducing parallel difficulties in the animal.
2. The human symptoms should be reflected in the behavior of the animal.
3. Techniques or procedures that ameliorate or correct the disorder in humans should be effective in producing improvement in animals.

One of the earliest animal models of a depressive disorder illustrated the fact that the tripartite cause–symptoms–cure criteria can be met. That disorder is known as the anaclitic depression syndrome (Harlow & Suomi, 1974; Rajecki et al., 1978, for reviews). Briefly, this syndrome has been reliably, if not universally observed in human children and young monkeys. In terms of its *cause,* prolonged, involuntary separation of the infant from the mother elicits the reaction. In humans, such separations take place, for example, when either the mother or the child has been hospitalized. In monkeys, separations are carried out experimentally. Regarding *symptoms,* the first phase of the syndrome is characterized in both children and monkeys by active distress, agitation, and protest. This phase is followed by a period wherein the infant is socially withdrawn, listless, and overtly miserable. In other words, depressed. If the separation is long enough, something like spontaneous recovery occurs, and the youngster brightens and becomes sociable again. Finally, in terms of a remedy or *cure* for the syndrome, the return of the mother herself, or in some instances the provision of substitute caretakers is known to alleviate the infants' distress in both humans and animals.

However, the range of applicability of the animal model of anaclitic depression is obviously limited, for it is doubtful that temporary (even if prolonged) separation from one's mother is at the root of much of serious adult depression in humans. Further, although there are other methods for inducing and relieving depression in nonhuman primates—such as extended social isolation and peer therapy—the utility of these particular models is extremely limited, if relevant at all. What is wanted is an animal model that has a more intuitive bearing on what we understand as day-to-day human depression.

Reactive Versus Process Depression

It will be helpful in this respect to distinguish between reactive depression on the one hand, and process depression on the other (Miller, Rosellini, & Seligman, 1977). Although the line between these two types of disorders is not always clear, process depression is characterized by a cyclic appearance with no identifiable external precipitant. One example of this kind of depression might be manic–depressive psychosis. The presumed causes of such process depression are biochemical or genetic, or a combination of the two. Reactive depression, for its part, is prompted by any of several sorts of traumatic life events, such as the loss of a loved one, reversal of fortune or career, or an incapacitating illness or injury. Here, the two forms of depression—process and reactive—may interact, in that an individual's predisposition to depression can mediate his or her reaction to traumatic events.

In separating these two conceptually distinct forms of depression, the possibilities for viable animal models become clearer. Although it may be ultimately possible to breed or otherwise produce animals with biochemical or genetic disorders and thus mimic human process depression, most animal laboratories are easily equipped with facilities to induce trauma into the lives of animals. Therefore, an animal model of reactive depression may be the most plausible for the time being. The catch here is to determine what trauma, and in what form, might induce something like reactive depression in nonhumans.

It happens that there have been hints of an answer to this question over the years. In working with wild rats, Richter (1957) discovered that handling the animals (which included clipping the rat's whiskers) and immersion in a swimming apparatus induced sudden death (but see Hughes & Lynch, 1978). It was Richter's view that such treatment denied the animals an opportunity to either fight or flee, which resulted in a situation of "hopelessness." Apparently, the rats simply gave up. But it remained for other researchers to come up with a reliable way to generate depressionlike states in animals in the laboratory. As we shall see, this procedure is said to induce learned helplessness. Then, once it was known that helplessness could be induced in animals via laboratory techniques, it became a tempting idea to try to induce it in humans in the laboratory. In tracing this progression we adhere to the cause–symptoms–cure criteria of McKinney & Bunney (1969) in seeking parallels in learned helplessness in animal and man.

Learned Helplessness in Animals: Cause, Symptoms, and Cure

Cause. Most of us are under the impression that what we *do* is responsible for what we *get* (see Lefcourt, 1973). That is, we have a sense that our reinforcements are under the control of our behavior, at least to a certain extent. Although this may be true, one wonders what would happen if we convinced an organism, animal or human, that this was simply not so, and that its behavior could not control important events or consequences. Seligman and Maier (1967) attempted to give this impression of helplessness to certain dogs in their laboratory by exposing them to trauma in the form of intense electrical shock. But it has already been noted that trauma per se may not be sufficient to induce helplessness, and this can be fully documented. Accordingly, let us first look at the procedures of the Seligman–Maier study, and then assess and discuss their consequences.

In one of the Seligman–Maier treatments, labeled *escapable* shock, a dog was placed in a harness so that it could not move around freely. The animal was then wired up so that within a 2-hour period it could be given a schedule of 64 painful shocks, each of 30 seconds' duration. This means that the animal was programmed to experience an enormous total of 1,920 seconds of anguishing pain. But in this escapable shock condition the dog could *do* something about the pain. If, during the presentation of any of the 64 shocks, the dog touched a panel located near its head, that shock was terminated immediately. Dogs in this escapable condition coped with the situation rather well. They "escaped" a considerable amount of discomfort they might otherwise have experienced. The actual amount of shock they received was *not* the potential 1,920 seconds, but rather only 226 seconds, on average.

Contrast the foregoing experience with that of animals in a learned helplessness condition labeled the *inescapable* shock treatment. Here, the dogs were prepared in the same way as in the preceding passages, with one major exception. While they were suspended in the harness there was nothing they could do to rid themselves of the pain of the 64 scheduled shocks. This does not mean that they experienced 1,920 seconds of pain, for their pattern of stimulation had been yoked (matched) to that of their counterparts in the escapable condition, so they too received only 226 seconds of shock, on average. Still, none of these 226 seconds of stimulation could be influenced by the dogs themselves. Hence, they were made to be literally helpless while in the harness situation.

Finally, there was a third treatment in which dogs were placed in the harness but were not shocked at all. This was a control condition for the effect of shock per se.

Symptoms. Would the psychological condition of the dogs in the escapable shock, inescapable shock, and control treatments differ? To evaluate this possibility, Seligman and Maier shifted the individual dogs to a new situation called a shuttle box. The box consisted of two cubicles with a shoulder-high barrier in between, and a wire grid floor through which shock could be delivered. The point of shocking dogs in this box was to determine if they would jump over the barrier, thus escaping the shock. In fact, a signal preceded the onset of shock so that an animal might avoid pain in the shuttle box situation altogether.

Based on past research, Seligman and Maier knew that normal dogs had little difficulty in mastering an escape or avoidance response in the shuttle situation. Normal dogs would quickly learn to jump over the barrier to safety as soon as the shock, or the cue signaling the shock came on. Seligman and Maier (1967) gave their experimental and control dogs 10 such trials (tries), each having a duration of 50 seconds. Therefore, if a dog learned to jump quickly it could escape or avoid pain, but if it failed to jump it would suffer nearly a minute's worth of pain on each failed trial. Indeed, it was the case that their control (normal) and escapable shock animals did well in learning to cope with the shuttle box problem. For example, the animals that had been able to escape shock in the harness were able to jump the barrier with an average latency of 27 seconds over all 10 trials, and they failed to escape on an average of only 2.63 trials out of 10. If anything, the controls did a little better. On the other hand, the shuttle box behavior of the dogs that had been given *in*escapable shock in the harness was inferior by comparison. Their average latency to jump was 48 seconds (50 seconds was the maximum!) and they failed utterly to jump on an average of 7.25 trials.

Recall that if a dog failed to jump on a given trial it received all 50 seconds of painful stimulation. The behavior of the inescapably shocked animals during these failures is interesting, if a bit pathetic. Initially, these animals were agitated in reaction to the pain, but unlike normals, they did not continue to run around until they stumbled across the barrier to safety. Rather, they stopped running and sat or lay down, quietly whining. That is, these particular animals acted as if they were helpless and incapable of escaping, which, of course, they were not. But, psychologically speaking, something had happened during the earlier experience with the uncontrolled shock that made them unlikely to cope with their present difficulty. In other words, they had learned to be helpless.

Cure. One way to explain the learned helplessness effect is to say that the organism has acquired the information that its outcomes or reinforcements are no longer contingent or dependent on its behavior. Presumably, this is what dogs learn when they are given inescapable shock in a harness. If this is true, one way to alleviate learned helplessness might be to give the dogs counterinformation, namely that their reinforcements could be contingent on their behavior.

To try this, Seligman, Maier, and Geer (1968) selected four of the most helpless dogs from Seligman and Maier (1967), and tried to show them that their behavior *could* produce results. All four selected animals had never escaped in any of the 10 trials in the shuttle box testing after experiencing inescapable shock in the harness. Retraining began with the lowering of the barrier, and attempts to coax the dog (''Here, boy'') to cross to the safe side during subsequent shock applications. If coaxing alone failed to persuade the animal to act, the dog was dragged to safety by its leash to give it the impression that escape was possible. As the dogs began to move on their own, the barrier was gradually raised to its original height. On the basis of these procedures, one dog recovered from its helplessness after 20 coaxing trials, whereas the others required from 20 to 50 leash tugs before they began responding on their own. All were thus ''rehabilitated.''

Summary. Animals who had been given the impression that their traumatic experience of electrical shock could not be controlled by their behavior while in the harness subsequently acted in a helpless fashion when confronted anew with the problem of dealing with shock. Given that they were literally helpless in the first place, they acted as if they had learned to be helpless by the time they reached the second situation. However, further experiences that indicated to them that their behavior *could* influence their outcomes seems to have reversed this development, and their learned helplessness was reduced.

Learned Helplessness in Humans: Cause, Symptoms, and Cure

Following the publication of the Seligman–Maier paper in 1967, there appeared a number of reports of something like experimentally induced learned helplessness in humans. For example, Glass and Singer (1972) reported task performance deficits following the application of unpredictable noise to people, and improved performance when subjects were merely given the impression that they could control noxious stimulation if they wished to. But it remained for Hiroto (1974) to try to replicate directly the Seligman–Maier dog results in humans. And not only did Hiroto try to duplicate the original results, he also sought to employ the general basic procedure of the original study!

Cause. I should hasten to relate that Hiroto (1974) did not strap people in a harness and jolt them with electrical shock to induce learned helplessness. Rather, he had college students wear a headset via which he could present a series of 30 very loud (100 db) 5-second-long tones. Some subjects were told that there was something they could do to terminate a tone after it came on. They were not told exactly what they could do, but as the only apparatus available to the subject was a button mounted on the table in front of them, most people figured out that if they pushed the button the noise would stop. Thus, these people were in an *escapable* noise treatment. A second group of subjects heard the loud tones and were also told that there was something they could do to turn them off. Here, however, the experimenter was lying, for pushing the button did not terminate the noise for this second group. Therefore, these subjects were in an *inescapable* noise condition. Finally, still other subjects (controls) were not exposed to this sort of auditory stimulation.

Symptoms. Like the canine subjects in Seligman and Maier (1967), the human subjects in Hiroto (1974) were then confronted with a second opportunity to cope with aversive stimulation, this time in a shuttle-response situation. Rather than leaping over a barrier, the humans were required to shift a lever (like a gearshift) from one side to another to control a second series of loud noises. As in the first situation, subjects were again told that there was ''something'' they could do to terminate each individual blast, even if they were not told precisely what that something was. But this time the lever shift response was truly effective for *all* subjects. In fact, a warning light preceded each tone by 5 seconds, and any of the subjects, if they learned to cope at all, could respond to the light with a lever shift, thus avoiding the noxious noise altogether. The question is, would subjects in all three conditions use the lever with equal effectiveness?

The answer is that people in the control, and in the escapable noise conditions made excellent use of the lever. By the end of 18 or so trials most of them were avoiding the noise completely by moving the lever in reaction to the warning signal of the light. The inescapable noise subjects, on the contrary, were relatively awful at escaping the noise. They were slower to use the lever and used it with less consistency than did the other subjects. Whereas the escapable-noise and control subjects failed to escape the noise on only 12% of the test trials, people in the prior inescapable treatment failed on fully 50% of their trials. As was the case with dogs, inescapable aversive events in one place rendered these people relatively helpless in another.

Cure. Of course, there is little to be gained from tugging people around on leashes, but, like the treatment for dogs' helplessness, there might be something one can do to change the state of relative helplessness induced in humans. For the dogs, Seligman et al. (1968) did something to restore their impression that their responses could lead to successful outcomes. A procedure very much along these lines was used to ''rehabilitate'' experimentally helpless humans.

To evaluate such a procedure, Klein and Seligman (1976) induced helplessness by following Hiroto's (1974) method and assessed helplessness using a shuttle arrangement also like Hiroto's. Of interest to us is the shuttle response latency of a group of control subjects who were never exposed to inescapable noise, and two separate groups of people given inescapable noise in a prior session. The thing that distinguished the two inescapable treatments is that subjects in one of the groups received a kind of rehabilitation session between the two tasks. These subjects were given a series of a dozen puzzles to solve, and they did quite well by solving nearly all of them. To put it another way, this group was given interpolated information that their behavior could indeed lead to success. In fact, when the latency to escape noise in the shuttle situation was measured, the rehabilitated subjects performed like normal controls. On early trials these two groups took about 3 seconds to escape a tone, and by the end of a series of some 20 trials this latency had improved to about 1 second. In contrast, the helpless, unrehabilitated subjects needed over 4 seconds to escape early in testing, and showed almost no improvement as testing progressed. Therefore, the helplessness effects of inescapable noxious stimulation can be reversed in humans as well as in dogs by using similar procedures. When a person is given information that his or her behavior can lead to success, helplessness effects dissipate.

Learned Helplessness: Summary and Conclusion

There has been controversy whether learned helplessness and clinical depression really have that much in common. But whatever the outcome of that debate we are still in a position to say that, at the very least, we have an excellent animal model of learned helplessness in humans. The three criteria of McKinney and Bunney (1969)—similarity in cause, symptoms, and cure—all seem to have been met when we examine the literature on dogs and college students.

Alas, despite their valuable contributions in the past, the place of dogs in future learned helplessness research is doubtful. The original theoretical formulation for helplessness effects was that a lack of contingency between response and outcome led to a reduction in the incentive for operant responding (Miller et al., 1977). By now, this relatively simple hypothesis has been supplanted by a more complicated formulation that incorporates ideas of (for example) universal versus personal helplessness, general or specific helplessness, whether future helplessness will be chronic or acute, and whether it will lower self-esteem or not (Abramson, Seligman, & Teasdale, 1978). Many features of the newer formulation rest on cognitive and attributional processes that are not usually associated with dog behavior. Indeed, Abramson et al. (1978) state that: "investigators of human helplessness . . . have become increasingly disenchanted with the adequacy of theoretical constructs originating in animal helplessness for understanding helplessness in humans. And so have we [p. 50]." Well, fine. But if Abramson et al. (1978) ever again need a good animal model of learned helplessness in humans, at least they know where they can find one.

CLOSING THOUGHTS

The late humorist P. G. Wodehouse sometimes referred to nonhuman animals as our "dumb chums" (Wodehouse, 1975, p. 47). Although engaging, this label is not altogether fitting, for it seems unconsciously to malign animals for either meager intellectual capacity, or an inability to speak, or both. On the contrary, the four selected case histories outlined here indicate that information provided by research on animal behavior can be valuable, one way or another, for promoting advances in our understanding or modification of human behavior.

The work on infantile attachment can be viewed as a search for behavioral analogies across species, and the comparisons of dominance patterns approach the criteria for the discovery of homologies in behavior (see Rajecki & Flanery, 1981). On the other hand, the areas of learned aversions and learned helplessness are implicit or explicit attempts to establish animal models of human behavior. The successes in these areas at least support a claim that it is worthwhile to seek homologies, analogies, or models, depending on the aims of the researcher. It is doubtless correct to say that organisms living in a common environment can only behave in a finite number of ways, and that the behavior of any of them has some potential for informing us about the behavior of others. We humans would be at something of a loss in building a science of psychology if we did not have access to the behavior of our dumb chums.

Still, people who are skeptical about the value of animal research vis-à-vis human behavior may wish to emphasize some of the shortcomings of various animal models, such as the one pointed out in relation to learned helplessness (p. 102). Notwithstanding, it could be argued that an animal model (or analogy, or homology), even if later proven inappropriate, can direct a variety of researchers toward a common problem that results in an earlier resolution of the problem compared with a situation in which every researcher works in a different framework. Related to this line of thought is the question of inappropriate models and their frequency of appearance. The skeptic who may grant the value of the comparisons in the four cases described here could probably offer four (or more) cases in which models (or analogies, or homologies) did not prove appropriate or valuable. What does the "successful comparative psychologist" say to the person who simply counts *failures* in comparative applications? Only that in the long run it is probably better to have compared and erred, than never to have compared at all. If one knows why a particular comparison is a failure, then one knows something important about the organisms to be compared.

ACKNOWLEDGMENTS

I am especially indebted to Dr. Timothy B. Baker of the University of Wisconsin for his assistance in providing both background and up-to-date information on learned taste aversions. Parts of this chapter were written when the author was on the faculty of the University of Northern Iowa.

REFERENCES

Abramovitch, R. The relation of attention and proximity to rank in preschool children. In M. R. A. Chance & R. P. Larsen (Eds.), *The social structure of attention*. New York: John Wiley & Sons, 1976.

Abramovitch, R., & Strayer, F. Preschool social organization: Agonistic, spacing, and attentional behaviors. In L. Krames, P. Pliner, & T. Alloway (Eds.), *Aggression, dominance, and individual spacing*. New York: Plenum Press, 1978.

Abramson, L. Y., Seligman, M. E. P., & Teasdale, J. D. Learned helplessness in humans: Critique and reformulation. *Journal of Abnormal Psychology,* 1978, *87,* 49–74.

Ainsworth, M. D. S. Object relations, dependency and attachment: A theoretical review of the infant–mother relationship. *Child Development,* 1969, *40,* 969–1025.

Ainsworth, M. D. S., Blehar, M. C., Waters, E., & Wall, S. *Patterns of attachment*. Hillsdale, N.J.: Lawrence Erlbaum Associates, 1978.

Allen, W. A giant step for mankind. *The New Yorker,* 1980, *56*(16), 36–38.

Austin, W. T., & Bates, F. L. Ethological indicators of dominance and territory in a human captive population. *Social Forces,* 1974, *52,* 447–455.

Baker, T. B., & Cannon, D. S. Taste aversion therapy with alcoholics: Techniques and evidence of a conditioned response. *Behavior Research & Therapy,* 1979, *17,* 229–242.

Bateson, P. P. G. The characteristics and context of imprinting. *Biological Reviews,* 1966, *41,* 177–120.

Beach, F. A. The snark was a boojum. *American Psychologist,* 1950, *5,* 115–124.

Bernstein, I. L. Learned taste aversions in children receiving chemotherapy. *Science,* 1978, *200,* 1302–1303.

Bernstein, I. S. Primate social hierarchies. In L. A. Rosenblum (Ed.), *Primate behavior: Developments in field and laboratory research* (Vol. 1). New York: Academic Press, 1970.

Blurton Jones, N. G. An ethological study of some aspects of social behaviour in children in nursery school. In D. Morris (Ed.), *Primate ethology*. Chicago: Aldine, 1967.

Boland, F. J., Mellor, C. S., & Revusky, S. Chemical aversion treatment of alcoholism: Lithium as the aversive agent. *Behavior Research & Therapy,* 1978, *15,* 167–175.

Bowlby, J. The nature of the child's tie to its mother. *International Journal of Psychoanalysis,* 1958, *39,* 350–373.

Bowlby, J. *Attachment and loss, Vol. 1. Attachment*. New York: Basic Books, 1969.

Camras, L. A. Facial expressions used by children in a conflict situation. *Child Development,* 1977, *48,* 1431–1435.

Cannon, D. S., & Baker, T. B. Emetic and electric shock alcohol aversion therapy: Assessment of conditioning. *Journal of Consulting and Clinical Psychology,* 1981, *49,* 20–33.

Chance, M. R. A. Attention structure as the basis of primate rank orders. In M. R. A. Chance & R. R. Larsen (Eds.), *The social structure of attention*. New York: John Wiley & Sons, 1976.

Deag, J. M. Aggression and submission in monkey societies. *Animal Behaviour,* 1977, *25,* 465–474.

Deutsch, R. D., Esser, A. H., & Sossin, K. M. Dominance, aggression, and the functional use of space in institutionalized female adolescents. *Aggressive Behavior,* 1978, *4,* 313–329.

Dollard, J., & Miller, N. E. *Personality and psychotherapy*. New York: McGraw–Hill, 1950.

Easterbrooks, M. A., & Lamb, M. E. The relation between quality of infant–mother attachment and infant competence in initial encounters with peers. *Child Development,* 1979, *50,* 380–387.

Elkins, R. L. Aversion therapy for alcoholism: Chemical, electrical, or verbal imaginary? *International Journal of the Addictions,* 1975, *10,* 157–209.

Fedigan, L. M. A study of roles in the Arashiyama West troop of Japanese monkeys (*Macaca fuscata*). In F. S. Szalay (Ed.), *Contributions to primatology* (Vol. 9). Basel: S. Karger, 1976.

Feshbach, S. Aggression. In P. H. Mussen (Ed.), *Carmichael's manual of child psychology*. New York: John Wiley & Sons, 1970.

Franks, C. M. Conditioning and conditioned aversion therapies in the treatment of the alcoholic. *International Journal of the Addictions,* 1966, *1,* 61–98.

Freud, A. The psychoanalytic study of infantile feeding disturbances. *Psychoanalytic Study of the Child,* 1946, *2,* 119–132.

Garcia, J., Clarke, J. C., & Hankins, W. G. Natural responses to scheduled rewards. In P. P. G. Bateson & P. H. Klopfer (Eds.), *Perspectives in ethology*. New York: Plenum Press, 1973.
Garcia, J., Ervin, F. R., & Koelling, R. A. Learning with prolonged delay of reinforcement. *Psychonomic Science*, 1966, *5*, 121–122.
Garcia, J., Hankins, W. G., & Rusiniak, K. W. Behavioral regulation of the milieu interne in man and rat. *Science*, 1974, *185*, 824–831.
Garcia, J., & Koelling, R. A. A relation of cue to consequence in avoidance learning. *Psychonomic Science*, 1966, *4*, 123–124.
Gartlan, J. S. Structure and function in primate society. *Folia Primatologica*, 1968, *8*, 89–120.
Gellert, E. The effect of changes in group composition on the dominant behaviour of young children. *British Journal of Social and Clinical Psychology*, 1962, *1*, 168–181.
Gewirtz, J. L. Attachment, dependence, and a distinction in terms of stimulus control. In J. L. Gewirtz (Ed.), *Attachment and dependency*. Washington, D.C.: Winston, 1972.
Glass, D. C., & Singer, J. E. *Urban stress*. New York: Academic Press, 1972.
Hamburg, D. A. Crowding, stranger contact, and aggressive behaviour. In L. Levi (Ed.), *Society, stress and disease*. London: Oxford University Press, 1971.
Harlow, H. F. The nature of love. *American Psychologist*, 1958, *13*, 673–685.
Harlow, H. F., & Suomi, S. J. Induced depression in monkeys. *Behavioral Biology*, 1974, *12*, 273–296.
Harlow, H. F., & Zimmerman, R. R. Affectional responses in the young monkey. *Science*, 1959, *130*, 421–432.
Hartup, W. W. Peer interaction and social organization. In P. H. Mussen (Ed.), *Charmichael's manual of child psychology*. New York: John Wiley & Sons, 1970.
Hiroto, D. S. Locus of control and learned helplessness. *Journal of Experimental Psychology*, 1974, *102*, 187–193.
Hodos, W., & Campbell, C. B. G. *Scala Naturae:* Why there is no theory in comparative psychology. *Psychological Review*, 1969, *76*, 337–350.
Hoffman, H. S., & Ratner, A. M. A reinforcement model of imprinting: Implications for socialization in monkeys and men. *Psychological Review*, 1973, *80*, 527–544.
Hughes, C. W., & Lynch, J. J. A reconsideration of psychological precursors of sudden death in infrahuman animals. *American Psychologist*, 1978, *33*, 419–429.
Kaufmann, J. H. Social relations of adult males in a free-ranging band of rhesus monkeys. In S. A. Altmann (Ed.), *Social communication among primates*. Chicago: University of Chicago Press, 1967.
Keverne, E. G., Leonard, R. B., Scruton, D. M., & Young, S. K. Visual monitoring in social groups of Talapoin monkeys (*Miopithecus talapoin*). *Animal Behaviour*, 1978, *26*, 933–944.
Klein, D. C., & Seligman, M. E. P. Reversal of performance deficits and perceptual deficits in learned helplessness and depression. *Journal of Abnormal Psychology*, 1976, *85*, 11–26.
Kolata, G. B. Primate behavior: Sex and the dominant male. *Science*, 1976, *191*, 55–56.
Lamon, S. Wilson, G. T., and Leaf, R. C. Human classical aversion conditioning: Nausea versus electric shock in the reduction of target beverage consumption. *Behavior Research & Therapy*, 1977, *15*, 313–320.
Lefcourt, H. M. The function of the illusions of control and freedom. *American Psychologist*, 1973, *28*, 417–425.
Lieberman, A. F. Preschoolers' competence with a peer: Relations with attachment and peer experience. *Child Devlopment*, 1977, *48*, 1277–1287.
Lockard, R. B. Reflections on the fall of comparative psychology: Is there a message for us all? *American Psychologist*, 1971, *26*, 168–179.
Maslow, A. H. The role of dominance in the social and sexual behavior of infra-human primates. I. Observations at Vilas Park Zoo. *Journal of Genetic Psychology*, 1936, *48*, 261–277.

Maslow, A. H. Dominance-feeling, behavior, and status. *Psychological Review*, 1937, *44*, 404–429.

McGrew, W. C. *An ethological study of children's behavior*. New York: Academic Press, 1972.

McKinney, W. T., & Bunney, W. E. Animal models of depression: Review of evidence and implications for research. *Archives of General Psychiatry*, 1969, *21*, 240–248.

Miller, W. R., Rosellini, R. A., & Seligman, M. E. P. Learned helplessness and depression. In J. D. Maser & M. E. P. Seligman (Eds.), *Psychopathology: Experimental models*. San Francisco: W. H. Freeman, 1977.

Montagu, A. The new litany of "innate depravity," or original sin revisited. In M. F. A. Montagu (Ed.), *Man and aggression*. New York: Oxford University Press, 1968.

Rachman, S. Aversion therapy: Chemical or electrical? *Behavior Research & Therapy*, 1965, *2*, 289–299.

Rachman, S., & Teasdale, J. *Aversion therapy: An appraisal*. In C. M. Franks (Ed.), *Behavior therapy: Appraisal and status*. New York: McGraw–Hill, 1969.

Rajecki, D. W. Imprinting in precocial birds: Interpretation, evidence, and evaluation. *Psychological Bulletin*, 1973, *79*, 48–58.

Rajecki, D. W. Ethological elements in social psychology. In C. Hendrick (Ed.), *Perspectives on social psychology*. Hillsdale, N.J.: Lawrence Erlbaum Associates, 1977.

Rajecki, D. W., & Flanery, R. C. Social conflict and dominance in children: A case for a primate homology. In M. E. Lamb & A. Brown (Eds.), *Advances in developmental psychology* (Vol. 1). Hillsdale, N.J.: Lawrence Erlbaum Associates, 1981.

Rajecki, D. W., Lamb, M. E., & Obmascher, P. Toward a general theory of infantile attachment: A comparative review of aspects of the social bond. *Behavioral and Brain Sciences*, 1978, *1*, 417–464.

Richards, S. M. The concept of dominance and methods of assessment. *Animal Behaviour*, 1974, *22*, 914–930.

Richter, C. P. On the phenomenon of sudden death in animals and man. *Psychosomatic Medicine*, 1957, *19*, 191–197.

Rowell, T. E. The concept of social dominance. *Behavioral Biology*, 1974, *11*, 131–154.

Rozin, P. The significance of learning mechanisms in food selection: Some biology, psychology and sociology of science. In L. M. Baker, M. R. Best, & M. Domjan (Eds.), *Learning mechanisms in food selection*. Waco, Tex: Baylor University Press, 1977.

Sanderson, R. E., Campbell, D., & Laverty, S. G. An investigation of a new aversive conditioning treatment for alcoholism. *Quarterly Journal of Studies on Alcohol*, 1963, *24*, 261–275.

Savin-Williams, R. C. An ethological study of dominance formation and maintenance in a group of human adolescents. *Child Development*, 1976, *47*, 972–979.

Savin-Williams, R. C. Dominance in a human adolescent group. *Animal Behaviour*, 1977, *25*, 400–406.

Seligman, M. E. P., & Maier, S. F. Failure to escape traumatic shock. *Journal of Experimental Psychology*, 1967, *74*, 1–9.

Seligman, M. E. P., Maier, S. F., & Geer, J. H. Alleviation of learned helplessness in the dog. *Journal of Abnormal Psychology*, 1968, *73*, 256–262.

Sluckin, W. *Imprinting and early learning*. Chicago: Aldine, 1965.

Smith, P., & Daniel, C. *The chicken book*. Boston: Little, Brown, 1975.

Smith, P. K. Aggression in a preschool group: Effects of varying physical resources. In J. de Wit & W. W. Hartup (Eds.), *Determinants and origins of aggressive behavior*. The Hague: Mouton, 1974.

Smith, P. K., & Green, M. Aggressive behavior in English nurseries and play groups: Sex differences and response of adults. *Child Development*, 1975, *46*, 211–214.

Southwick, C. H., Siddiqi, M. F., Farooqui, M. Y., & Pal, B. C. Effects of artificial feeding on aggressive behaviour of rhesus monkeys in India. *Animal Behaviour*, 1976, *24*, 11–15.

Stott, L. H., & Ball, R. S. Consistency and change in ascendance-submission in the social interaction of children. *Child Development,* 1957, *28,* 259–272.

Strayer, F. F., Chapeskie, T. R., & Strayer, J. The perception of preschool social dominance. *Aggressive Behavior,* 1978, 183–192.

Strayer, F. F., & Strayer, J. An ethological analysis of social agonism and dominance relations among preschool children. *Child Development,* 1976, *47,* 980–989.

Strayer, J., & Strayer, F. F. Social aggression and power relations among preschool children. *Aggressive Behavior,* 1978, *4,* 173–182.

Sundstrom, E., & Altman, I. Field study of territorial behavior and dominance. *Journal of Personality and Social Psychology,* 1974, *30,* 115–124.

Tinbergen, N. *The study of instinct.* London: Oxford University Press, 1951.

van Lawick-Goodall, J. The behaviour of free-living chimpanzees in the Gombe Stream Reserve. *Animal Behaviour Monographs,* 1968, *1,* 161–311.

van Lawick-Goodall, J. A preliminary report on expressive movements and communication in the Gombe Stream chimpanzees. In P. Dolhinow (Ed.), *Primate patterns.* New York: Holt, Rinehart, & Winston, 1972.

Vaughn, E. D. Misconceptions about psychology among introductory psychology students. *Teaching of Psychology,* 1977, *4,* 138–141.

Vogler, R. E., Lunde, S., Johnson, G., & Martin, P. Electrical aversion conditioning with chronic alcoholics. *Journal of Consulting and Clinical Psychology,* 1970, *34,* 302–307.

Wickler, W. Socio-sexual signals and their intra-specific imitation among primates. In D. Morris (Ed.), *Primate ethology.* Chicago: Aldine, 1967.

Williams, T. A., Friedman, R. J., & Secunda, S. K. *The depressive illnesses.* Washington, D.C.: National Institute of Mental Health, 1970.

Wodehouse, P. G. *Full moon.* New York: Ballantine Books, 1975.

Wrangham, R. W. Artificial feeding of chimpanzees and baboons in their natural habitat. *Animal Behaviour,* 1974, *22,* 83–93.

Zivin, G. Facial gestures predict preschoolers' encounter outcomes. *Social Science Information,* 1977, *16,* 715–730.

Reader's guide to Section 2
THEORETICAL CONCERNS: OPPORTUNITIES AND CONSTRAINTS IN A COMPARATIVE APPROACH

In Chapter 5, Sober announces that "since the work of comparative psychologists indicates that the field is beginning to tentatively emancipate itself from the strictures of behaviorism, it may be well to review the philosophical issues concerning human perception and cognition that have convinced many philosophers that the science of mind may postulate the existence of representations and information processing devices." Insofar as mentalistic concepts are admissible and have special utility in understanding human psychology, what are the possibilities that such concepts might also be admissible and have utility where the study of nonhumans is concerned? The author first endeavors to clear the path of mentalism of barriers raised by behaviorists, namely that mentalistic concepts are teleological, untestable, or unfalsifiable; mentalistic states are immaterial; and that other explanations for cause and effect relationships in behavior are to be preferred. Sober attacks the behaviorist position in turn by arguing that in many cases an appeal to the existence of beliefs or representational structures (mental states) rather than a focus on the physical properties of stimulus situations makes possible an understanding of why physically different conditions can produce the same behavior, while physically similar conditions can result in quite different behavior. What is critical is not so much the relation of physical stimuli to one another, but the relation of the stimuli to the perceiver. Finally, by way of showing how an analysis of nonhuman psychology can be compatible with mentalistic concepts, the author broaches the important point of how we are to learn of these states. "Our grounds for ascribing mental states to nonhuman animals are pretty much the same as our grounds for ascribing mental states to each other: We postulate the existence of such states to account for and predict what others will do. Since the postulated stories work so well, this offers us some reason . . . for thinking that they are true." Sober then concludes with a philosophical discussion of the terms "belief," "concept," and "consciousness," suggesting avenues for their use in comparative psychology, along with precautions against their misuse.

Demarest, in Chapter 6, wishes to "explore how individuals have speculated about what is perhaps the most fundamental question in comparative psychology, 'What is the nature of evolutionary change in behavior?' " He begins by outlining how notions of change, progress, and continuity found in various philosophical and scientific systems of thought are seemingly wed to characteristics of the sociopolitical climate in which they emerge. For example, the author suggests how the disparate theories of evolution produced by Lamarck and Darwin were influenced by the nature of the social revolutions experienced in their respective countries: the relatively violent, rapid French Revolution for Lamarck, and the relatively nonviolent, extended Industrial Revolution for Darwin. Products of different lines of such theories are that changes in the natural world may or may not have been in a certain direction (perhaps progressive, perhaps not), and at a certain tempo (perhaps continuous, perhaps not). The upshot for comparative psychology is that, in terms of attempts to trace possible lines of mental evolution, there is a four-fold catalog of theoretical positions available concerning the direction and tempo of differences in the animal kingdom. Regarding direction of these differences, there may have been *anagenic* (directional) versus *steady state* (nondirectional) mental evolution. As for tempo, there may have been *continuous* (slow, gradual) versus *discontinuous* (rapid, punctuated) mental evolution. The catalog, therefore, includes theories of four types: anagenic/continuous; anagenic/discontinuous; steady state/continuous; and steady state/discontinuous. Demarest identifies certain schools of thought in comparative psychology—for example, classic behaviorism, classic ethology, and other, more recent camps—and shows where each fits into the typology of theoretical stances, thus satisfying his aim to organize the way in which fundamental questions in comparative psychology have been asked.

In Chapter 7, Zivin draws attention to some possibilities concerning the intellectual or cognitive processes researchers employ when actually doing comparative work. In particular she suggests that "when a puzzling or previously unnoticed behavior in one species is found to be related significantly to a better-known behavior in another species, the researcher's understanding of the better-known behavior tends to condition his or her understanding of the behavior of the other species. That is, the researcher's model (or 'notion' or 'concept') of the behavior in the better-known example—subsuming how the behavior works and the factors that influence it—will be the model by which the researcher initially understands the similar behavior in the less-well-studied case." To illustrate these propositions, she points to the emergence of a hybrid model of dominance behavior in children. Earlier in this century psychologists generally conceived of human dominance as the result of a personality trait, of which individuals were hypothesized to have a certain amount. Thus the fact that some people were ascendant while others were submissive was located squarely *in* the individual.

Meanwhile, animal behaviorists were developing broadly-based conceptualizations of dominance that identified the phenomenon not as a property of individuals, but rather of the group's social nature, where one's degree of dominance could only be known in reference to one's place in a hierarchy. Indeed, the well-developed ideas of the social-structure (animal) theory proved to be compelling, and by the late 1960s developmentalists were enthusiastically applying concepts and observational techniques from the study of nonhuman primate behavior to that of social patterns among children. According to the author, "a revolution against nonsocially oriented, nonbiologically informed, nonsystem oriented, prematurely experimental, and motivationally inferential traditions occurred in human studies. Implications of traits, as in older personality psychology, were thus avoided with the new, more behavioral techniques." Hence the researchers' understanding of the better-known behavior (that in the animal literature) tended to condition their understanding of the behavior in the other species. Zivin further suggests why such borrowing will lead to hybrid models rather than outright adoptions, and discusses limits in borrowing models across evolutionary grades.

Suomi and Immelmann (Chapter 8) stress the fundamental importance of "generalization" for the very existence of science. They describe the traditional biological approach to generalization, which is taxonomic and is based on phylogenetic relatedness between species: the categories of homology, convergence, and analogy. They then put forward an additional basis for establishing generalizations between species, namely the specification of the *ultimate* factors that "cause," and the *proximate* factors that "control" the feature in question. "The 'ultimate-proximate' dichotomy arises from the fact that those factors responsible for the appearance or cause of a given characteristic or phenomenon in a species may be quite different from those factors that actually control its appearance in individual members of the species." For example, in the case of habitat selection, ultimate factors—which result in greater reproductive success for organisms living in one sort of environment than another—might be adequate food supply, protection from predators, favorable conditions for reproduction, and others. Proximate factors, on the other hand, are the particular features of the environment (or environment-organism mechanisms) by which a single individual is able to recognize the most suitable habitat. "Ultimate factors are concerned with survival value, proximate factors with adaptations in physiology and behavior." The authors go on to provide a scheme of comparison that yields *degrees* of generalizations based on similarity of appearance, common ancestry, sensitivity to the same ultimate factors, and comparable influence by common proximate factors (while taking into account considerations of level of analysis, and ontogeny). In their view, the greater the similarity over one or more of these four dimensions, the more compelling and valid the generalization. Finally, the authors discuss practical applications of knowledge about cross-species generalities based on their scheme, with a special emphasis on comparisons between human and nonhuman subjects.

5 Mentalism and Behaviorism in Comparative Psychology

Elliott Sober
University of Wisconsin at Madison

INTRODUCTION

A salient lesson of recent work in the philosophy of mind is that there can be no methodological objection to a psychology that is mentalistic. Because the work of a number of comparative psychologists indicates that that field is beginning to emancipate itself tentatively from the strictures of behaviorism, it may be well to review the philosophical issues concerning human perception and cognition that have convinced many philosophers that a science of the mind may postulate the existence of representations and information-processing devices. These philosophical considerations have not arisen in a vacuum, but coincide with the development within human psychology of mentalistic models and theories that appear to be fruitful and suggestive.

Mentalism says that mental states are *inner*. They are the *causes* of behavior and therefore are *not identical with* behavior. Mentalism and behaviorism are contrary doctrines, because behaviorism claims that insofar as there really are such things as beliefs and desires, these items must somehow be reducible to—be "nothing more than"—items of behavior. Besides claiming that mental states cause behavior, mentalism goes on to say how these inner states manage to do so. Mental states have representational properties; cognizing beings have beliefs *about* the world they inhabit. Not only is this a truism of common sense; additionally, mentalism asserts that the science of psychology will accord a fundamental place to the information-bearing and transforming characteristics of the inner states it postulates.

Mentalism is quite compatible with another philosophical *ism*—namely, *materialism*. It is entirely possible, according to mentalism, that for an individual to be in some mental state—say, that of wanting food—is for it to be in some physical state of the central nervous system. This is the so-called "mind/brain identity theory" advanced by Smart (1959) and Place (1962). Equally well, mentalism is compatible with the idea that there is no one physical state that individuals must be in when they are in some particular psychological state. When I wanted food yesterday, I was in some physical state. When you want food today, you'll be in some physical state too. But, according to what philosophers nowadays call "functionalism," we needn't be in the same physical state. So your wanting food today will be identical with your being in some particular physical state, but the general property of wanting food will not be a physical property. According to this view, psychologists abstract from the physical differences that obtain between individuals who are in the same phychological state and try to describe that psychological state in terms of its functional role in connecting stimuli, behavior, and other mental states. Functionalism is currently being explored as an alternative to the materialist thesis of mind/brain identity (Block, 1978; Block & Fodor, 1971; Dennett, 1971; 1978a; Fodor, 1968b, 1975; Lycan, 1981; Putnam, 1967; Sober, 1982).

So we can identify mentalism in terms of what it does and does not say about the *reducibility* of mentalistic constructs, and in terms of the characteristic feature that mental states possess: The mental is not reducible to observable behavior, although mental states may or may not be reducible to physical states of the central nervous system. Additionally, the central feature of mental states is their representational—information-bearing—character. Philosophers use a technical term to allude to the fact that mental states can be *about* things—that they can be representations *of* things in the world. Mental states have *intentionality*. Note that this technical sense is broader than the commonsense idea of an action's being "intentional."

The current proliferation of information-processing models in theories of human perception and cognition makes mentalism seem to be a promising, if not inevitable, research program (Neisser, 1967; Norman, D., Rumelhart, D., Abrahamson, A., Eisenstadt, M., Gentner, D., Kaplan, R., Kareev, Y., Levin, J., Linton, M., Munro, A., Palmer, S., Scragg, G., & Stevens, A., 1975). Yet, if behaviorist criticisms like those offered by Skinner (1964, 1966, 1971, 1974) are correct, these models are fundamentally flawed. In what follows, I examine and try to undermine these systematic objections. Of course, clearing these problems from the path of mentalism does not show that that outlook is correct. In particular, it does nothing to delimit the domain in which mentalistic constructs find their application. This is particularly relevant to determining the role of information-processing models in comparative psychology. Although scientists interested in human perception and cognition are often quite comfortable with ideas of inner representation and inference, mentalism is a far less common orientation in comparative psychology. As Donald Griffin brings out in his interesting book *The Question of Animal Awareness* (1976), the fear of anthropomorphism, and the idea that hypotheses about such inner mental states in the case of nonhuman animals are untestable, continue to exert considerable influence. In order to bring these issues into clearer focus, it is important to say something about the difference between "real" cases of organisms that have mental states and those cases in which it is just a fiction—possibly a useful one, and possibly an unhelpful one—to talk this way. The data offered by Griffin are tantalizing, in that the pattern of investigation in case after case seems to be that the behavioral repertoire of many species is far more intricate than investigators first imagined. But what does this greater complexity imply about the issue of mind? Although no solution is offered, I try here to clarify the questions involved.

Against Teleology

One of the main ideas underlying movements within psychology that are broadly sympathetic with behaviorism involves a view about how progress has been achieved in the rest of science. Here I have in mind the idea that a science progresses by replacing teleological concepts with ones that are untainted by ideas of goals, plans, and purposes. In Aristotelian physics, the motion of physical objects was understood in terms of objects seeking their natural places. Heavy objects in the sublunar sphere have as their goal locating themselves where the center of the Earth now is (Lloyd, 1968). This teleological outlook was replaced by the Newtonian conception in which ideas of seeking and purpose no longer appeared. Newton's laws, of course, had their natural states, in that the laws say what an object would do in the absence of "interfering" forces; the first law of motion asserts that if no forces are applied, an object will remain at rest or in uniform motion. But, just as in the Newtonian conception of gravity, certain traditional, further questions about the causes behind the regularities were not addressed. *Hypothesi non fingo* was in part a slogan announcing that certain questions would not be addressed. Just as the Newtonian theory of gravitation simply does not tackle the question of what makes physical objects obey the inverse-square law, so the laws of motion do not say why constant velocity is the state of motion that occurs when all forces are equal to zero. Conjectures about purpose were not part of the picture. It appears that the progress involved in moving from Aristotle to Newton in part consisted of forswearing "occult qualities" like teleology.

So, in a certain sense, Newtonian physics was a more modest undertaking than the Aristotelian tradition it superseded. Aristotle tried to answer a question about causation that Newtonian theory forswore. It is part of the standard positivist historiography of physics that what Newton chose to ignore was not a "real" question to begin with. Within this philosophical tradition, it is customary to claim that the teleological ideas of Aristotle were not explanatory. It is even suggested that the very question these teleological ideas were intended to answer—the question of why *these* regularities rather than others—is really a pseudoquestion. Within this positivist orientation, the purpose of science is simply to record regularities; to ask for deeper explanation is to risk finding one's self babbling metaphysical nonsense.

This philosophical orientation received further impetus from the Darwinian revolution in biology. Newton had proscribed purposefulness from the realm of physics, but it still survived in the science of biology. Just as Aristotelian teleology was intended to answer real questions about causal mechanism, so the argument from design was supposed to be a respectable scientific argument. The idea that species were specially created by God, which found systematic development in Paley's *Natural Theology* (1819), was developed as an inference to the best explanation. The observed adaptation of organisms to their environments could best be explained, so it was suggested, by the hypothesis that an intelligent designer had created organisms to fit their environments. Darwinism swept this argument aside, replacing the teleological mechanism postulated by Creationists with the nonintelligent activity of natural selection. Although it might appear that adaptations are the handiwork of an intelligent creator, this appearance is unmasked—explained away—by the Darwinian hypothesis. Here again, progress consists in the elimination of teleology.

Psychology has regularly looked to the other sciences for some indication of how to progress, in the hope that the recipes that have worked elsewhere can be applied successfully at home. Behaviorism, as exemplified by the work of B. F. Skinner, has apparently extrapolated from these other sciences and reached the conclusion that progress in psychology will consist in the unmasking of teleology. Human behavior, according to this view, only *appears* to be under the causal control of the beliefs, desires, and intentions of individuals. A teleological conception of human behavior is embedded in common sense, but if we are to move beyond uncritical "folk wisdom," we must do what the other sciences have done. Behaviorists claim that we must abandon the idea that human behavior is explicable in terms of the rational, intelligent deliberation of agents.

The behaviorist's insistence that we need to go beyond commonsense idioms of rational deliberation must have more behind it than this simple extrapolation from physics and biology. Inductive arguments of this kind are very weak. Why suppose that the transformations observed in other sciences should be emulated by psychology? Maybe the difference between the motion of the planets and the growth patterns of plants, on the one hand, and our cognitive and perceptual activity, on the other, is precisely that the latter, unlike the former, really is controlled by purposeful, intelligent agents. The inductive argument, unless supplemented with some further explanation, invites this simple counterclaim. Skinner attempts to address this issue by saying what specifically is wrong with teleology *in psychology*.

His defense of behaviorism has two parts. The negative strategy is to demonstrate the bankruptcy of mentalism. More positively, he has tried to develop a research program whose actual results and future promise give us a substantive alternative to the position he wishes to discredit. Both of these lines of argument have been challenged. The mentalism Skinner attacks has been claimed to be a straw man. And the standards he applies to psychology have been seen as arbitrarily restrictive impediments to scientific theorizing. As for the positive program, critics of behaviorism have claimed that Skinner's program fails to characterize some of the most interesting and important aspects of human psychology, that is, our capacity to generalize and to act creatively. Although these criticisms have been formulated in the context of human psychology, they in fact have a broader application. In what follows, I try to bring out their relevance to the psychology of other species.

Although Skinner's own brand of behaviorism is no longer as influential as it once was, reviewing the errors and limitations of his position has more than historical interest. "Behaviorism" has turned into a rather sloppy term. For some, to be a behaviorist is simply to think that psychological hypotheses ought to be tested by observing behavior. This formulation, it seems clear, turns "behaviorism" into a truism. Often, it is the formulation that one hears defended when behaviorism is under attack. But when the argument subsides, and psychologists return to their theorizing, allegiance to behaviorist ideas sometimes takes on a much more restrictive and nontrivial form. Behaviorist tenets of this more contentful variety place severe constraints on the sorts of questions one can ask and the sorts of hypotheses one can take seriously in the explanation of behavior.

MENTALISM AND HOMUNCULI

My discussion here is very much indebted to the work of Dennett (1978b) and Keat (1972), who have painstakingly assembled and analyzed the diverse remarks that Skinner has made concerning what is wrong with mentalistic explanations. One standard comment that behaviorists make about the attribution of beliefs and desires—what philosophers have called "intentional" states (of which garden variety intentions are just one)—is that they presuppose that some sort of nonphysical stuff—some immaterial soul—is involved. Although this is a correct interpretation of Cartesian dualism, there is no particular reason why beliefs and desires have to be understood as immaterial items. Mind/body identity theorists, as I have mentioned, hold that mental states are identical with states of the central nervous system. For them, the use of mentalistic locutions is perfectly compatible with the rejection of Cartesian dualism; our mentalistic vocabulary does, as behaviorists fear, purport to name something *inner,* but not, as they allege, something immaterial (Place, 1962; Smart, 1959). Skinner's attitude toward this traditional complaint against mentalism has been ambivalent; sometimes he appears to think it is important (Skinner, 1971, p. 11; 1974, p. 31);

at other times, he accords it a peripheral role in his argument (1974, pp. 12 and 191).

Another standard Skinnerian (1974) criticism, from which he also occasionally backs off, is that mentalism is unsatisfactory because it involves attributing states "which can only be inferred [p. 14]." Chomsky (1959/1964) takes this to be the central defect of Skinner's outlook and correctly points out that every science is in the business of postulating the existence of states that cannot be directly observed. Although there can be little doubt that this sort of operationalist bias has been important in Skinner's thought, it is nevertheless true that he sometimes allows that science may make claims about what scientists cannot see (1964, p. 84). At other times, Skinner objects to mentalism on the broad ground that mental states are *internal* ones, and he seems to be quite untroubled by the fact that this indictment condemns physiological investigation just as much as it censures mentalism (1966, p. 16). It is odd that a psychology bent on emulating physics should eschew internal states as the causes of external behavior.

Dennett (1978b) has argued that these various remarks are somewhat peripheral. For Skinner, the main problem with mentalism is that it is *too easy* (Skinner, 1964, p. 80, 1971, p. 160, p. 195). For almost any behavior, says Skinner, one can make up a story according to which the behavior is a consequence of the interaction of beliefs and desires. This story will be untestable and therefore will be empty of any explanatory content. Mentalism, for Skinner, is committed to the self-defeating strategy of postulating homunculi; it is not illuminating to try to explain perception, memory, or intention by positing the existence of a little man in the head who looks at an image, retrieves a memory trace, or forms a resolve.

To assess this line of criticism, we need to disentangle two ideas. First, there is the idea that mentalism involves postulating a little man in the head—a homunculus—and that this postulation is explanatorily empty. Second, and separable, is the idea that the broad theoretical framework of mentalism—that actions are to be explained as the result of beliefs interacting with desires—is unfalsifiable, because if one mentalistic ascription doesn't work, another can always be dreamed up to do the trick.

Let's begin with those little men in the head. What, exactly, is methodologically wrong with the claim that they are responsible for the behavior of the individuals in whom they are allegedly housed? It is customary to compare homunculi in psychology with the joke in Molière's (1964) "The Imaginary Invalid," in which the good doctor "explains" the fact that opium puts people to sleep by saying that it possesses a *virtus dormitiva* (that is, a dormative virtue). Although there is nothing so unfunny as trying to explain why a joke is funny, let's persist with this humorless question.

One striking difference between homunculi and dormative virtues concerns the issue of *truth*. It seems absolutely clear that none of us have little men living in our heads who, like the homunculi at the end of Woody Allen's movie

Everything You Always Wanted To Know About Sex, control behavior. Yet, it seems patently obvious that opium puts you to sleep because it has certain properties that endow it with that capacity. So it is just plain wrong to claim that every homunculus explanation is unfalsifiable. Not only are they sometimes falsifiable; we can have the best reasons for thinking that they are false. If anything is unfalsifiable, it is the claim that opium has the power to put you to sleep; given that it does put people to sleep, it is hard to see how the hypothesis of a dormative virtue might be disconfirmed. But if this is right, what is supposed to be the common flaw of homunculus and dormative virtue explanations?

It seems clear that both are rather short on details. Granted, opium does have the power of putting you to sleep, and its having that power causes you to go to sleep when you ingest it. But what Molière's doctor could not explain was *how* the opium produces sleep. The causal story he tells is so impoverished that it leaves us no better off in our search for explanation. What needs to be done is that we supplement his incredibly trivial (but true) claim with more details; the story he told was not false, just grossly incomplete.

But does this pinpoint the defect of homunculus explanations? If one provided details, would this remedy the difficulty? In the Woody Allen movie, lots of hilarious mechanical details were supplied in the form of pulleys, levers, and computer terminals. Did this make the homunculus explanation any less absurd or methodologically suspect? In addition, it is important to remember that very good scientific theories are often silent on crucial details concerning mechanism. Newtonian physics doesn't tell us why objects obey the laws of motion or how they come to feel the effects of gravitational influence. And Darwinian theory was not, in Darwin's lifetime, equipped with an account of the origins of life. The fact was just taken as a given. In order to develop the idea that *virtus dormitiva* and homunculus explanations are flawed by their radical incompleteness, we would have to provide a more precise account of when incompleteness is a damning flaw. This is harder to do than might first appear.

Another issue muddying the waters is that the pattern exhibited by homunculus explanations seems to be exploited in other sciences, but without conjuring up the same impression of explanatory vacuity. Suppose a physicist explained the motion of the planets around the sun by appealing to the fact that subatomic particles revolve around atomic nuclei. I take it that this explanation would not be accused of emptiness simply because the parts exhibit the same property as the whole. Similarly, suppose a physiologist tried to explain the digestive capacities of an organism by claiming that the organism contained an intestinal parasite that had the very same digestive powers. Although this explanation might be wrong for any number of reasons, I doubt that it would be ruled out in principle as incapable of accounting for the behavior of the host organism. These two explanations, and homunculus explanations as well, might be called *Chinese box* explanations. In them, a capacity or property of an object is explained by describing an object inside of the original object that has the very same capacity or property. It is hard to see why such explanations must always be wrong or empty.

The resolution of this difficulty can be found in distinguishing two sorts of explanatory problems that a science confronts. We need to distinguish *type* from *token*. Types are properties or kinds, whereas tokens are individual objects. The nine planets are nine tokens of the type *planet*. Chinese box explanations may explain why the planets are now revolving around the sun, why this organism is now capable of digesting sugar, and why (yes) this subject is now seeing the piece of paper. A (token) physical object may have some particular property (type) because some of its subparts do too. But even if this were true, we would still be completely in the dark about another level of question. What is the nature of revolving? What is digestion? What is seeing? Chinese box explanations are incapable of answering these questions about types, even though they are, in principle, quite capable of addressing the more specific questions about tokens. For if a system is in fact a Chinese box, then the question about the type is found to subsume both the system and its part. If I can see and digest because some of my constituents can too, then the question of what seeing and digestion are applies to my parts just as much as it does to me. A Chinese box explanation doesn't help answer this kind of question; it merely broadens the class of objects whose shared property requires elucidation.

So we now can understand what is wrong with mentalism, *if* mentalism is committed to postulating the existence of little men in the head who have the very same mental characteristics as are possessed by the individuals whose psychology we set out to explain. But is mentalism committed to this line of argument? Several psychologists and philosophers have argued that there is nothing wrong with homunculi, provided that the postulated homunculi have skills that are more rudimentary than those possessed by the containing organism. In short, the homunculi must be *stupid,* or, at least, stupider than we are (Attneave, 1960; Dennett, 1971; Fodor, 1968a, 1968b). The larger the gap between the intelligence we have and the intelligence that our postulated homunculi have, the more powerful a homunculus explanation will be—the more illumination it will afford of what intelligence, perception, memory, and so on, are all about. It is plausible to expect a mentalistic psychology to reduce gradually the level of intelligence it requires of constituent homunculi, rather than do so all at once. In the end, we hope to be able to explain the nature of mental states by appeal only to homunculi that are purely mechanical and mindless in their operations.

The impetus for this construal of mentalism, of course, comes from the computer conception of mind. A computer might be quite capable of playing a respectable game of chess. As we watch it play, or perhaps try to beat it, we find ourselves attributing desires and beliefs to it. We perhaps tell ourselves that such attributions are just useful fictions; it isn't as if the computer really *wants* to capture my rook, or really *believes* that its bishop is in danger. Rather, it is simply useful for the purposes of predicting its behavior to pretend that it has beliefs and desires. The important point is that attributing intentional states to the computer is perfectly consistent with expecting that such ''beliefs'' and ''desires'' can be further analyzed, perhaps by looking at the program. The program would show how these ''sophisticated,'' ''intelligent'' capacities of the computer are simply the upshot of lots of very stupid homunculi behaving in perfectly mechanical ways.

The kind of mentalism that animates the creation of information-processing models in psychology is of the same form. The attribution of mental states to human beings, or indeed to animals, is quite compatible with such states being further analyzed into simpler constituents. But there is an important difference between what researchers might feel about the upshot of this process of analysis in the human case and what it reveals in the computer case.

We might imagine ourselves examining a very simple machine, one capable, let us say, of adding large columns of figures. Before we learn very much about how it works, we might find ourselves telling a story about this machine according to which the machine *looks* at the first two numbers, *figures out* what the sum is, *remembers* this sum, and then *figures out* what the sum is of this and the third item of the list, and so on. Once we find out the program by which the machine works, we would confidently assert what we might already expect—namely, that the machine has none of the mental states just italicized. This was just a useful fiction. We perhaps feel the same way about the chess-playing computer—that an analysis of its component capacities shows that our initial attributions of beliefs and desires were *false* (though useful).

But when it comes to the human case, researchers in cognitive science often have a different reaction. If we could break down the ''learning program'' by which we form and transform our beliefs and desires into constituents each of which was a mindless mechanical computational device, we would not, I expect, conclude that people really are not intelligent, that they really don't have beliefs and desires at all. We would more likely say that we had discovered the basis of our very real capacities to form beliefs and act intelligently to reach our desired goals.

That is, the kind of mentalism at work in cognitive science seems to be devoted to *explaining* rather than *explaining away* our mental capacities. By showing what underlies our apparently purposeful and intelligent behavior, we show that the appearance is a reality and how this manages to be so. The underlying story that researchers try to reconstruct *supplements* and *elaborates* our initial attributions of intentional states rather than *refuting* them.

Dennett (1978b) quotes a vivid example from Woodridge (1963) of how the mechanical story of why an organism engages in a behavior can *unmask* the behavior, *explaining away* its apparent intelligence:

> When the time comes for egg laying the wasp *Sphex* builds a burrow for the purpose and seeks out a cricket which she stings in such a way as to paralyze but not kill it. She drags the cricket into her burrow, lays her eggs alongside, closes the

> burrow, then flies away, never to return. In due course, the eggs hatch and the wasp grubs feed off the paralyzed cricket, which has not decayed, having been kept in the wasp equivalent of deep freeze. To the human mind, such an elaborately organized and seemingly purposeful routine conveys a convincing flavor of logic and thoughtfulness—until more details are examined. For example, the wasp's routine is to bring the paralyzed cricket to the burrow, leave it on the threshold, go inside to see that all is well, emerge, and then drag the cricket in. If, while the wasp is inside making her preliminary inspection the cricket is moved a few inches away, the wasp, on emerging from the burrow, will bring the cricket back to the threshold, but not inside, and will then repeat the preparatory procedure of entering the burrow to see that everything is all right. If again the cricket is removed a few inches while the wasp is inside, once again the wasp will move the cricket up to the threshold and re-enter the burrow for a final check. The wasp never thinks of pulling the cricket straight in. On one occasion, the procedure was repeated forty times, always with the same results [p. 65].

Here, the discovery that the wasp's behavior is so rigid and mechanical makes us think that our initial attributions of intelligence, and of beliefs and desires as well, were probably just manners of speaking. Dennett explains how behaviorism has sought to achieve a similar discrediting of mentalism:

> In "Behaviorism at Fifty," he [Skinner] gives an example almost as graphic as our wasp. Students watch a pigeon being conditioned to turn a clockwise circle, and Skinner asks them to describe what they have observed. They all talk of the pigeon *expecting, hoping* for food, *feeling* this, *observing* that, and Skinner points out with glee that they have observed nothing of the kind; he has a simpler, more mechanical explanation of what has happened, and it *falsifies* the students' unfounded *inferences* [p. 65].

Dennett argues that it isn't the mechanical character of Skinner's preferred explanation, but its *simplicity,* that justifies one in concluding that attributing beliefs and desires to the pigeon is misguided. If instead of the simple story of how the reinforcement proceeded, we were handed some incredibly intricate story of why the pigeon turns in a circle, we might be more inclined to stick with our initial claim that, indeed, the pigeon was intelligent and did have mental states after all.

There is a major puzzle here, which we should not assume to be solved by the word "complexity." It is quite possible to tell a very complicated story indeed about the pigeon and about any rudimentary adding machine. Just start talking about the interaction of elementary particles. I noted previously that those who are not inclined to reject mentalistic claims as *always* mistaken want to pick and choose; some objects really do have mental states, whereas others do not, even though it may be convenient occasionally to treat them as if they do (the chess-playing computer discussed before may be of this latter kind). One of the major tasks that a successful mentalism should be able to discharge is the illumination of the basis of this distinction. ''Complexity'' probably has something to do with it, if we only knew what ''complexity'' means and what kind of complexity is relevant.

Another large issue involved here is that of saying why the rigidity of the pigeon's or the wasp's behavior shows that the organisms are not really acting on their beliefs and desires. In the wasp case, we perhaps want to say that the insect is not acting intelligently; if it had any smarts, it would realize that there is a better way of achieving its ends. But this answer raises the main question about this example: What does the wasp's lack of intelligence have to do with the question of whether it has beliefs and desires? Dennett (1971/1978), Davidson (1975) and other philosophers have argued that the attribution of beliefs and desires makes sense only on the assumption that the being is ''rational,'' where ''rationality'' for them includes the idea of intelligence. I find these arguments unconvincing (for details see Sober, 1978, 1981). Why not, alternatively, say that the wasp is unintelligent precisely because it is incapable of *realizing* that there is a better way to achieve its *desires*? Here one retains the claim of mentality while giving up the claim of intelligence.

Rather than saying that the pigeon is not acting on beliefs and desires because its behavior is so rigid, why not conclude that its behavior is *extremely intelligent?* After all, what would *you* do if you were in some comparable, total environment subject to conditioning? We might imagine some horrendous regimen of rewards and punishments that would make any of us turn in a cicle. Would this show that we really weren't intelligent, or that we really didn't have beliefs and desires? Clearly not; indeed, it might be precisely because of our intelligent analysis of the best way for us to attain our desires, given our beliefs, that we turn in little circles just as the experimenter wishes.

The behavior of the pigeon in the experiment, *by itself,* is inconclusive. We don't know yet whether the behavioral regularity is caused by the organism's mental states. Arguably, the further consideration that is involved here, which supplies the motivation for Skinner's experimental methods in general, is some extrapolation about what would happen ''in the wild.'' If pigeons *always* behave in such regimented, mechanical ways, then perhaps they don't have mental states; they are just like simple adding machines. If, however, our behavior in our natural environment were much more variegated and flexible (which it presumably is), then we might be disposed to view our behavior in the weird laboratory setup as perfectly consistent with the fact that, even there, we were acting intelligently on our own beliefs and desires. In short, the relevant issue appears to be the possibility of extrapolating from the laboratory to the natural environment. Pigeon behavior in the laboratory shows that pigeons are unintelligent only if the same rigidity would be exhibited in less contrived environments. So it needn't always be true that demonstrating stable stimulus–response regularities somehow unmasks—explains away—the attribution of intelligence, beliefs, and desires.

Our discussion so far has focused on the behaviorist's claim that mentalism is committed to an explanatorily empty postulation of homunculi. Certainly the decomposing of sophisticated abilities into more rudimentary ones plays a prominent role in the research program of cognitive psychology, and if one wishes to call these subcomponents homunculi, there can be no harm in doing so. But the fact that these homunculi are stupid—the fact that the explanations constructed are *not* Chinese box explanations—vindicates mentalism from the charge that its postulates are without explanatory power. In addition, we have made a distinction that is of some importance in determining the applicability of mentalistic constructs to other species. There is a difference between *explaining* intelligent, purposeful behavior, and *explaining away* the appearance of such. This difference has to do with the complexity of the underlying mechanisms involved in producing the behavior, but it is not at all clear just how this issue of complexity is to be understood. Complex behavior makes it useful for us to describe an organism mentalistically, but this does not answer the theoretical question of what it is to have mental states.

MENTALISM AND FALSIFIABILITY

I now turn to another theme in Skinner's indictment, one that also argues that mentalism is empty, but does not do so by seeing smart little men in the head as part of mentalism's inevitable conceptual baggage. Here I have in mind the idea that mentalism is unfalsifiable, because for any action that is observed, one can always cook up a story of the required kind. According to this criticism, the broad framework of mentalism is consistent with any possible behavior and so is unfalsifiable and inherently unscientific.

Philosophers nowadays are not as confident as they used to be that the idea of falsifiability is well understood, or that it can play the role that Popper (1959) and others had assigned it—namely, that of providing an answer to the demarcation problem of separating science from nonscience. In science, however, the use of the charge of unfalsifiability continues to enjoy a vigorous life within debates relating to behavior. Hinde (1974) argues that the school of ethology most associated with the work of Lorenz and Tinbergen was too quick to postulate instincts, and that such free and easy claims about instincts are unfalsifiable and empty of explanatory power. A similar charge has been made by Lewontin (1979) and others against sociobiology, charging that it is a caricature of Darwinism in which any trait can, one way or another, be made to fit into the overarching theory. In what follows I try to illuminate some of the delicate questions that surround the concept of falsifiability, focusing mainly on Skinner's charge against mentalism, but with an eye to extracting more general conclusions.

To begin with a rather *ad hominem* remark, it is well to note that the same charge that Skinner levels against mentalism can be made against behaviorism.

Consider an action I perform that appears to be novel and unprecedented (the example is Dennett's): I am held up at the point of a gun and, hearing "Your money or your life," I hand over my wallet. Now I have never been threatened with a gun or with any other weapon before. Perhaps the closest I have come to a stimulus situation resembling this in its physical details is that when I was a kid, I played cops and robbers. I do not remember what I habitually did then, but maybe I just giggled and ran away. Certainly this is not what I do in this all too real encounter. How are we to explain my behavior?

The data a behaviorist regards as important in this task are not accessible; we have no record of my history of reinforcements. Yet, the behaviorist will feel confident that some (unknown) history of conditioning would provide the needed explanation within the constraints of behaviorism. The behaviorist may even hazard a guess as to what my history of conditioning might have been like. But one might then ask whether there is any possible behavior that would be intractable from the behaviorist's point of view. Isn't every possible behavior in principle explicable by the schedule of reinforcements that preceded it? As Dennett points out, the behaviorist's invoking some sort of behavioral disposition to hand-over-the-money-if-threatened is very much like a *virtus dormitiva;* it is a something-I-know-not-what that must be there, although the details are largely unknown.

Although Dennett intends his criticism of Skinner to be just as devastating of behaviorism as Skinner's criticism was meant to be of mentalism, it is important to see that both of them are a bit off the mark. The charge of unfalsifiability is being bandied about too cavalierly. In a certain way, every general theoretical orientation, especially one in a field that is still developing, will contain enough flexibility so that the failure of one particular explanation can be interpreted as simply calling for another explanation also consistent with the general perspective. Although specific claims about the particular beliefs and desire that caused a person to act can be disconfirmed, it is much less easy to disconfirm the more general thesis that the person acted on the basis of some beliefs and desires or other. This is hardly unique to mentalism, though. A mind/body identity theorist—someone who holds that each mental state or event is identical with some physical state or event—need not be troubled by the discovery that pain is not invariably correlated with the firing of C fibers. Although this finding refutes one identification of pain with a physical state of the nervous system, the possibility still remains that some other physical trait is to be identified with the mental property of being in pain. Claims like those embodied in behaviorism, mentalism, and the mind/brain identity thesis are so general that it is difficult, if not impossible, to refute them by a single observation.

Often, it is the way people use theories that renders their beliefs unfalsifiable, and rarely if ever the theories themselves that are intrinsically unfalsifiable. Here I depart from the guiding assumption of Popper (1959), who thought that the logical properties of a theory determine whether or not it is falsifiable. My point can be illustrated by taking an example of the sort discussed in the sociobiology debate. Suppose I claim, as E. O. Wilson (1975, 1978) has done, that homosexuality is adaptive, the postulated genetic predisposition for it having been selected for in ancestral populations by a process of kin selection. Now when various arguments are presented to challenge this adaptationist account, I merely retreat to another conjecture about the trait's adaptive significance. Regardless of the evidence provided, I am determined to find some adaptive significance in this, and every, trait. Now my point is that this kind of behavior shows something about me, and nothing in particular about Darwinism, or about the theory of natural selection. Darwinian theory may or may not be falsifiable. But the kind of attitude I have taken about the adaptive significance of homosexuality—one in which I will give up other beliefs in order to preserve my allegiance to this favored one—I could well have taken with respect to perfectly good theories in any science. One does not show that relativity theory is unfalsifiable by finding people who are unwilling to allow any possible evidence to count against *their* belief in the theory. What the existence of such intellectual attitudes shows is something about the people involved and nothing special about the theories they claim allegiance to.

My final point about falsifiability concerns an observation due to Pierre Duhem (1954), which Quine (1960) has generalized and articulated. Typically, a theory will have observational consequences only when taken in conjunction with other theoretical assumptions. What this means is that, in the short run, the failure of an observational prediction leaves logically open the question of which of the starting assumptions was to blame. Some one or more of these must be false, but it is a matter for further determination to find out which. I do not want to give this fact the reading that Quine has given it, according to which there is no such thing as a prediction's having particular bearing on the confirmation or disconfirmation of some one among the many starting assumptions that generate it (Glymour, 1980). But what Duhem's point does mean is that theories can sometimes plausibly be preserved from refutation by revising one's background assumptions. This is precisely what led Leverier to his discovery of Neptune. The Newtonian prediction of the position of Uranus was off and, instead of giving up Newtonian theory, which was immensely well confirmed, Leverier decided that the initial conditions had been misunderstood. Leverier conjectured that some as yet unobserved planet was present that would account for the failure of the prediction. It turned out, of course, that Leverier was right, in that Neptune was the culprit. When Leverier pursued the same line of reasoning to explain the fact that Mercury's perehelion differed from its predicted value, he was wrong. Rather than there being an interMercurial planet (Vulcan) as Leverier thought, the problem this time was with Newtonian theory itself. The precession could not be accounted for until relativity theory was developed.

One can imagine someone jeering at Leverier's reaction to the failure of Newtonian theory to correctly predict the position of Uranus. One can imagine the claim being made that Newtonian theory is unfalsifiable, as any observation can be rendered consistent with the theory by suitable juggling of other beliefs. In a sense, the observation that juggling is always possible is correct, although the conclusion that something is wrong with Newton's theory or with Leverier's scientific standards fails to follow. It is a delicate matter of judgment to say when an apparent refutation of a theory is to be taken seriously; it is a matter of judgment to know what hypotheses to blame when a prediction fails. The single result of a failed prediction does not tell one what to do; only against the background of more systematic considerations can a reasonable decision be made. Although it is always *possible* to hold on to a beloved hypothesis in the light of recalcitrant experience, it is not always *plausible* to do so. The reader will perhaps already have noticed that the appeals to "plausibility" and to "judgment" in this paragraph are mere placeholders for ignorance. In my opinion, philosophers of science do not understand very much about how such plausibility considerations are to be formulated and judged.

The upshot of this discussion is that I find nothing very compelling in the behaviorist's claim that mentalism is "unfalsifiable." In commonsense explanation, we have a multitude of ways of confirming and disconfirming hypotheses about what other people believe and desire. In cognitive psychology, hypotheses about internal representations and processes can be tested, although not with the absolutely knockdown character that one might (mistakenly) think a proper science should aspire to. So, although commonsense psychology is not the same thing as a science of the mental, I see no reason for thinking that commonsense concepts of belief and desire cannot be tuned and refined so as to yield a science of the mind. This appears to have already begun, and the fruitfulness of the results is perhaps more of a testimony to the legitimacy of the methods than any abstract philosophical argument could contribute.

HOW BEHAVIORISTS MUST GENERALIZE OVER STIMULUS CONDITIONS

In the previous section, I mentioned the behaviorist's postulating a disposition to hand-over-my-wallet-when-threatened in order to account for my behavior in the face of a robber. Now it is time to be a bit fussier about exactly what the disposition is that I exhibited on that occasion, and whether it can be analyzed in conformity with the requirements of behaviorism.

One might take the line that I exhibited no such disposition on that unhappy occasion. It would be more accurate to say that I had a disposition to hand over what I believe is my money, when I believe that I have been threatened. Or perhaps, even more circumspectly, we ought to say that I had the disposition to do what I thought was handing over what I believed to be my money, when I believed that I had been threatened. It is this highly mentalistic disposition that accounts for the fact that in a wide variety of physically different situations, I will "behave in the same way." Whether I see a man, a woman, or just a piece of paper, I may give up my cash. If the acoustical wave packets projected into the air by my assailant had not been interpreted as English, I would simply have exhibited "confusion-indicating behavior." The point I am making is that one can describe what I did in behavioristic-sounding language, but that should not obscure the fact that the kinds of states involved must crucially make reference to mental items, preeminently to beliefs and desires, or their scientifically more sophisticated replacements (information storage and motivational structure, perhaps). So a psychologist who studies such money-giving-when-threatened behavior will probably turn out to be not very behavioristic, once one looks carefully at what is going on. I want to emphasize this point by looking, not at the kinds of dispositions that the behaviorist ascribes, but at the way the behaviorist describes stimulus situations. This point deserves emphasis because it seems to be an occupational hazard of behaviorists that one veers away from behaviorist tenets without really realizing it.

Let us begin with a truism. There is *something* about the stimulus situation of the robbery that, given my constitution at the time, caused me to hand over the money. If this stimulus situation had been different, I would have acted differently. This truism is not something that a mentalist wishes to deny. The stimulus situation caused the behavior, just as striking the match caused it to light. Causes are not by themselves sufficient for bringing about their effects, but causes plus initial conditions are (we ignore thorny and irrelevant issues concerning quantum mechanical indeterministic causality).

The task is to find out what properties of the stimulus situation were causally relevant in the production of the behavior. We must move beyond saying that there is *something* about the stimulus situation that was causally efficacious, and say what that something is. One experimental way of doing this is to manipulate the stimulus situation and see whether the original behavior follows. This is a familiar technique in the study of animal behavior. One of Tinbergen's three-spined sticklebacks attacked another fish when that fish was placed in the first one's tank (Tinbergen, 1951). The hypothesis was that some property or other of the introduced fish caused the aggressive behavior in the original resident. By using models instead of real fish, Tinbergen discovered that *having a red belly* was the causally efficacious property. To simplify a bit, one can vary many other properties of the fish at will, but this would be irrelevant; the resident fish attacks when and only when the intruder has a red belly.

The essential point about this investigation, which makes it consistent with the constraints of behaviorism, is that the property of the stimulus claimed to be causally efficacious is a physical property. If Tinbergen had, instead, announced that the property of the stimulus situation that caused aggressiveness was that the introduced fish was believed by the original fish to be an intruder, this would hardly count as a behaviorist explanation.

Let us be clear on the structure of this experimental technique and on the requirements that it imposes on what can count as an adequate explanation. First, we try to find out which property of the stimulus is causally efficacious in a given situation by considering what happens in *other* (similar but different) stimulus situations. Second, the other stimulus situations considered are ones in which we hold the internal state of the organism constant. If we had blinded the resident stickleback, it would not have behaved aggressively even if the intruder had had a red belly. And third, the efficacious property must be sufficient for producing the effect in the other stimulus conditions considered.

The third assumption involves a certain simplification. We assume that it is a *single property* (or a single conjunction of properties) that brings about the effect. But suppose sticklebacks are a bit more complicated. Suppose that there are six properties affecting aggressive behavior, and that a stickleback will attack when an intruder exhibits any five of them. If we set out to find a single invariant property that is present when the effect occurs and is absent when the effect is too, we will fail. There is no such property. It is not too difficult to imagine how Tinbergen's experimental paradigm might be modified to take this possibility into account.

Let's imagine that we can carry out the same sort of experiment on me, to find out what property of the stimulus situation caused me, in that circumstance, to hand over my cash. It isn't the color of the gun, we find. It isn't the physical property of the noise emitted by my assailant. Robbers can work silently, and, besides that, there is no physical property of acoustical signals that explains what unites all the various noises that I will interpret as instances of the English sentence "Your money or your life" (Fodor, Bever, & Garrett, 1974). And, of course, the use of that particular sentence isn't essential either; others will do the trick. The physical differences one finds in stimulus situations that all miraculously manage to end in my handing over my wallet are quite staggering. And equally startling is the fact that two stimulus situations can be physically quite similar, and yet produce wholly different effects (suppose the robber says "Your bunny has a knife"). We are a far cry from the stickleback's red belly here.

It is just barely conceivable that careful work by behaviorists will manage to uncover a common thread in this heterogeneous class of stimulus situations of the right kind. What is required is a description of the stimulus situation that is purely physical. It might be possible to uncover this sort of explanation, but whether it is worth searching for it is quite another matter.

The mentalist will begin with the commonsense assumption that it is probably the belief and motivational structure I have at the time that accounts for why I will respond the same way to some stimulus situations, and differently to others. Common sense tells us that I hand over my money because I believe that the robber will shoot if I don't, and I don't want to get shot. But this is only the beginning of the story, because science, unlike common sense, will strive to understand what sorts of structures are involved in belief and desire, and how these produce behavior.

The behaviorist tries to find a common thread in the various stimulus situations by comparing the physical properties of stimulus situations *with each other*. A mentalist, on the other hand, will try to solve this problem by postulating internal states that make the comparison of different stimulus situations more complicated. What unites two stimulus situations and makes them both capable of causing me to hand over my wallet, is the fact that each bears certain sorts of relations to my hypothesized mental states. What is up for comparison is not the intrinsic, nonrelational physical properties of stimulus situations, but the *relations* that stimuli bear to me.

In numerous instances, investigations of human perceptual and cognitive capacities have exhibited this pattern. What properties of a familiar face influence your capacity to recognize it? What are the characteristics of an inscription of the letter ''b'' that make it recognizable as an instance of that letter? What acoustical properties of a noise determine whether or not English speakers will say that it is an instance of the word ''horse''? It certainly is empirically possible that we should have turned out to be like the stickleback is alleged to be. But we didn't. The requisite physical properties that are supposed to underlie all the stimulus conditions simply don't seem to be there.

Progress has been made in the aforementioned areas of investigation by postulating the existence of representational structures (see, for example, Fodor et al., 1974; Norman *et al.,* 1975). Faces, written letters, and spoken words are recognized by formulating representations and operating on them in accordance with specified rules. It is assumed that these rules, and the representational system they are part of, are stored in the brain, in much the same way that a program is physically stored in a computer. Such theorizing is a far cry from Skinner's caricature of the vacuous homunculus. Indeed, one discouraging thing about such theorizing is its vulnerability to empirical refutation. It isn't that such models are unfalsifiable; on the contrary, the present problem is that so many models are known to be empirically inadequate.

One last question needs to be raised about the methodology of Tinbergen's experiment and what it reveals when applied to the example of my being robbed. I mentioned before that Tinbergen varies the stimulus situation, but *keeps the state of the organism constant*. Why is this appropriate? To begin with, one might quite modestly say that Tinbergen was interested only in those causal factors located in the stimulus situation. As a matter of interest, he simply didn't investigate other, internal factors. But if the decision to run the experiment in this way was based simply on one's interest in one question rather than another, one cannot conclude from the experiment that a completely satisfying explanation of the stickleback's aggressiveness was simply that the introduced fish had a red belly.

Let me illustrate the problem here by a fanciful example. Suppose I decided that I would attack any red object that is placed in front of me. A scientist notices my aggressiveness and does an experiment by varying the stimulus situation while holding my internal state constant. The scientist discovers that a red stimulus object will cause aggressiveness, and no aggressiveness occurs if the object is not red. The scientist concludes that my behavior can be explained just by appeal to physical properties of the stimulus object. Is this a victory for behaviorism? The point here is the same one made about Skinner's pigeons. The fact that an organism obeys a certain behavioral regularity (specified in purely physical terms) leaves entirely open the question of what makes it obey the regularity. The stickleback and the pigeon, I suppose, don't conform to the regularity in virtue of their beliefs and desires; but, in the case just described, I do.

We might summarize the two points of this section as follows: Certain characteristic forms of human behavior seem to defy behaviorist analysis. Their complexity has made it impossible for us to detect physical properties of the stimulus situation that account for whether or not the behavior ensues. This sort of complexity opens the way for mentalism as a research strategy. But behavioral complexity seems to be at best only a sufficient condition for proceeding with this methodology. Even quite mechanical behavior—behavior that is the behaviorist's dream of an easy research problem—can be mediated by mental states. The problem this points to is that, although we have a fairly good grasp of the reasons that can motivate mentalism, we have only a tenuous hold on what the precise connection is between behavioral complexity and mental states.

THE CONCEPT OF BELIEF AND THE CONCEPT OF CONCEPT

So far I have characterized some in-principle objections to mentalism; they are "in principle" because they do not challenge the adequacy of some particular mentalistic model, but rather argue in advance that any possible mentalistic construct will be defective. Rejecting such global criticisms of course leaves open the hard question of in what positive direction mentalistic theorizing ought to proceed, and also leaves open the possibility that many such models will be defective in various ways. In this and the following two sections, I discuss some basic problems that arise in the context of mentalism. The questions are not meant to show that mentalism is impossible or incoherent, but simply to mark some of the fundamental issues that need to be sorted out. It would be absurd to demand that central theoretical concepts be made clear *before* major theories about those concepts have been developed.

Let's begin with the concept of belief. In everyday life and in science as well, it is often predictively useful to ascribe beliefs and desires to nonhuman animals.

Although it seems clear that many of the beliefs that human beings can have are not available to dogs and cats, some of them appear to be. With respect to these, our grounds for ascribing mental states to nonhuman animals are pretty much the same as our grounds for ascribing mental states to each other: We postulate the existence of such states to account for and predict what others will do. Because the postulated stories work so well, this offers us some reason (though not a conclusive one) for thinking that they are true. Although it is conceivable that science will show that this view of ourselves and of certain other animals is a mass delusion, there is no very strong reason to think that this is in the offing. What we might more plausibly expect of a comparative psychology is that it illuminate and explain how numerous species are capable of internally representing their environment and acting on the basis of those representations.

Several philosophers (e.g., Davidson, 1975; Stich, 1979) have argued that this commonsense picture, shared to some extent by explanations offered in ethology and other sciences, is defective, at least when it comes to the concept of belief. The conclusions they reach are quite radical; they deny, for example, that nonhuman animals ever have beliefs. Stich goes even further, holding that even people lack this kind of characteristic, because he thinks that the concept of belief ought to be abandoned as incoherent. Davidson, on the other hand, allows that language users such as ourselves do have beliefs, but that the having of beliefs requires a host of other properties that nonhuman animals lack. Although I think that neither of these philosophers has offered good reason for accepting these radical conclusions, the argument they exploit does pose an important challenge to researchers in comparative psychology. The task is to formulate a concept of belief that does the required explanatory work and avoids the difficulties these philosophers have unearthed.

The problem is very simple. Suppose, on the basis of observing your dog Fido, you claim that Fido believes that there is a bone buried under the tree in your yard. We might suppose that this hypothesis, plus numerous others, makes Fido's behavior predictable and intelligible to you. The success of your "theory" is grounds for thinking it true. So mentalism with respect to Fido receives some support. We then ask the question: Exactly what properties does Fido possess by virtue of which he has the belief in question?

Presumably, to have the belief that there is a bone buried in the yard, one must have various concepts. One must have the concept of *bone,* of *yard,* of *tree,* and of *burying*. But what is it to "have" these concepts? To have the concept of *bone,* does one need to know that bones are found inside a variety of living creatures? To have the concept of *yard,* does one need to know that yards are units of property surrounding certain kinds of dwellings? To have the concept of *tree,* does one need to know certain biological facts about trees?

To these kinds of questions, both Davidson and Stich say *yes*. To have a concept, according to them, one must know *a lot*. One must know numerous things about the objects to which the concept applies. But Fido, so they argue, has no such knowledge. Fido doesn't know what a tree is, or what a bone is, or anything of the sort. But because Fido lacks these concepts, we conclude that Fido does not have the belief we offhandedly attributed to him.

Davidson and Stich conclude that Fido doesn't have any beliefs at all. The idea, presumably, is that we could repeat this style of argument for any belief ascription we might be tempted to make. They have tried to offer a recipe for refuting any such suggestion. There is, however, a fairly simple point, and a much more complicated issue, that we should examine before we follow them in concluding that mentalism is a mistake with respect to nonhuman animals.

First, we should note that what this and other similar examples show (at best) is that the beliefs we can correctly ascribe to dogs and cats must be conceptually impoverished in certain ways. The concepts enshrined in the vocabulary of human languages, however, appear to be rich. The words "bone," "yard," and so on seem to have a great deal built into them. Perhaps we ought not to use *them* in describing Fido's mental states. But, instead of giving up on mentalism altogether, perhaps what we need is a set of concepts that are appropriate for the species we are talking about; what we need is to characterize certain concepts that are impoverished in the right way. This is something that comparative psychology might undertake; if ordinary English doesn't furnish the required tools, we may just have to invent them ourselves.

So, instead of ascribing to Fido the belief that a bone is buried in the yard, we might ascribe to him the belief that a*bone*is*buried*in*the*yard*, where the starred terms pick out appropriately circumscribed canine concepts. This strategy would permit us to retain our mentalism, both in common sense and in science. However, we have so far only put a name on the approach involved. We now need to consider the more fundamental question of what it is to have a concept.

To begin with, we should note that most of the rhetorical questions we asked earlier about Fido could well be asked about ourselves. Does your neighbor know enough biology to say what a *tree* is? Can't someone doubt that bones are found within many or all living creatures and still have the concept of *bone?* It is far from clear exactly what knowledge is required for *anyone* to have a concept. Certainly, knowing the whole truth about *X*'s isn't needed for one to have the concept of an *X*. And if it is said that to have the concept of *X*, one must know what an *X* is, then it should be replied that there are many equally correct but unequally informative answers to questions of the form "What is an *X*?" What is a bone? A bone is a thing, an object smaller than a house, a thing that is inside the bodies of certain animals, a thing made of calcium, and much else. The problem we are facing in trying to describe what it is to have a concept seems to issue from the fact that we are trying to distinguish between two kinds of facts. First, there are the facts about bones that one needs to know in order to have the concept *bone*. Second, there are the facts about bones that aren't needed to have the concept, but rather are facts we may or may not come to know. Most discussions of this issue tend to assume that this dichotomy is real, although some philosophers (Quine, 1960) have argued that there is no principled distinction here. What seems clear, however, is that the whole subject of what it is to have a concept is presently a morass.

One feature of the basic approach we have taken to this issue is worthy of notice. For an organism to have a *belief,* it must possess *concepts,* and this, in turn, requires *knowledge*. The argument described previously, which criticizes the ascription of mental states to nonhuman animals, makes this assumption. And my discussion of how one might retain mentalism in comparative psychology by defining certain impoverished concepts likewise presupposes this proposition. But there is another, somewhat attractive way of thinking about conceptual competence that may have its uses within comparative psychology. The rough idea is that one can have the concept *X* if one has a state that is differentially sensitive to *X*'s. Here, having a concept involves a relation (perhaps a causal one) between a state of the organism and a state of the world. It needn't involve *knowledge*.

The behavioral evidence for saying that Fido has the concept of *bone* is very roughly this: He has the capacity to recognize bones. Granted, he can be tricked into treating as a bone something that is not a bone (but just looks like it is); and, granted, he can fail to treat something as a bone when it is a bone. Still, he is pretty good at discriminating. Let's imagine that there is a state that Fido goes into when and only when there is a bone present. Such a state would mean that Fido is a perfectly reliable detector of bones. Fido fails to live up to this ideal, but let's ignore that for the moment. If Fido has a state of this kind, and if it plays a certain role in his psychology, we might then conclude that Fido has the concept of *bone*.

Notice that it is perfectly possible that an organism might have a state responsive to *X*'s without having very many beliefs that characterize what an *X* is. A case that is even more plausible than our example of Fido's bone is perhaps certain simple color words. We might imagine an organism that is sensitive to *red* things in the required way and thereby is said to have the concept of redness. However, this being might not believe anything at all like a definition of what it is to be red. Conceptual competence, in this scenario, is a much less heavily cognitive achievement than it was according to the earlier view.

The kind of proposal I have just sketched is developed in considerable detail by Dretske (1981). It has promise for articulating a view of conceptual competence that can be used in comparative psychology. In fact, it seems to capture a great deal of the kinds of evidential considerations that inform the tentative mentalism already at work in the study of animal behavior. In spite of its appeal, I want to note certain problems with this approach that need to be worked out if it is to be acceptable.

First, there is the issue, already noted, that we are not perfectly reliable detectors of anything, and neither are other organisms. Fido does not infallibly respond to bones. But we probably could describe in a quite complicated way a set of properties that Fido does respond to. We might have to use fairly complicated geometrical ideas and language involving theoretical physical properties to circumscribe what Fido will treat as a bone. Yet we don't want to say that Fido has the complicated concept described in this scientific language. We want to say that he has a much more impoverished concept—namely that of a bone (or even something more impoverished than that, perhaps). So the first problem is reconciling the thesis that possessing a concept of *X* is having a state that is differentially sensitive to *X*'s with the fact that no organism is a perfect detector.

The second question involves the fact that an organism can be sensitive to environmental properties without having any concepts at all. I get a rash when and only when I touch poison ivy, but this doesn't show anything about my conceptual abilities. What is required to flesh out the picture is that the state, besides being differentially sensitive, also plays certain kinds of causal roles. The state must be *used* in certain ways, if it is to represent a concept for an organism. Belief requires conceptual competence, but it appears that we are not able to say what conceptual competence involves without also having a theory of how concepts are used in the formation of beliefs.

The third issue involves a somewhat recondite phenomenon that philosophers treat under the rubric of inten*s*ionality (spelled with an *s;* this is different from, though connected with, the idea of inten*t*ionality so far discussed). It is presumably possible for someone to have the concept of *water* without having the concept of H_2O. The latter is acquired by learning a little chemistry, whereas the former is standardly assimilated as part of ordinary common experience. People have had the concept of water for a very long time, but possessing the concept of H_2O is a relatively recent historical phenomenon. The problem is that something is made of water if, and only if, it is made of H_2O. Yet, on the assumption that we have here *two* (different) concepts, a theory of what it is to have a concept must explain how it is possible to acquire one but not the other of this pair. The problem, of course, is that any state that an organism goes into when and only when water is present will also be a state that the organism goes into when and only when H_2O is present. Much more must be said about conceptual competence than the mere assertion that to have the concept *X* is to possess a state that reliably indicates the presence of *X*'s, if we are to be able to draw distinctions like the one involved here between the concept of water and the concept of H_2O.

One natural maneuver is to distinguish simple from complex concepts. The concept of H_2O is complex, in that an organism possesses this concept only if it possesses the concept of hydrogen and the concept of oxygen. But nonhuman animals don't appear to have these concepts; they don't have states that are reliable indicators of the presence of oxygen or of the presence of hydrogen of the appropriate kind (the fact that they pass out when there is no oxygen isn't the right sort of connection; see the preceding discussion). So it follows that dogs and cats don't have the concept of H_2O. One would then go on to argue that the concept of water works a bit differently. But at this point, one must be prepared to say what *is* required to have *this* concept, and, as we have noted, this is a tall bill to fill. It is no good just stipulating that having the concept of water requires a lot less than having the concept of H_2O, One must also give a non-ad hoc explanation of what this difference consists in.

The problem we have been considering arises whenever one wants to attribute beliefs to an organism. It also arises when the term *belief* is not used, but other, connected, ideas are involved. When one attributes perceptual states of certain kinds to organisms, the same problem arises. What does it mean to say that the bird *sees that* a predator is approaching, or *hears that* its offspring are hungry? These locutions indicate states of perceptual knowledge; to see that *P* is to know that *P* is true by visual means (here, *P* is a proposition like "the predator is approaching"). And because knowledge implies belief, one again has to give an account of what is involved in having beliefs. The point of these remarks is not to discourage comparative psychology from trying to make use of mentalistic constructs in the explanation of behavior. Rather, the point is to be alert to certain theoretical problems that work in psychology can potentially illuminate.

THE FORM OF REPRESENTATIONS

The questions raised in the previous section concerned the *content* of representations; assuming that Fido has beliefs, the problem was to say exactly what it is that he believes. An equally central set of issues concerns the form that representations take; if Fido believes that there is a bone in the yard, there must be some particular way in which he represents this belief. Although psychologists of human cognition have been relatively untroubled by the first set of problems, the latter has received considerable attention, and a range of experimental techniques has been developed for testing hypotheses about how information is represented and structured.

You know the names and phone numbers of certain people. How do you represent this information to yourself? Have you encoded an alphabetical list by first name? Almost certainly not. When I ask you for Zelda's number, you take no longer to answer than when I ask you for Aaron's. Although this piece of evidence is by no means foolproof, the use of *reaction time* experiments can shed considerable light on questions about the form of representation. A hypothesis about the form that a representation takes can have behavioral consequences, when it is conjoined with assumptions about the mode of accessing and manipulation. We can get behavioral evidence not just for whether or not you know certain people's phone numbers, but also for how the information (or misinformation) you possess is encoded.

Let me give a more realistic example of how this technique has been deployed in psychological theorizing about human visual perception. Psychologists have devoted considerable attention to refining and exploring the following crude question (see Paivio, 1971, and Pylyshyn, 1981, for discussion): Does the brain encode information in a way similar to the way that sentences encode information, or does the brain form representations that are similar to pictures? This question is sometimes expressed by asking about the role of "propositional" versus "imagistic" modes of representation. Shepard and Meltzer (1971) approached an aspect of this rather vast problem by asking how it is that subjects figure out whether two line drawings like the ones seen in Fig. 5.1 are different views of the same shape. Our introspective inclination might be to say that we solve this problem by rotating mental images, but is this "anecdotal" report to be taken seriously? Shepard and Meltzer discovered that the larger the rotational angle separating the two drawings, the longer it takes subjects to solve the problem. In fact, it takes about twice as long to analyze a pair of figures separated by 100 degrees as it takes for a pair separated by 50 degrees. This reaction time data, Shepard and Meltzer concluded, confirms the hypothesis that subjects represent this problem to themselves pictorially, not propositionally.

This kind of experimental paradigm has been used quite fruitfully in psycholinguistics (Fodor et al., 1974) and in other areas of human psychology. My purpose in describing it is not to claim that it automatically solves the problems associated with identifying the form that representations are given. Indeed, other experimental arrangements are available, and this is just one among many. My point is to highlight one of the tasks that will confront a mentalistic comparative psychology. For any object that encodes information—whether it is a human being, a computer, or a nonhuman animal—there must be an answer to the question: *How* does it do it? Representations must have form as well as content. Studies of human cognition may help comparative psychologists to construct models for investigating this crucial issue.

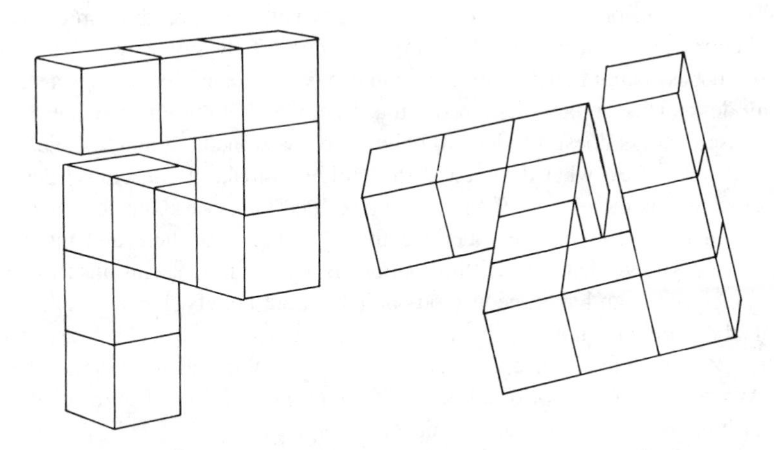

FIG. 5.1. Examples of line drawings used to test hypotheses regarding "propositional" versus "imagistic" models of mental representation (see text).

CONSCIOUSNESS

Those comparative psychologists who feel inclined to allow more scope for mentalistic constructs than has traditionally been allowed very often feel that the main issue to be considered is that of consciousness. The very title of Griffin's (1976) provocative book is significant in this respect; his plea for mentalism gives central importance to the question of animal *awareness*. It is interesting that this issue should be thought of as central by psychologists. Among philosophers who are sympathetic with mentalism, at least within the context of human psychology, consciousness is often regarded as a special issue, one on which the case for mentalism does *not* stand or fall. This perspective is quite consistent with thinking of consciousness as the mystery of mysteries, so far as the study of the mind goes. But there are numerous fundamental issues that can be tackled while putting consciousness to one side.

This philosophical attitude in some ways takes its cue from the information-processing models currently being explored in cognitive psychology (see, e.g., the introduction to Fodor, 1981). It would be a mistake to think that we now really understand human perception, language acquisition, problem solving, and memory. Still, it does seem plausible to think that information-processing models provide the broad framework in which these psychological phenomena may eventually be understood. We at least seem to possess a vocabulary that makes it possible to construct and test models in these domains. However, even though current psychological theorizing may tell us important things about perception, memory, and problem solving, the problem of consciousness remains as ever a formidable and apparently intractable obstacle. The difficulty seems to be that we don't even possess concepts that would allow us to talk about the phenomenon very well.

What is consciousness? Part of the difficulty is that it is hard to circumscribe the phenomenon in a noncircular way. I'll start off by saying what it *is not*. Consciousness is not belief. I am now conscious of the fact that there is a stapler on my desk. This implies, I suppose, that I believe that there is a stapler there. But consciousness doesn't reduce to belief; there is something more. Beliefs can be *un*conscious. So what do we add to a belief to make it conscious? Perhaps consciousness is merely belief about one's beliefs: I am conscious that there is a stapler on the desk because I believe that the stapler is there and moreover believe, or know, that I have this belief. But this, too, seems insufficient,: presumably one can have *un*conscious beliefs about beliefs. Neither belief, nor belief about beliefs, nor any iteration thereof seems to get to the heart of the matter. We give this concept other names—''phenomenal awareness'' being one. We find the metaphor of a light bulb going on appealing. But the character of this transparent mental state is entirely opaque.

Fortunately for cognitive psychology, there is a lot more to mentality than consciousness. As intractable as the problem of awareness appears to be, progress looks possible in numerous other areas. Although mentalism appears to be going full steam ahead on numerous aspects of human psychology, I am not aware of any research that is pursuing the issue of consciousness, properly so called. What this means is, I think, that if comparative psychology is interested in going mentalistic, it ought to turn to the problem of consciousness last, not first.

CONCLUDING REMARKS

The versions of behaviorism and mentalism I have surveyed see each other as *competitors*. Each focuses on the behavioral repertoire of some organism and views the other as having nothing to contribute to its elucidation. I have tried to show how Skinner's objections to mentalism can be defused, and also to point to some of the main concepts that a developing mentalism will have to clarify.

In spite of the defects of the behaviorist critique of mentalism and in spite of the inadequacy of a strict behaviorism in accounting for some behaviors, it is not difficult to see why comparative psychology should be a hospitable home for behaviorism. If one is interested in finding characteristics that organisms in different species can share, it may be well to define these characteristics nonmentalistically. If ''aggression'' and ''infantile attachment'' are to be properties of species that lack mental states as well as of those that have them, then these terms must be defined so as to be neutral concerning mental capacities. It is *neutrality* that is required here—one doesn't want to define the properties so as to *rule out* the possibility that beings who are aggressive are endowed with minds of various sorts. The theoretical terms should be silent on this question. A behaviorism of this variety will not compete with mentalism. Mentalism might provide a characterization of the mechanisms by which this or that species is capable of certain behaviors, whereas a behaviorist theory might generalize over these diverse cases by abstracting away from the mechanisms involved. There is no a priori reason why theorizing at this level of generality should be impossible.

Studies of behavior within evolutionary theory often approximate this behaviorist model. A theory of parental investment, for example, is concerned with the selective advantages of certain behaviors, and not with the particular devices—whether mentalistic or not—that one or another species uses to implement them. Although such models do not presuppose uniformity of mental mechanism, they do seem to presuppose another sort of uniformity with respect to causation. Critics of sociobiology (Allen, E., Beckwith, B., Beckwith, J., Chorover, S., Culver, D., Duncan, M., Gould, S., Hubbard, R., Inouye, H., Leeds, A., Lewontin, R., Madansky, C., Miller, L., Pyeritz, R., Rosenthal, M., & Schreier, H., 1975/1978; Allen, E., Beckwith, B., Beckwith, J., Chorover, S., Culver, D., Daniels, N., Dorfman, E., Duncan, M., Engelman, E., Fitten, R., Fuda, K., Gould, S., Gross, C., Hubbard, R., Hunt, J., Inouye, H., Kotelchuck,

M., Lange, B., Leeds, A., Levins, R., Lewontin, R., Loechler, E., Ludwig, B., Madansky, C., Miller, L., Morales, R., Motheral, S., Muzal, K., Ostrom, N., Pyeritz, R., Reingold, A., Rosenthal, M., Mersky, M., Wilson, M., & Schreier, H. 1976/1978) have argued that sociobiological explanations (like those of Wilson, 1975) lump together behaviors that are functionally quite distinct. One example they discuss is "slavery." If "slavery" in human beings has an entirely different cause and function from "slavery" in the social insects, then there will be no such behavioral category as "slave making." The point is not that human slavery is mediated by mental structures not found in ants. Darwinism is founded on the insight that artificial and natural selection have a common causal structure in spite of their differing at the level of mental causation. Rather, the criticism at issue finds sociobiology guilty of a kind of mistake that potentially confronts any behaviorist explanation in comparative psychology. Cross-species categories of behavior abstract from differences in mental mechanism; if they abstract from causal structure completely, they are left with nothing to talk about. Sociobiology is committed to the assumption that categories of social behavior that apply across a range of species have a common causal structure at the level of the genetic process of natural selection. It is an empirical question as to which behaviors and which species can be treated in this fashion.

As we look to the past, we see behaviorism and mentalism at odds with each other. Behaviorism found mentalism deficient in rigor, and mentalism replied in kind. It will be interesting to see whether the future of comparative psychology recapitulates this opposition. Perhaps a division of explanatory labor is possible. Only time and hard work will tell.

REFERENCES

Allen, E., Beckwith, B., Beckwith, J., Chorover, S., Culver, D., Duncan, M., Gould, S., Hubbard, R., Inouye, H., Leeds, A., Lewontin, R., Madansky, C., Miller, L., Pyeritz, R., Rosenthal, M., & Schreier, H. (And cosigners from The Science for the People Study Group on Sociobiology.) Against "sociobiology." *New York Review of Books*, November 13, 1975, pp. 182–186. Reprinted in A. Caplan (Ed.), *The sociobiology debate*. New York: Harper & Row, 1978.

Allen, E., Beckwith, B., Beckwith, J., Chorover, S., Culver, D., Daniels, N., Dorfman, E., Duncan, M., Engelman, E., Fitten R., Fuda, K., Gould, S., Gross, C., Hubbard, R., Hunt, J., Inouye, H., Kotelchuck, M., Lange, B., Leeds, A., Levins, R., Lewontin, R., Loechler, E., Ludwig, B., Madansky, C., Miller, L., Morales, R., Motheral, S., Muzal, K., Ostrom, N., Pyeritz, R., Reingold, A., Rosenthal, M., Mersky, M., Wilson, M., & Schreier, H. (And cosigners from The Science for the People Study Group on Sociobiology.) Sociobiology—Another biological determinism. *Bioscience*, 1976, *3*, pp. 182–86. Reprinted in A Caplan (Ed.), *The sociobiology debate*. New York: Harper & Row, 1978.

Attneave, F. In defense of homunculi. In W. Rosenblith (Ed.), *Sensory communication*. Cambridge, Mass.: MIT Press, 1960.

Block, N. Troubles with functionalism. In C. Savage (Ed.), *Perception and cognition: Minnesota studies in the philosophy of science* (Vol. 9). Minneapolis: University of Minnesota Press, 1978.

Block, N., & Fodor, J. What psychological states are not. *Philosophical Review*, 1971, *81*, 159–181.

Chomsky, N. A review of B. F. Skinner's *Verbal behavior*. *Language*, 1959, *35*, 26–58. Reprinted in J. Fodor & J. Katz (Eds.), *The structure of language*. Englewood Cliffs, N.J.: Prentice–Hall, 1964.

Davidson, D. The method of truth in metaphysics. In P. French, T. Uehling, & H. Wettstein (Eds.), *Midwest studies in philosophy* (Vol. 2). Minneapolis: University of Minnesota Press, 1975.

Dennett, D. Intentional systems. *Journal of Philosophy*, 1971, *68*, 87–106. Reprinted in D. Dennett (Ed.), *Brainstorms: Philosophical essays on mind and psychology*. Montgomery, Vt.: Bradford Books, 1971.

Dennett, D. Artificial intelligence as philosophy and as psychology. In D. Dennett (Ed.), *Brainstorms: Philosophical essays on mind and psychology*. Montgomery, Vt.: Bradford Books, 1978. (a)

Dennett, D. Skinner skinned. In D. Dennett (Ed.), *Brainstorms: Philosophical essays on mind and psychology*. Montgomery, Vt.: Bradford Books, 1978. (b)

Dretske, F. *Knowledge and the flow of information*. Montgomery, Vt.: Bradford Books, 1981.

Duhem, P. *The aim and structure of physical theory*. New York: Atheneum, 1954.

Fodor, J. The appeal to tacit knowledge in psychological explanation. *Journal of Philosophy*, 1968, *65*, 627–643. (a)

Fodor, J. *Psychological explanation*. New York: Random House, 1968. (b)

Fodor, J. *The language of thought*. New York: Thomas Crowell, 1975.

Fodor, J. *Representations: Philosophical essays on the foundations of cognitive science*. Cambridge, Mass.: MIT Press, 1981.

Fodor, J., Bever, T., & Garrett, M. *The psychology of language*. New York: McGraw–Hill, 1974.

Glymour, C. *Theory and evidence*. Princeton, N.J.: Princeton University Press, 1980.

Griffin, D. *The question of animal awareness*. New York: Rockerfeller University Press, 1976.

Hinde, R. A. *The biological bases of human social behavior*. New York: McGraw–Hill, 1974.

Keat, R. A critical examination of B. F. Skinner's objections to mentalism. *Behaviorism*, 1972, *1*, 17–32.

Lewontin, R. Sociobiology as an adaptionist program. *Behavioral Science*, 1979, *24*, 5–14.

Lloyd, G. E. R. *Aristotle: The growth and structure of his thought*. Cambridge, Mass.: Cambridge University Press, 1968.

Lycan, W. Form, function, and feel. *Journal of Philosophy*, 1981, *78*, 24–50.

Molière, J. *Le Malade Imaginaire*. Paris: Bordas, 1964.

Neisser, U. *Cognitive psychology*. New York: Appleton–Century–Crofts, 1967.

Norman, D., Rumelhart, D., Abrahamson, A., Eisenstadt, M., Gentner, D., Kaplan, R., Kareev, Y., Levin, J., Linton, M., Munro, A., Palmer, S., Scragg, G., & Stevens, A. *Explorations in cognition*. San Francisco: Freeman, 1975.

Paivio, A. *Imagery and verbal processes*. New York: Holt, Rinehart, & Winston, 1971.

Paley, W. *Natural Theology*. London: Rivington, 1819.

Place, U. T. Is consciousness a brain process? In V. C. Chappell (Ed.), *The philosophy of mind*. Englewood Cliffs, N.J.: Prentice–Hall, 1962.

Popper, K. *The logic of scientific discovery*. London: Hutchinson, 1959.

Putnam, H. Psychological predicates. In W. H. Capitan & D. D. Merrill (Eds.), *Art, mind, and religion*. Pittsburg: University of Pittsburg Press, 1967. Reprinted as The nature of mental states. In Putnam, H. (Ed.), *Mind, language, and reality*. Cambridge, Mass.: Cambridge University Press, 1975.

Pylyshyn, Z. The imagery debate: Analog media vs. tacit knowledge. In N. Block (Ed.), *Imagery*. Cambridge, Mass.: MIT Press, 1981.

Quine, W. V. O. *Word and object*. Cambridge, Mass.: MIT Press, 1960.

Shepard, R., & Meltzer, J. Mental rotation of three-dimensional objects. *Science*, 1971, *171*, 701–703.

Skinner, B. F. Behaviorism at fifty. In T. W. Wann (Ed.), *Behaviorism and phenomenology*. Chicago: University of Chicago Press, 1964.

Skinner, B. F. Operant behavior. In W. K. Honig (Ed.), *Operant behavior: Areas of research and application*. New York: Appleton–Century–Crofts, 1966.

Skinner, B. F. *Beyond freedom and dignity*. New York: Knopf, 1971.

Skinner, B. F. *About behaviorism*. New York: Random House, 1974.

Smart, J. J. C. Sensations and brain processes. *Philosophical Review*, 1959, *68*, 141–156.

Sober, E. Psychologism. *Journal for the Theory of Social Behavior*, 1978, *8*, 165–191.

Sober, E. The evolution of rationality. *Synthese*, 1981, *46*, 95–120.

Sober, E. Panglossian functionalism and the philosophy of mind. In P. Machamer (Ed.), *Naturalistic epistemology*. Berkeley: University of California Press, 1983.

Stich, S. Do animals have beliefs? *Australasian Journal of Philosophy, 1979, 57,* 15–34.

Tinbergen, N. *The study of instinct*. Oxford: The Clarendon Press, 1951.

Wilson, E. O. *Sociobiology: The new synthesis*. Cambridge, Mass.: Belknap Press, 1975.

Wilson, E. O. *On human nature*. Cambridge, Mass.: Belknap Press, 1978.

Woodridge, D. *The machinery of the brain*. New York: McGraw–Hill, 1963.

6 The Ideas of Change, Progress, and Continuity in the Comparative Psychology of Learning

Jack Demarest
Monmouth College

When Hodos and Campbell (1969) announced that there was no theory in comparative psychology some dozen years ago I was a bit puzzled. How could such a statement be true when the very essence of research in this field was concerned with questions about the nature of mental evolution? I've come to realize that Hodos and Campbell were referring to the highly refined mechanistic model of theory used in the physical sciences, that is, a formal set of quantitative principles, systematically organized for the purpose of empirical verification and prediction. This restricted notion of theory certainly has never been particularly true of comparative psychology, nor of any other discipline within psychology for that matter. On the other hand, a broad conception of theory, to include a set of loosely related assumptions, speculations, and suppositions, has always had its place in this field and in other biological disciplines concerned with broad questions about the principles of macroevolution. Unfortunately, most people in psychology do not explicitly state what their assumptions are when reporting research or formulating hypotheses. This can present problems because many times the implicit assumptions underlying a research program provide the theoretical foundation for interpreting and generalizing the results. Ironically, although most scientists are well aware that their observations of nature are determined in large part by tacit theoretical assumptions, few ever stop to consider what these assumptions are or how they came to hold them. It is one thing to know that a system of viewpoints directs our attention to certain phenomena at the expense of others, and entirely another to understand why we hold these points of view.

The purpose of this chapter is to explore how individuals have speculated about what is perhaps the most fundamental question in comparative psychology, that is, "What is the nature of evolutionary change and variation in behavior?" In the process we compare various unspoken assumptions that have followed from this question and form the basis for many of the theories that have been proposed as answers. The chapter begins by examining the genesis of three concepts in the history of thought: change, progress, and continuity. The concepts are all interrelated, as the idea of progress is a statement of belief about the direction of change, and the idea of continuity is a statement of belief about the tempo of change. The prevailing ideas of change, progress, and continuity at different points in history were profoundly influenced by contemporary social and political forces, and therefore some of these influences are described.

Next, the chapter explores the relationship of these ideas, in the form of metaphors, to the establishment of a comparative psychology. I have chosen to use the metaphor in this analysis because it has often been used in scientific thinking to convey an abstract idea in concrete terms, because specific metaphors represent historical signposts in the evolution of ideas, and perhaps most importantly, because the vivid images provided by these metaphors influenced the way people thought about the world around them—often serving as an explanatory and descriptive principle of nature. In this part I trace the idea of continuity in the writings of Aristotle and Leibniz, and in the works of naturalists like Bonnet and later the German biological school of thought called *Naturphilosophie,* and its impact as demonstrated in the writings of Charles Darwin and the earliest comparative psychologists. The third part of the chapter dissects the Darwinian tree metaphor into two of its assumptions, the direction and the tempo of evolutionary change, and explores how each idea is related to mental evolution—in particular to the evolution of learning. Under the aegis of evolutionary tempo, the Mental Continuity hypothesis and Discontinuity hypothesis are reviewed, and their histories traced within comparative psychology. A similar treatment is given to three hypotheses devised to account for the direction of mental evolution, two of which imply progress (cladogenesis, anagenesis) and one that does not (steady-statism). Finally, I show that any broad theory of the evolution of learning can be distinguished according to the set of assumptions its adherents hold concerning the direction and tempo of mental evolution. Failure to appreciate these distinctions can lead to a narrow assessment of the nature of comparative psychology.

ON CHANGE, PROGRESS, AND CONTINUITY

Conventionally, science is thought to be a march of continuous progress toward a final truth through the gradual accumulation of facts. A scientific theory, according to this point of view, explains these facts with accuracy and reliability, is testable, and leads to predictions about enduring features of the real world. A theory that is not supported by the facts must be abandoned, lest it threaten to obstruct the search for truth (Popper, 1959). This model of scientific enterprise has come under considerable criticism, some might even say discredit, from historians of science (Gould, 1977b; Kuhn, 1962). According to Gould: "Science is no inexorable march to truth, mediated by the collection of objective information and the destruction of ancient superstition. Scientists, as ordinary human beings, unconsciously reflect in their theories the social and political constraints of their times [p. 15]."

We see the ideas of change, progress, and continuity develop and entwine in 18th- and 19th-century social and political thought (Bury, 1920; Lovejoy, 1936). Science reflected these ideas and comparative psychology grew out of this science. In this section I examine how scientists consciously or unconsciously attempted to reconcile the assumptions on which they based their work with prevailing social and cultural attitudes.

On Change and Variation. It is generally agreed upon among historians that nature, before the 19th-century, was viewed as static and unchanging. The Greeks are said to have invented the concept of change, and the question of the reality of the flux versus the forms of nature (i.e., change versus permanence) became the central problem of pre-Socratic philosophers. The most important conclusions reached in this analysis were that whatever has real being—whatever really is—must be eternal and unchanging, and that such real existence can never be discovered by the senses (Robinson, 1981). Even philosophers who made change the focus of their systems of thought did so by referring to change as a permanent substance or thing. Thus Heraclitus, who saw everything in constant flux, identified the source of the flux with the eternal substance fire. The existence of change was a property of fire, and those things that exhibited substantial change were simply composed of more of this element than things that changed or varied only a little. Change as a dynamic process was not an idea entertained by the Greek mind.

The idea that change is not a process, but that it is a static characteristic of a thing or collection of things can be traced from the earliest Greeks, to Plato and Aristotle, to the Medieval religious thinkers throughout the western world. Anaxagoras, the pre-Socratic cosmologist and believer in the existence of unchangeable forms of Being, explained change and variation as the result of differences in the elements whose mixture makes a thing. The philosopher's interest in this system was centered on understanding the nature of this mixture and its cause, and not on change itself. The path of development from this idea to that of a universal Reason in Socratic thought should be obvious. Plato's belief in the reality of ideas and the unreliability of sense impressions is based on the attitude that ideas have an unchanging permanence that transcends time and individual thinkers. Change and variation, to Plato, are imperfect reflections of eternal, unchanging ideas in the illusory world of sensory appearances. Even Aristotle, who reintroduced the study of nature and its diversity into philosophical speculation, failed to appreciate the dynamic basis of change. His emphasis on classification and specification of types, coupled with the influence of Plato's teaching about the reality of the unchanging idea of a thing (i.e., its form), reduced the Aristotelian model of change to a hierarchical classification system in which each class was conceived solely in terms of a set of defining properties.

As a result of the Greek emphasis on static conceptions of the universe, change remained a vague and largely ignored concept until the late 18th century. It is not the purpose of this section of the chapter to trace the complete history of the idea of change, but only to show that the dynamic view of the concept did not emerge until relatively recently. From classical antiquity to the Medieval and Renaissance periods change was considered to be a static concept. Perhaps the most important element in this sequence of events was the steady demise in importance of day-to-day concerns with problems of living and an increased interest in spiritual, eternal ideas. The demise of Athenian power after the Peloponnesian War, the gradual depletion of economic trade and intellectual activity after the deaths of Alexander and Aristotle, and the growing corruption of society and law under Roman rule led to a disillusionment with materialistic speculation, and the rise of humanistic–idealistic philosophies. As people became more concerned with existence beyond the disappointments of the material world, faith came to supersede reason, and Platonic concerns with eternal unchanging forms of Being became the dominant intellectual force of the period. Even as Thomas Aquinas reintroduced Aristotelian thinking into religious dogma in the 13-century, and with it a renewed interest in the natural world, change and diversity in nature continued to be viewed as a fixed entity. The machinelike regularity and precision in the systems of Descartes and Newton provided new but essentially static conceptions of a changing universe, and even biological systems were conceived in terms of rigid class hierarchies. For example, the taxonomic system of binomial nomenclature devised by Linnaeus in the 18th century is fundamentally Aristotelian. Later it became apparent that this system made it very difficult for biologists to indicate what a species really was because species exhibit variability and change (Ghiselin, 1969; Gould, 1977b; Mayr, 1970).

In one sense it may be easier to understand the failure of Medieval intellectuals to include change in their speculations about the world by recognizing that there was little in the way of genuinely creative effort in literature, art, music, or architecture during this 1000-year period of time. With such inconsiderable change in the sensory world, should one expect to find *Homo sapiens* speculating zealously on the nature of change?[1] We must instead turn to the 18th and 19th centuries to examine scientific assumptions concerning change. Mendelsohn

[1]We should be wary of attributing this disappointing outcome simply to intellectual or aesthetic impoverishment, or to religious orthodoxy or authoritarianism (Robinson, 1981). For even as the 12th and 13th centuries ended an era of literary, artistic, and architectural restraint, another 400 years would pass before change became the focus of social, political, economic, and scientific institutions.

(1980) has described how prior to this time 17th-century naturalists were influenced by the clockwork world view of Descartes and other mechanists—a view founded on principles of stability, uniformity, and lawfulness. Voltaire, for example, in *The Ignorant Philosopher* (cited in Dampier, 1966), remarks: "it would be very singular that all nature, all the planets, should obey eternal laws, and that there should be a little animal, five feet high, who, in comtempt of these laws, could act as he pleased, solely according to his caprice [p. 197]." Hill (1974) has even tried to draw a connection between the scientific theories of this period and the structure of the several absolute monarchies that governed most of the western world, both systems based on consummate and interminable laws. But as the 18th century drew to a close, stability and order came under attack. Social, political, economic, and religious upheaval presented the common man with his first taste of dramatic change from the immediate past. Theories of change and variation appeared hand in hand with sharp, violent revolution.

On Progress. In France and England radical change took different paths. The industrial revolution in England was a gradual, cumulative struggle. Although occasionally violent, change could be understood only as the sum of many seemingly insignificant events (Mendelsohn, 1980). By contrast the French Revolution was sudden, dramatic, and catastrophic. Scientific theories were colored by these images of gradual versus cataclysmic change. For example, the French naturalist Lamarck, a sympathizer with the revolution, offered the world its first full-scale theory of organic diversity and transmutation by introducing sharp breaks from the past (spontaneous generation), environmental dislocation and struggle (adaptation), and progressive improvement (evolution). But the revolutionary metaphor was not restricted to a theory of evolution. George Cuvier, the indefatigable leader of the Academie des Sciences and proponent of the immutability of species, advocated "catastrophism," a progressionist view of the world in which geological catastrophes allowed God to re-create living beings several times. Each creation was a distinct improvement over the last, leading ultimately to the modern creation dominated by man. In England, on the other hand, the revolution metaphor was not a violent concept. Lyell's theory of uniformitarianism in geology painted a picture of slow, steady change through roughly equivalent periods of time. He saw natural law as invariant and timeless, despite the literal appearance of catastrophies in the fossil record (Gould, 1977a, 1977b).

Despite their differences, all these points of view shared a common thread of belief in the unidirectional, progressive nature of change. Actually, Lyell (1832) originally asserted that nature existed in a state of dynamic stability. Species came into existence and disappeared, mountain ranges rose and fell, but the world remained fundamentally the same. But by the mid 1850s, even Lyell had come to accept progression among the vertebrates. Some say that his conversion to Darwin's brand of evolution was not due to his association with Darwin, but rather it was a "minimum retreat" from his earlier views once he had modified his stand on progression (Gould, 1970).

The period from the American and French Revolutions to the outbreak of World War I was imbued with progressionist thought. Belief in progress became not just a hypothesis, but a dogma (Bury, 1920; Wagar, 1969). It "dominated the European mind to such an extent," writes Christopher Dawson, "that any attempt to question it was regarded as a paradox or a heresy [1960, p. 7]." Thus the idea of change and the idea of progress became one.

On Continuity and Discontinuity. The social and political forces that shaped the progressionist thinking of Cuvier and Lyell also fostered a conceptual dichotomy that has been the focus of scientific controversy ever since. Cuvier's catastrophism was imbued with the violent overtones of 18th century political events in France. Indeed, his viewpoint was that the whole of earth's history was punctuated by revolutions. Lyell's theory of uniformitarianism, on the other hand, rejected revolutions, violence, and catastrophe in favor of a system of gradual, steady, orderly change. According to Gould (1977a), the debate between uniformitarianism and catastrophism (i.e., between gradual versus cataclysmic change) was the cornerstone of scientific activity in the early 19th century. Geology became: "the queen of the sciences. The greatest minds of Europe flocked to it . . . the relationship between geological change and biological change became a major subject and source of contention [Gould, 1977a, p. 9]."

In view of Lyell's conversion to progressionism it is ironic that progress and the direction of change were not particularly important issues for Darwin. Natural selection refers to nothing more than successive adaptation to changing local environments, it is not a statement of directed trends or inherent improvement. Like his British colleague Lyell, Darwin seems to be saying that nature had much time and required little effort to produce the broad panorama of past events. What seemed to be most important to Darwin's way of thinking was not the direction of change so much as the tempo of change. Does it proceed gradually in a continuous and leisurely fashion, or is it episodic (Gould, 1977a)? This was the crux of the debate between Lyell's uniformitarianism and the catastrophists, and Darwin followed the gradualist argument on this issue. In a sense the industrial revolution in England provided the perfect backdrop for Darwin's theory of natural selection. The tempo of change in each was slow and steady, both relied on continuous competition—a "struggle for existence" that led to the "survival of the fittest," and both emphasized adapting oneself to the conditions that prevailed rather than advocating violent change. It is not surprising that in the course of the 19th century, evolution and revolution came to have opposite meanings. Evolution became the equivalent of slow, continuous change, whereas revolution meant rapid, abrupt, discontinuous change.

The historian Everett Mendelsohn (1980) has argued that attitudes shaped by political and social events regarding continuity and discontinuity were the main forces governing the structure of scientific discourse in the 19th century. Attitudes toward the stability of nations were reflected in scientific debate; continuist epistemologies were linked with political conservatism; discrete, discontinuist epistemologies were associated with more radical social change. Moreover, we are not without modern examples. Lysenko's brand of Lamarckism, for example, was fostered on Russian biology in the wake of the violent and rapid revolution of 1917. As class systems were overturned in one swift movement, so too could the characteristics of food staples be altered in a single generation of selective breeding. Another example, one closer to home, representing the conservative attitude of progressive capitalism in the United States, is provided by Richard Lewontin (1977):

> There are still rich and poor, powerful and weak . . . How is this to be explained? We might suppose that . . . society . . . has inequality built into it and even depends on that inequality for its operation. But that supposition, if taken seriously, would engender yet another revolution. The alternative is to claim that inequalities reside in properties of individuals. . . . This is the claim that our society has produced about as much equality as is humanly possible and that the remaining differences in status and wealth and power are the inevitable manifestations of *natural* inequalities in individual abilities [p. 7].

Lewontin argues that this assertion is the basis of racism, sexism, class superiority, and the ideology of "human nature" and that science, in the shadow of this ideology, has fostered deterministic theories sympathic to this view. The eugenics movement, the IQ controversy, the controversy over women's rights, and the challenge of sociobiology to comparative psychology all stem from this issue.

Although not one of our main concerns here, we must remember that these attitudes continue to form the foundations of contemporary systems of thought. The lively controversy known as the Kuhn–Popper debate in the history of science is perhaps the best example of this. Kuhnian science proceeds for long periods of time without any fundamental change, guided by an interest in showing that an orthodox theory (a paradigm) is true rather than a compulsion to refute what is generally accepted. When change does occur (a paradigm shift), it is sudden and dramatic. Popper's (1959) view of scientific discovery, on the other hand, portrays science as a long, tedious march from hypothesis to relative certainty. Appearing as it did a century after Darwin's major work, it quite clearly reflects a conservative, evolutionary epistemology. It is no coincidence that Kuhn's influential work was entitled *The Structure of Scientific Revolutions* (1962), whereas Popper's latest contribution is called *Objective Knowledge: An Evolutionary Approach* (1972). Similar distinctions exist in modern evolutionary biology (e.g., the current debate between Eldredge and Gould's [1972] notion of "punctuated equilibria" and Gingerich's [1976] defense of phyletic gradualism) and in modern comparative psychology (following section).

METAPHORS OF CHANGE AND THE ORIGIN OF COMPARATIVE PSYCHOLOGY

In the previous section I argued that the idea of change is a relatively modern concept in science, that it has been linked historically with the idea of progress, and that conservative science has typically viewed change as a slow continuous process. Comparative psychology, like its sister disciplines comparative anatomy, physiology, and embryology, is a product of this view of nature and its theoretical assumptions have developed along similar lines.

About a dozen years ago the concept of continuity in comparative psychology was attacked for being atheoretical and antievolutionary (Ghiselin, 1969; Hodos & Campbell, 1969). It was equated with a belief in a "unitary, graded, continuous dimension known as the *Scala naturae* or Great Chain of Being [Hodos & Campbell, 1969, p. 338]." My purpose in this section of the chapter is to show that there are several different versions of this scale and that the systems differ in their adherence to the idea of a graded continuum of features upon which organisms may be ranked, on the progressive nature of the scale, and in the number of dimensions involved. Some of these systems (e.g., the Aristotelian latticework, the Leibnizian chain, the transcendentalist ladder) have been intuitively appealing but scientifically unproductive as models of change in nature. Other systems (e.g., cladogenesis, anagenesis, steady-statism) may not be so easily dismissed.

The Lattice. The idea that organisms can be arranged on a graded scale of complexity with respect to certain structural and functional features has its roots in Aristotelian taxonomy. Given a set of defining characteristics, Aristotle believed that any organism could be ranked according to each of its features along a continuum of similar traits graded from simple to complex. Aristotle did not believe that each animal occupied a distinctly different place on the continuum, rather that grades existed within which many different animals might fall. Most importantly, and what is often overlooked by critics of an Aristotelian classification system, is that Aristotle never argued for a single graded scale with man at the top, but for many scales, in only some of which man held the ascendent position. He recognized that organisms differ in many ways—in external appearance, in anatomy, in development and complexity of particular organs, in habitat, in sensory abilities and intelligence—and that there are no necessary relationships among these diverse categories (Lovejoy, 1936). Though he never used such a metaphor, the lattice comes to mind as an appropriate model of the Aristotelian system. A given point in the lattice may at the same time be the beginning, middle, or end of a configuration of similar points. An organism 'superior' in one respect may be 'inferior' in another.

The Chain. In spite of the stress Aristotle placed on multiple patterns of classification, Medieval writers came to embrace only two of these arrangements. One, based on a vague notion of 'perfection' (defined variously in terms of developmental complexity at birth, physical self-sufficiency, or metaphysical completeness and closure), satisfied the difficult problem of ordering mortal and immortal, material and immaterial entities on the same scale, with God at the apex and man cast humbly in the middle. A second hierarchical arrangement was based on the possession of differential 'powers of soul', and emphasized the superiority of the rational soul over the nutritive (plants) and sensitive (animal) souls. A merging of these two ideas, the hierarchies of reason and perfection, came to be known as the *Scale naturae* or Great Chain of Being. This was the metaphor that dominated thought in philosophy and natural history through the Middle Ages and down to the late 18th century (Lovejoy, 1936).

The linked-chain metaphor and the assumptions of continuity and unidirectional gradation were made explicit in the philosophic systems of Locke and Leibniz in the 17th century. Leibniz in particular (quoted in Lovejoy, 1936) developed the idea with clarity:

> All the different classes of beings which taken together make up the universe are, in the ideas of God who knows distinctly their essential gradations, only so many ordinates of a single curve so closely united that it would be impossible to place others between any two of them. . . . And, since the law of continuity requires that when the essential attributes of one being approximate those of another all the properties of the one must likewise gradually approximate those of the other, it is necessary that all the orders of natural beings form but a single chain, in which the various classes, like so many rings, are so closely linked to another that it is impossible for the senses or the imagination to determine precisely the point at which one ends and the next begins [p. 145].

Note the departure of Leibniz from Aristotle in declaring that when the essential attributes of one being approximate those of another, *all* the properties of the one must approximate the other. In Leibniz we have for the first time a unitary scale of nature based on an unchanging but graded continuum of characteristics.

The tremendous impact that Locke and Leibniz had on thinkers of the 18th century can be seen in the number of philosophers, scientists, poets, essayists, and theologians of that period who wrote about, speculated upon, and drew implications from the unitary, graded Great Chain of Being. Bolingbroke, Pope, Bonnet, Diderot, Kant, Schiller, and Goethe, to name just a few, all accepted the general concept with its political, religious, and social implications. The metaphor provided a strong image and was an effective means of supplying order to a chaotic universe. The religious and political symbolism of a Chain of Being was also ''pleasing to the mind [Mendelsohn, 1980].'' It had a parallel in the absolute monarchies of the age that gave it a kind of worldly relevance (Hill, 1974), and as a declaration of biological and social relations, the concept provided a sense of certainty and justification in the Christian belief in man's dominion over nature, and in the European conquest of undeveloped, ''primitive'' nations. We must also not overlook the fact that Leibniz's allusion to ''essential gradations'' cast in terms of the new calculus imparted to this scheme a quantitative flavor suited to the European trend to mathematicize nature. The Chain of Being was more than a metaphor, it became the model of the universe in the 18th century and the stimulus for virtually all scientific speculation.

The Ladder. As the idea of the Great Chain of Being with its assumptions of continuity and unilinear gradation became widely accepted, its implications were explored and problems began to arise. Nowhere was this more apparent than in the static clockwork version of the universe known as preformation theory. A preformationist argument held that all organisms were created by the Christian God in a single act. Coupled with the idea of a Chain of Being it implied a gradation of structure and intelligence from the minutest animal, too small to see even with a microscope, up to man and to beings unknown to man because of their superior place in the universe. The Church had always placed man in a similar position: inferior to countless holy entities above him in the cosmic hierarchy but superior to the living creatures below him in that scale. When examined carefully, however, this belief implied that man must differ physically and intellectually from his nearest nonhuman species only infinitesimally. Jenyns, an 18th-century essayist, tried to diminish the heretical impact of this deduction by emphasizing the many degrees of intelligence within the human species. His point was that although the psychological distance between the highest animals and the lowest men was indeed infinitesimal, between either of these and the intellectual giants of the age a vast chasm existed. According to Jenyns (1790), as quoted in Lovejoy (1936):

> Animal life rises from this low beginning in the shellfish, through innumerable species of insects, fishes, birds, and beasts, to the confines of reason, where, in the dog, the monkey, and chimpanze, it unites so closely with the lowest degree of that quality in man, that they cannot easily be distinguished from each other. From this lowest degree in the brutal Hotentot, reason, with the assistance of learning and science, advances, through the various stages of human understanding . . . till in a Bacon or a Newton it attains the summit [p. 197].

We begin to see the germ of evolutionary thinking even within this static world view. Yet, despite attempts to rescue the metaphor of the chain, there remained the implication that human nature could not be unique in the eyes of God. Furthermore, the idea of each individual as a fixed link in the Great Chain of Being evoked a theology in which hope for a better life was denied. Each individual's position in the chain was established at once and for all time in the creation.

By the late 18th century this static view of nature began to change. Bonnet's (1764) theory of palingenesis invested preformationist ideas with the concept of progressive change (Gould, 1977c) and, soon after, the hypothesis of transformation of species began to emerge. The ideas of continuity and progress became fused in a dynamic view of the fluctuating nature of the universe (Bury, 1920). In Germany, a group of biologists combined the progressive view of nature with romanticist ideas in vogue at the time to produce the school of *Naturphilosophie*. Its accent on the inevitability of beauty and order, the primacy of man, and continuous progressive change became the ''religion of nature'' in the early 19th century. The Chain of Being was reworked from a rigid set of fixed links to a ladder that organisms might climb (Lovejoy, 1936). Man—the Faust ideal—was, by the will of his Maker, insatiable in mounting every rung.

Evolutionary Metaphors. The naturalistic point of view spread rapidly throughout Europe, into England, and to America. The ''natural history method'' became the featured means of exploring the world. To understand civilization, one studied ''savages''; to understand human nature, one explored the rest of nature (Robinson, 1981). The influence of this perspective is plentiful in the books and essays of the early 19th century, and even the general periodicals of the day were filled with accounts of the life and ways of ants and bees, fish and fowl. Darwin's studies of climbing plants, of the habits of earthworms, and of the behavior of ''higher'' organisms including his infant son proceeded from the same naturalistic perspective (Ghiselin, 1969, 1973). There was one crucial exception, however. Like the Naturphilosophen he thought along developmental lines, with continuity and gradual divergence marking the tempo of nature's change. But where the Naturphilosophen saw a single, progressive direction in nature, Darwin recognized a multiplicity of possibilities. Where the former saw a ladder, Darwin saw a tree.

It would be incorrect, however, to say that Darwin denied the possibility of directional change as an outcome of natural selection (Gould, 1977a). Evolution is adaptation to local environments, but there are many ways in which this can be accomplished. One way may involve progressive ''improvement'' through the advancement of structural organization conferring greater evolutionary success. In Darwin's own words (1859/1959): ''I do not doubt that this process of improvement has affected in a marked and sensible manner the organization of the more recent and victorious forms of life, in comparison with the ancient and beaten forms [Line 196].'' And again: ''The inhabitants of each successive period in the world's history have beaten their predecessors in the race for life, and are, in so far, higher in the scale of nature; and this may account for that vague yet ill-defined sentiment, felt by many palaeontologists, that organization on the whole has progressed [Line 263].'' Of course Darwin noted that organization can also become regressive, and in this sense his views might be better termed directionalist than progressionist.

We continue to see the directionalist influence on contemporary paleobiology: According to Huxley (1958a):

> As regards *higher and lower* in the biological sector, for one thing it is obvious that some organic types have a higher (more complex, more efficient, and more integrated) organization than others; and for another, paleontology and comparative anatomy have demonstrated that . . . the acquisition of higher organization in this sense has in fact conferred evolutionary success, the more highly organized types having become more abundant and dominant at the expense of less highly organized types [p. 452].

In matters pertaining to the origin and nature of the mind, we find that Darwin's (1871) views were very similar. In his words:

> It is therefore, highly probable that with mankind the intellectual facilities have been mainly and gradually perfected through natural selection. . . . Undoubtedly it would be interesting to trace the development of each separate faculty from the state in which it exists in the lower animals to that which exists in man; but neither my ability or my knowledge permits the attempt [p. 128–129].

Moreover, it appears from his letters and notebooks that he equated higher intelligence with the relative degree of efficiency of the ''lines of transmission'' in the brain (de Beer, 1963; F. Darwin, 1903). Although characteristically vague when it came to psychological topics, he seemed to believe that greater efficiency of transmission, complexity of brain organization, and the possession of higher mental faculties were correlated phenomena. As this was the position taken by the earliest comparative psychologists (Morgan 1894; Romanes, 1883; Washburn, 1908), and general statements about the psychological propensities of animals were typically framed with qualifications concerning neural complexity (James, 1890; Spencer, 1872) it would seem logical to conclude that comparative psychology was initiated, not as an attempt to trace probable lines of mental evolution as Hodos and Campbell (1969) suggest, but as a means of assessing ''improvement'' in the various modes of adaptation that animal species inherit with successive grades of neural organization (Yarczower & Hazlett, 1977).

Darwin's belief in the continuity of mental abilities referred to the slow, gradual modification of modes of adaptation with advances in brain organization and can be contrasted with a discontinuity view that predicts the sudden, punctuated appearance of new and qualitatively distinct modes of adaptation under conditions of routine variation in successive grades of neural complexity. There is nothing atheoretical or antievolutionary in the discontinuity view, and it is as valid today as it was 100 years ago (Gould, 1976). There are other equally valid versions of the tree metaphor, however, and these become apparent when we reestablish the distinction between theoretical assumptions about the directional nature of change and assumptions about the tempo of change.

THE TEMPO OF MENTAL EVOLUTION

With Darwin's *Origin of Species* (1859/1959), a theoretical base was established for exploring the evolution of mental abilities. *The Descent of Man* (1871), published 12 years later, provided the framework for the comparative psychology of learning in the next century. Darwin and virtually every subsequent investigator of learning and the "higher mental processes" in animals and man were led to a position that advocated the directional nature of mental evolution as it pertains to modification in behavior. Higher mental processes developed out of simpler ones, presumably because of increased neural complexity. One also gets the impression that a unitary, directional evolutionary process is implied when reading textbooks that deal with topics in learning listed from simple (e.g., habituation, conditioned reflex) to complex (e.g., selective learning, insight, reasoning) (Dethier & Stellar, 1961/1970; Maier & Schneirla, 1935/1964; Watson, 1914). But conceptual theories on the evolution of mental abilities did differ in one important respect—their assumption about the tempo of mental evolution. Darwin, for instance, saw evolution as a slow, steady, gradual process leading to a diversity of species that should not differ dramatically from one another. Others viewed the discontinuity in the way different animals learn to solve problems as evidence for sudden, episodic changes in mental evolution (Bitterman, 1965a; Schneirla, 1959/1972). The first point of view has come to be known as the Mental Continuity Hypothesis and the second will be called the Discontinuity Hypothesis.

Mental Continuity Hypothesis. In the *Descent of Man* (1871), Darwin spelled out in detail the Mental Continuity Hypothesis: Psychological faculties such as courage, temperament, and general intelligence were the products of a long, slow process of natural selection:

> If no organic being excepting man had possessed any mental power, or if his powers had been of a wholly different nature from those of the lower animals, then we should never have been able to convince ourselves that our high faculties had been gradually developed. . . . My object in this chapter is to show that there is no fundamental difference between man and the higher mammals in their mental faculties [p. 287].

From the opening paragraphs of the chapter on the comparison of the mental powers of man and the lower animals, one realizes why this view was of such importance to Darwin's theory of natural selection. Although many of his contemporaries were willing to accept his evidence for structural similarities through common descent, most argued that because man differed so greatly in mental ability from all other animals, some auxilliary force must be invoked to account for this discontinuity. A small minority of biologists, such as Broom (1933), felt impelled to invoke "spiritual agencies" to account for change of this sort, but it was generally realized that an *elan vitale* was unnecessary. Most critics were satisfied with the Lamarckian explanation of mental evolution, which explained the human mind as the result of the progressive inheritance of acquired talents (Spencer, 1855, 1872). This point of view postulated that the social context of human mental development was far more important than a purely mechanical process of heredity could permit (Lewes, 1879/1978).

And so Darwin's task was set. In order to argue that mind is a product of natural selection, he had to show that a continuity of mental faculties existed between man and beast. One full chapter in *The Descent* was devoted to intelligence and problem-solving abilities of animals. As Lockard (1971) has pointed out, his listing included many of the research topics that would collectively come to be known as comparative psychology. Imitation of parental song in birds (Darwin, 1871, p. 291), taste aversion in mammals (p. 295), "shaping" of prey-catching behavior in young hawks (p. 291), avoidance learning in fish (p. 292), modifying "set" responses and progressive improvement in tasks (p. 294) were all touched upon. Even tool usage and language were discussed. It was Darwin's contention that intelligence, abstraction, and reasoning evolved gradually through the development and combination of simpler forms of memory, association, and attention, a position taken up on a somewhat more ambitious scale in his book on worms (1881). With the hypothesis of mental continuity and the model of experimental analysis provided by his worm research, the way was opened for others. The followers of Darwin were concerned with uncovering proof of the similarity of mental processes of men and animals. Typically they tended to minimize the gulf separating human and animal performance by writing anthropomorphic anecdotes. Romanes, the most celebrated of the anecdotalists, published a series of books with evidence for the hypothesis of continuity of behavioral characteristics among animal species (1883, 1884, 1888). So important were his writings in the field that Boring (1950) and other historians describe Romanes as the father of comparative psychology. Romanes said that one can observe gradual quantitative increments in mental capacity as one moves up a hypothetical phylogenetic scale of complexity from single-celled organisms, through the invertebrates to the vertebrate classes. At no point was there a sudden change in the quality of mental abilities.

Although Romanes has been rightly accused of too uncritically accepting stories of somewhat questionable character for inclusion in his books, it is only fair to point out that he wrote in much the same style as had Darwin (1871) and Spencer (1855) before him, and the short accounts of animal behavior that he described and discussed were typical of natural history essays popular in the 19th century (Mountjoy, 1980; Young, 1973). As it is, he was selective about those anecdotes that he subsequently published, establishing a criterion related to the qualifications of the observer and the frequency of corroboration. He also anticipated the critics who objected to the inferential nature of his method. In Romanes' (1883) words: "skepticism of this kind is logically bound to deny evidence of mind, not only in the case of lower animals, but also in that of the higher, and even in that of men other than the skeptic himself [p. 6]." Like many writers of the era, he believed that the mentality of animals could be inferred in much the same way as philosophers inferred mental states in others. He also reasoned that the more similar the neural machinery of man and animal, the more confidence one could have in the inferences drawn about their mental states. Thirty years later a similar argument was still being advanced. According to Holmes (1911):

> With beings much like ourselves our inferences may be fairly accurate. When thrown amid the people of other nations or races our judgements are more apt to be erroneous. And when we try to infer what goes on in the mind of a cat or a dog the difficulties are very greatly increased. If we try to imagine what sort of psychic states are associated with the supra-oesophageal ganglion of an ant or a crayfish, analogy almost completely fails us [p. 3].

The thesis developed by Romanes was that the mind rose gradually through evolution from primitive reflexes to consciousness. His comparative psychology was formed in a period of clashing metaphors of chains and ladders and trees, at a time when experimental psychology was emerging as an independent discipline concerned primarily with consciousness and mental life, and at the height of the vogue of Spencer (Hofstadter, 1955). The result was a marriage of pre-Darwinian progressionist and structuralist thinking that shaped the nature of animal psychology for the next three decades (James, 1890; Washburn, 1908; Wundt, 1894/1907).

By the early 1900s Romanes' anecdotal approach to animal psychology had already become obsolete, replaced by systematic experimental methods. Yet the progressionist–structuralist reasoning that formed the basis for Romanes' comparative psychology lived on. Nowhere is this more apparent than in Washburn's book, *The Animal Mind* (1908). Written in the framework of Titchner's Structuralist school of thought, the book took the position that organisms having a nervous system similar to a human, upon reacting to a stimulus in a way that is similar to a human, must, by inference, have an inner experience that resembles our own. Where consciousness and mind arose in the animal world, no one could say, but by the method of introspection and the careful use of inference from experimental evidence, the mind of animals could be understood and the nature and structure of the consciousness of animals could be revealed. However, this approach was never a particularly popular one, despite the fact that the book went through four editions (i.e., 1908, 1917, 1926, 1936), and attempts to infer the mental life of animals from behavior were soon abandoned.[2]

With the transition from introspective psychology to behaviorism at the turn of the century the only thing that remained of significance in Romanes' system was the idea of mental continuity. The cast of characters instrumental in this change included Morgan (1894) and the Functionalist school, especially James (1890) and Angell (1909/1978). Thorndike (1911), a student of James, and Watson (1914), a student of Angell, made the transition complete. Their respective contributions to the comparative psychology of learning are well documented in virtually every textbook on the history of psychology (e.g., Boring, 1950; Robinson, 1981) and I will not detail them again here. It is important to note, however, that this transition did not represent a conceptual trend different from that already apparent in Romanes' writings. What changed was merely a methodology and the acceptable limits to the domain of animal psychology. In the words of Angell (1909/1978): "Our whole tendency nowadays is to recognize and frankly admit, that inasmuch as we must infer the psychic operations of animals wholly in terms of their behavior, we are under peculiar obligation to interpret their activities in the most conservative possible way. [p. 169]"

Conceptually, comparative psychology continued to adhere to the idea that the tempo of change in mental evolution is slow and steady. This was reflected in the view that became synonomous with learning psychology, that is, different species all learn in essentially the same way (Thorndike, 1911; Watson, 1914), and the differences between what we call intelligent behavior in man and the regulatory behavior of lower organisms is only a matter of degree; "one grades insensibly into the other" (Jennings, 1906, p. 335). As a consequence of this belief, all major learning theorists in the first half of the 20th century failed to consider in any significant sense the issue of species differences in learning. Thus Pavlov (1927/1960, p. 9), Tolman (1932, p. 415), Skinner (1938, p. 442), and Hull (1943, p. 29) all mentioned their belief in the essential continuity of mental processes in animals and man, but they did not examine their opinion in detail. Skinner, for one, was particularly fond of presenting cumulative response curves from three different species—a rat, a pigeon, and a monkey, for example—without telling his audience which was which, and asking them to identify the curve that belonged to each animal. Because the curves looked nearly identical, Skinner (1956) answered his own question: "It doesn't matter . . . once you have allowed for differences in the ways in which (these species) . . . make contact with the environment, and in the ways in which they act upon the environment, what remains of their behavior shows astonishingly similar proper-

---

[2]In recent years this approach has taken on renewed interest with the publication of Griffin's *The Question of Animal Awareness* (1976). The difficulties in arriving at an *accepted* method to address the subject have not disappeared, however, and so the subject of consciousness in animals continues to remain outside the domain of contemporary comparative psychology (see also Rensch, 1971).

ties. Mice, cats, dogs, and human children could have added other curves to this figure'' [p. 230].

The Discontinuity Hypothesis. Similar learning curves for different species suggested mental continuity because it implied that the minds of these organisms were similar and that evolution had altered their mental abilities only slightly. Presumably, evolution of mental faculties moved at the same rate in all these organisms. If mental evolution was a slow, gradual process then we should expect to find only slight differences in closely related species, and qualitatively different capacities should be found only at the broadest taxonomic comparisons. Because the general form of the learning curves for chicks, cats, and dogs was the same in Thorndike's studies (1911) the logical conclusion was that mental evolution is very slow indeed. However, continuity theory was soon challenged by hints of discontinuity. Kohler's (1925) studies of chimpanzees provided evidence of ''insight'' not exhibited by Thorndike's cats. Kohler argued that some animals learn in different ways than others, and he accused Thorndike of using tasks that prevented insightful solutions. Species differences were also found using complex learning problems such as the delayed response test (Hunter, 1912), the multiple-choice test (Hamilton, 1911) and the reversal learning test (Schneirla, 1946/1976).

By 1935 it had become apparent to some comparative psychologists that ''the nature of modification by experience expresses itself differently in various situations and in various animals; and these differences are both qualitative and quantitative in nature'' [Maier & Schneirla, 1935/1964, p. 335]. Groups of researchers, working independently, began to provide evidence for a theory of discontinuity in mental evolution. European investigators were among the first to express this view. Tinbergen (1951), for example, pointed out that the digger wasp excels in a naturalistic delayed response test—it can remember exactly how much food to bring to each of many larva nests even if delayed up to twenty-four hours. Most of the mammals in Hunter's (1912) delayed response test were unable to delay this long. Nissen in the early 1950's also argued that evolution had been a discontinuous process rather than a continuous one (1951). Basing his views largely on the morphological evidence of discontinuity presented by D'arcy Thompson (1917), he felt that large and abrupt variations characterized both structural and behavioral change, and that the idea that new capacities and new behavioral mechanisms emerged in the course of evolution was in keeping with the biological thinking of the day. He also argued that the emergence of these new capacities should coincide with the abrupt changes in structural characteristics differentiating the major taxonomic groups. Nissen's program of research was therefore very similar to the one begun by Maier and Schneirla (1946/1972): Trace the manner in which the most basic behavior mechanisms of simple tissue adaptation in the lowest organisms expand and become supplemented by new abilities in animals higher in the phylogenetic scale. A modern example of this approach can be seen in a passage from a recent review of habituation. According to Wyers, Peeke, and Herz (1973):

> It is often said that associative learning begins with the platyhelminthes, while learning by ''habituation'' characterizes all animal phyla. In this view, the former depends on the presence of synaptic differentiation of neurons, the ganglionic congregation of interneurons, and bilateral symmetry of bodily and neuronal organization, or all three together. On the other hand, learning by habituation is seen as depending only on the presence of neuronal, or neuroid, differentiation of transmissive capacity. Thus, the evolution of learning capacity is seen as involving a discontinuity and ''emergence'' of new learning properties dependent upon a new organization of neural processes [p. 50].

Discontinuity viewpoints argue against the idea of a slow, gradual rate of change in mental evolution, but what they argue *for* is less clear. There are two possibilities. The first, the position taken by Nissen (1951) and Wyers et al. (1973), is that the evolution of neural mechanisms and novel learning abilities proceed in unison. A sudden change in the complexity of the nervous system will yield a qualitatively distinct behavioral capacity. In this view both neural evolution and mental evolution are episodic and discontinuous. However, it is not necessary for the rate of change in structure and function to be linked in this way. A small change in one may result in a dramatic change in the other. If small variations in the differentiation of the nervous system could result in qualitatively distinct behavioral capacities, then the tempo of neurological evolution may proceed at a slow, steady pace whereas mental evolution would appear punctuated with sudden changes. This may be what Schneirla (1959/1972) had in mind when he wrote: ''learning can differ qualitatively at different phyletic levels. Even the sparse existing evidence indicates that differences of a fundamental nature exist, not to be understood in terms of ''complexity. [p. 301]''

So despite the widespread acceptance of the Mental Continuity Hypothesis among students of animal learning in the middle of this century, discontinuity theories of learning continued to surface at regular intervals. In addition to Nissen's work, the 1950s and 1960s were marked by the discontinuity writings of Dethier and Stellar (1961/1970) and, most importantly, Bitterman (1965a, 1965b, 1969; Bitterman, Wodinsky, & Candland, 1958). Simultaneously, however, Harlow (1958, 1959a, 1959b) was presenting comparative data from his learning set experiments that he said demonstrated mental continuity. A similar argument was heard from Warren (1965a, 1965b). Most of this work on comparative animal problem solving, especially the learning set and reversal learning studies of Harlow and Bitterman, came under heavy criticism in the late 1960s and 1970s for conceptual and methodological reasons. Objections were raised about the selection of species for examination and the small number of representatives of a species actually tested (Lockard, 1968; Rumbaugh, 1968; Warren, 1973), about the suitability of the test situation to the sensory and motor abilities of the animal (Hodos, 1970; Moore, 1973), about the lack of controls for motivational differences across species (Gossette, 1974; Hodos, 1970), and the lack of a significant phyletic relationship between the animals compared (Hodos & Campbell, 1969; Hodos, 1973). In fact, with the exception of the work of comparative neuroanatomists and a few psychologists (e.g., Masterton, Bitterman, Campbell, & Hotton, 1976), no one seems to be taking a broad comparative perspective today. The disappearance of this approach in the psychology of learning is in part due to an inability or unwillingness to deal with the criticism of the earlier work, partly because of a shift in interest to questions of adaptive significance and ecological determinism, and partly due to the fact that the continuity and discontinuity hypotheses were incomplete, were not structurally rigorous, and did not generate sound biological hypotheses in light of the modern synthetic theory of evolution (Mayr, 1970). In addition, I do not believe that anyone has been aware that statements of mental continuity and discontinuity are hypotheses about the tempo of mental evolution. Made aware that the construction of evolutionary scales is inappropriate to the search for the phylogenetic *origins* of learning ability (Hodos & Campbell, 1969), comparative psychologists seemed to have abandoned this approach altogether. The question of evolutionary *tempo* remains, however. Although we all make the assumption that intellectual abilities have changed in the course of evolution, we still have no idea whether the change has been slow and gradual, or sudden and discontinuous.

THE DIRECTION OF MENTAL EVOLUTION

Although all the theories of learning mentioned in this chapter have implicitly accepted the directional nature of mental evolution, none have been concerned with the means by which such change could take place. I think this was due to a lack of understanding of the several possible ways in which directional evolution can be studied. In evolutionary biology, for example, three levels of analysis can be discerned. Experimentalists, working with populations of organisms representing subdivisions of species, explore the fundamental principles of evolutionary processes. Modern systematics, on the other hand, is usually done at the species or genus level, and paleontology at higher levels (Mayr, 1969; Simpson, 1961). A similar generalization can be made about the study of learning in comparative psychology. Learning processes and their genetic basis are typically studied experimentally in a small number of domestic species (Beach, 1950; Ehrman & Parsons, 1976), the study of the phylogenetic origins of learning is properly accomplished with closely related groups of species (Hodos & Campbell, 1969); and historical questions in evolutionary psychology, concerned with the evolution of learning on a broad scale, are examined using categories above the family level (Yarczower & Hazlett, 1977). Questions concerning the directional nature of mental evolution may be treated by any of these methods, but traditionally only the last of these approaches has been employed to any significant extent in comparative psychology.

Anagenesis and Cladogenesis: The Directional Evolution Hypothesis. There appear to be at least three ways to account for directional evolution on a broad scale that may be applied to animal learning: cladogenesis, anagenesis, and steady-statism.

Cladogenesis, or branching evolution, is essentially a study of phyletic lineages (Rensch, 1960, 1971). Both evolutionary time and phyletic divergence are used as axes for comparisons of these lineages, and the branching diagram describing this relationship shows when and how rapidly the lineages split to form new species. A complete picture of cladogenesis would yield a phylogenetic tree. True monophyletic branches of an evolutionary tree are properly referred to as cladograms with each separate evolving line referred to as a clade (Mayr, 1969; Simpson, 1961; Wilson, 1975), keeping in mind that the *true* sequence may never be known (Gould, 1976). Cladogenesis implies that some trends in evolution are marked by progressive improvement in the structure, physiology, or behavior of groups of organisms, but it also assumes that degradation or recessive evolution will mark other trends. Directional change in this approach is therefore multidimensional, and progressive improvement is only one of these dimensions.[3] In point of fact it may be misleading to use progress in this context because the idea of progress is confounded with the fact of phylogeny—an animal is not "higher" simply because it has evolved later.

In contrast, anagenesis, or upward evolution, is unidimensional and concerned only with progressive improvement of structure or function in an evolutionary sequence of characteristics (Huxley, 1943, 1958b; Rensch, 1960). Central to anagenesis is the concept of an evolutionary grade. Evolutionary grades

---

[3]The criteria for defining progressive improvement vary considerably from person to person, and I have chosen to adopt Huxley's definition, which includes greater integration of biological systems, and an elevation of the upper limit of biological efficiency (i.e., increased control over and independence of the environment). It is important to realize that these trends are not universal and inevitable; in the words of Huxley (1943): "evolution may perfectly well include progress without being progressive as a whole [p. 558]." He also noted that this is not the same as specialization. In his words: "Specialization . . . is an improvement in efficiency of adaptation for a particular mode of life: progress is an improvement in efficiency of living in general [p. 562]." This is an important distinction because specialization means simply adaptation to local conditions, whereas progress, as used here, implies a gradual process of building upon simpler structure and function under the continuous influence of directional selection. Unfortunately the vagaries inherent in Huxley's phrase "improvement in efficiency of living in general" makes this sort of analysis a questionable enterprise (Simpson, 1944, 1953). However, because directional selection has been shown to produce greater cell integration, efficiency, and autonomy (Rensch, 1960, and Gould, 1976, have applied the concept of evolutionary progress to optimization of design in the engineering sense), I continue to use the term.

are distinct stages or levels of organization in an evolutionary series of morphological, physiological, or behavioral traits. By definition, higher grades exhibit greater complexity and efficiency of design than lower grades. It should be recognized, however, that grades or stages of progress may differ according to the criterion employed as the means of measurement (e.g., mammals analyzed as running machines form different grades than do the same mammals judged as problem solvers). We should also realize that convergence and parallel evolution can result in several different groups occupying the same grade. For these reasons, some attention to the phylogeny of the group under question is necessary, but this is less important in anagenesis than in cladogenesis. Anagenesis is concerned more with the hierarchy of evolutionary grades attained by various taxa, whereas cladogenesis is concerned more with describing lineages traversing various grades.

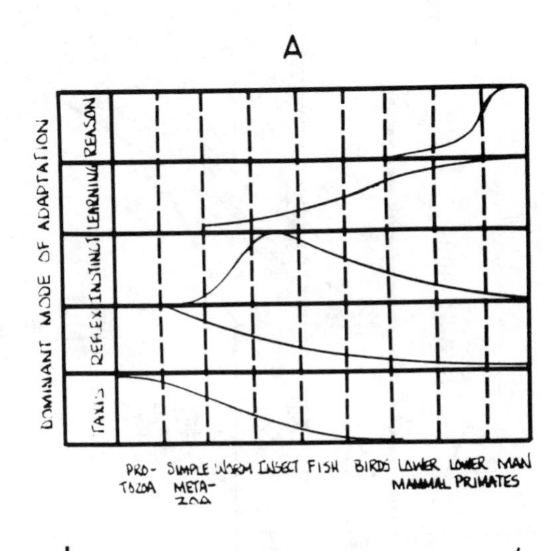

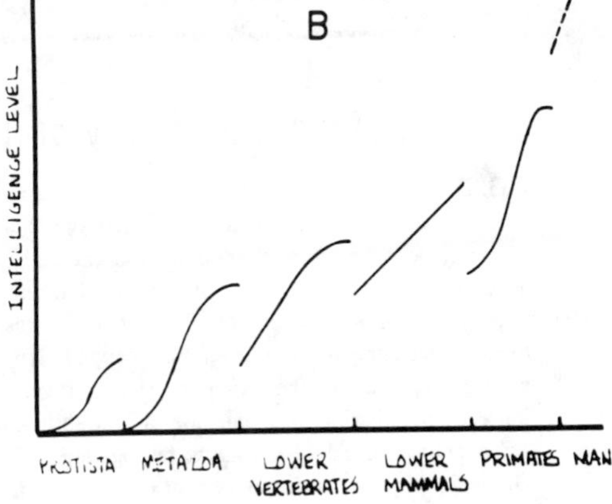

FIG. 6.1. Two examples of anagenic analysis in comparative psychology. Fig. 6.1A (from Dethier & Stellar, 1970) shows the relative dominance of different modes (grades) of adaptation at different levels of the phylogenetic scale. Fig. 6.1B (from Nissen, 1951) represents a tentative estimate of comparative intelligence in the animal kingdom. The vertical axis represents only the order or direction of differences, but grades of mental ability are implied.

Where cladogenesis and anagenesis differ most is in their concern with the temporal features of evolution. Evolutionary time is central to cladogenesis, but it is only incidental to anagenesis. Although anagenesis assumes that progressive improvement has occurred over time, it typically bypasses the time dimension and proceeds to describe the course of this directional change by ranking species related at least by parallel evolution according to a hierarchy of grades. The usual method of depicting anagenesis involves the construction of a table or figure with sequences of grades (anagenesis) plotted on one axis and phyletic diversification (cladogenesis) plotted on the other. Two examples of this method are shown in

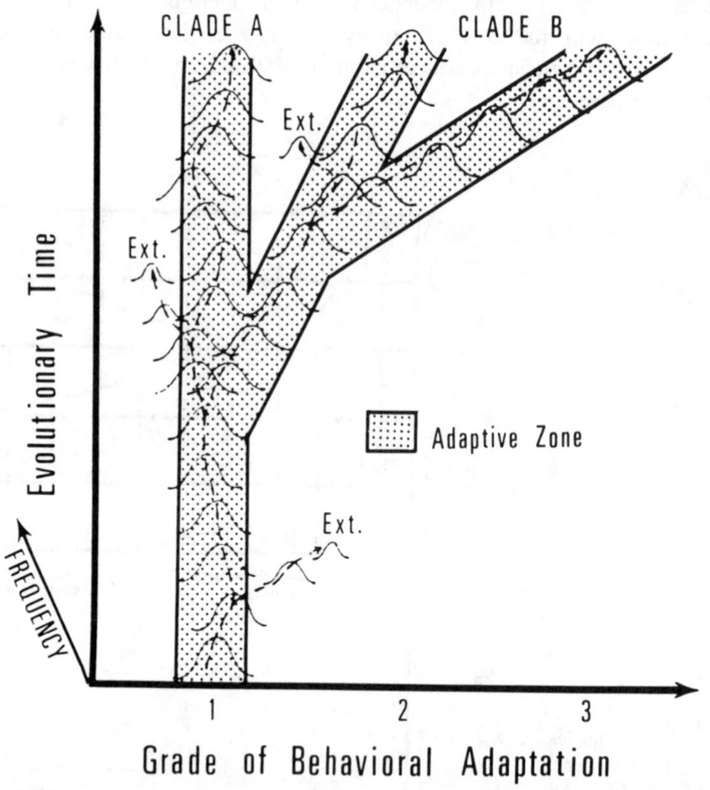

FIG. 6.2. The cladogram reflects several patterns of phyletic evolution including a population exhibiting one grade of behavioral adaptation well-adapted to a stable zone (Clade A), a bifurcating population gradually changing its dominant mode of adaptation in response to a shifting adaptive zone (Clade B), and several subpopulations whose change in behavior is not in response to selection pressures of the adaptive zone and who eventually become extinct (Ext.). A slice of evolutionary time representing the present age would reveal three populations each exhibiting a normal distribution of phenotypic variability around a different mode (grade) of adaptation. Grade 3 is a more biologically efficient mode of adjustment than Grade 2, which is an improvement over Grade 1.

Fig. 6.1, taken from Dethier and Stellar (1961/1970) and Nissen (1951). At best, both figures reveal only very abstract relationships between several grades of behavioral adaptation or intelligence level and diversification within taxonomic groupings. The diagrams do not include a temporal reference scale or detailed information concerning phyletic lineages, and they do little more than reflect the intuitive statements about the direction of mental evolution made by Darwin and his followers. But, despite all the restrictions and subjectivity, this procedure provides a baseline for the proper assessment and understanding of the selective force of functional efficiency of design (Gould, 1976; Huxley, 1958b).

There is another way of depicting anagenesis that maintains the advantages of assessing improvement in biological efficiency of design without eliminating the potential objectivity of the phyletic tree. This method incorporates both the anagenic and cladogenic procedures. For our purposes, a description of the direction of evolution can be drawn by fitting the cladogram to a graph of the relationship between evolutionary time and evolutionary grade (Fig. 6.2). Note that at any slice of evolutionary time there may be one, two, or more grades of the phenotype (e.g., brain organization, mode of adaptation). Another important advantage of this method of depicting anagenesis is that the variation in phenotype may be shown. Phenotypic variation within each grade can thereby reflect adaptation to the adaptive zone, which may remain stabile or may have a direction—in this case a progressive direction. An adaptive zone, it should be pointed out, is a concept Simpson (1953) used to refer to the mutual relationship between environment and organism, and not a place where life is led. Later in the chapter I contrast this gradual, continuity model of anagenesis with a punctuated, steady-state model of directional change.

Anagenesis in the Comparative Psychology of Learning. Although the methodologies of both cladogenesis and anagenesis are frequently used in paleontology, their application to the study of the evolution of learning would appear to be virtually useless because learning and other psychological characteristics of ancestral species do not fossilize and leave a record. We can infer very little about how extinct species behaved, much less how they learned. The research programs of Bitterman (1965a, 1965b, 1969) and Harlow (1958, 1959a, 1959b) represent preliminary attempts at the anagenic procedure in animal learning, but they contain serious violations of the assumptions of anagenesis, most notable of which is the hypothesis that ancestral fish, reptile, or primate grades of learning ability are represented by the performance of a few members of one or two living species (Hodos & Campbell, 1969; Warren, 1973). However, there are ways to deal with this problem. Hodos (1973), for example, has devised a system of comparative testing that is designed to abstract the essential nature of the learning ability of grades of ancestral taxa. He argued that with studies of many species at each taxonomic level, a prototype of the common ancestral traits can be generated. Those characters that do not appear consistently across a single taxonomic level could be excluded as derived characteristics. These would be related, of course, to the particular adaptations of the organisms in their differing habitats. Hodos called this approach "phylogenetic arithmetic." Its goal would be the idealization of the defining characters of a taxon, a "morphotype" as he called it, a term borrowed from the comparative biology of Nelson (1969, 1970). The problem with this approach, as Nelson (1969) himself has pointed out, is that exactly how well such a morphotype mirrors past reality is something we can never know, for we are never able to compare the two. Ziegler (1973) has found this criticism important enough to render Hodos' proposition useless, based primarily on the fact that individual characters often undergo divergence and convergence in evolution. Although this is an accurate criticism, the same has also been made of numerical taxonomy (Hull, 1970) without negating its worth. The point is, given enough characters to work with, the morphotype will be useful in generating hypotheses about directional trends in the evolution of learning ability.

There have been some moderately successful attempts to do just this. For example, Gossette (1970, 1974; Gossette, Gossette, & Riddel, 1966; Gossette, Kraus, & Spiess, 1968) has used a procedure that he calls "taxonomic calibration" to plot "control ranges" of learning abilities. Control ranges reflect performance differences across various motivational and procedural variables. By fractionating control ranges into performance units using response scaling techniques, Gossette found phyletic clumping effects on various learning set and discrimination reversal problems. Those species most closely related taxonomically tended to perform alike.

Using a somewhat different logic, Rensch (1956, 1963) reasoned that increased body size and learning capacity should be correlated because the final size of the cerebral cortex grows at a rate faster than the rest of the brain, and a larger number of cortical cells makes more elaborate learning possible. In studies of many closely related species differing in absolute size, this relationship held true. Moreover, the correlation between brain size and learning ability was found for closely related birds, mammals, and beetles, suggesting that we are not speaking about a random trend. This finding is of great interest for it suggests that learning ability itself may serve as a directional selection factor, and that this would account for the prevalence of evolutionary progressions in overall size among many different radiations of animals. Rensch also noted that as a general rule, organisms with a short infancy are less capable of complex learning than are closely related organisms with a longer childhood—a rationale for the directional evolution of neoteny (c.f. Gould, 1977c).[4]

---

[4]Neoteny is the retention of formerly juvenile characters by adult descendants produced by retardation of somatic development. It has been suggested that delayed maturation and its consequent prolonged parental investment permits increased amounts of learning and improvement in the ability to learn. In addition, retention of rapid fetal growth rates during postnatal development would allow for an increased brain size.

The evidence collected by Jerison (1973), however, raises problems for this analysis and for anagenic thinking in general. Using endocranial fossil casts of many extinct organisms, and comparing the volumes of these casts to the expected brain volume for an average organism of the same clade (i.e., an encephalization quotient), Jerison found that amount of brain has not simply increased progressively in various lineages but that brain size is related to specific requirements in the ecosystem of select groups for organisms. For example, both carnivores and their probable prey among ungulate herbivores have displayed a continual increase in brain size during their evolution, but at each stage (early, middle, and late Tertiary period, and modern mammals) the carnivores always had the larger encephalization quotient. In South America, however, where advanced carnivores are not found and marsupial carnivores have relatively small encephalization quotients, there was no increase in brain size among herbivores. Jerison speculated that in general predators require a larger brain than their prey in order to do all the things necessary for a successful hunt. In all areas except South America the intense selection pressure imposed by the big-brained predators placed a premium on herbivore ''intelligence,'' and their brain size grew in response to this selection pressure. The increased herbivore brain size in turn placed a premium on the evolution of a bigger carnivore brain to maintain the necessary edge required for successful hunting. In South America, where large-brained carnivores are absent, there was no directional selection pressure influencing herbivore brain size and therefore no ''progressive improvement'' is apparent.

The Steady-State Hypothesis. Anagenesis stops, according to Huxley (1953), when it has reached its limit, when ''natural selection cannot push it any further in that particular direction.'' Obviously, brain size in South American herbivores did not reach an upper limit in any sense, and the herbivore's ability to avoid being eaten seems to be more an adaptation to local conditions than a grade in the progressive improvement of more efficient predator evasion strategies. In contrast to models of directional change, a system called steady-statism (Gould, 1977a) seems better able to account for evolutionary trends in the modification of behavior of this sort. This theory does not argue against the possibility of change, but rather for the absence of environmentally determined directional trends toward increasing complexity, integration, and efficiency. Trends that appear to be directional according to this model are actually the result of random, fluctuating periods of intense disruptive selection followed by longer periods of stabilizing selection. The model is borrowed from and hence very similar to the punctuated equilibrium model of the diversity of species (Eldredge & Gould, 1972).

The steady-state model of change at first sounds like a contradiction: How can change occur in a steady state? The answer is apparent when we realize that selection is essentially a random, stochastic process that may proceed very rapidly over short ecological intervals of time. Darwin maintained that evolution is nothing more than successive adaptation to changing local environments. Local environments fluctuate stochastically, so change, at least on a quantitatively small scale, is inevitable. However, there is no directional trend through time in this system because change in one direction in one generation is just as likely to run in the opposite direction in the next generation. Over ecological time, natural selection will be steady-statist and will be governed by alterations in the physical environment. But what about all the evidence that has been mounted for directional change in mental evolution—is it to be denied by

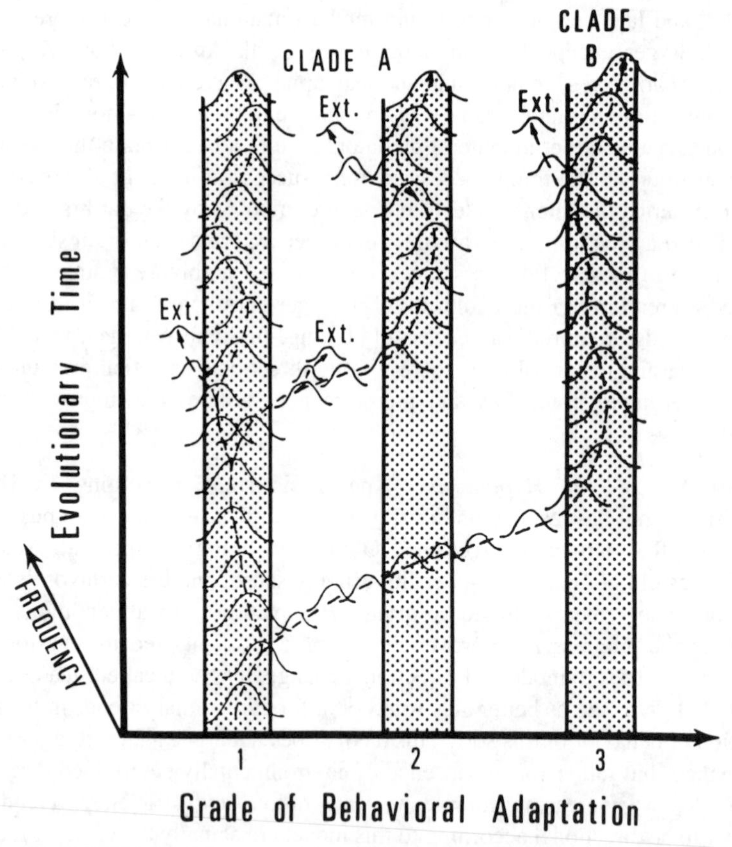

FIG. 6.3. *Steady—Statism*. The cladogram reflects the two main features of this model: (1) populations well-adapted to a stable zone with relatively little change in the dominant grade of behavioral adaptation; and (2) bifurcating populations whose evolutionary change in behavior is not in response to selection pressures of the adaptive zone. Some of the latter populations become extinct, but some change rapidly enough and endure long enough to achieve a new adaptive zone, and hence a new grade of behavior. Grades of behavioral adaptation are the result of random, chance factors, and are not necessarily a function of the recency of phyletic divergence (e.g., Clade B diverged early from Clade A, yet B displays a more advanced grade of behavior than the lineages in Clade A).

this model? The answer is no, because this model, based on the theory of punctuated equilibrium, couples the idea of general stability in the expression of behavioral characteristics with the idea that the tempo of evolution is sudden and discontinuous. Mental evolution is primarily nondirectional for it reflects only adaptation to local environments. But speciations may progress at different rates and have differential phyletic longevity, that is, some divergent lineages will endure through time longer than others (Fig. 6.3). In a random set of speciations, some lineages may become extinct because they were less capable than the ancestral population of competing in an adaptive zone. Occasionally, however, when random selection forces are intense, speciation will occur rapidly and endure long enough for achievement of a new adaptive zone. These stochastic invasions of new environments by peripheral lineages might result in new and improved neural mechanisms giving rise to a new grade of behavior, or there may simply be a change in the efficiency of mental processes through the interplay of the novel features of the new adaptive zone acting upon the inherited constitution of the species, even when it is fundamentally the same as the ancestral species. Gottlieb (1976) for example, has shown that certain environmental contingencies can exert an inductive effect on species-typical and species-atypical behavioral development by increasing the likelihood of occurrence of certain behaviors from the animal's response repetoire that ordinarily would not occur in the absence of this experience. There is nothing inherently directional about these random invasions, but when they occur as indicated, the effect will be the outward appearance of progressive, directional improvement in mental abilities.

Steady-Statism and Grades of Learning Ability. There is no overwhelming evidence to support either anagenesis or the steady-state model of mental evolution. For example, although we know very little about the relationship of brain structure and grades of learning ability, it seems apparent that the major lines of descent of the higher land vertebrates show a fairly general tendency to improve the efficiency of the brain. At the very least, an increase in structural complexity, especially an increase in the number of brain cells, has led to greater division of labor among the tissues, and therefore to more efficient functions. Presumably this has produced grades of improvement in sensation, perception, and learning. Gould (1977b, 1977c) has suggested that markedly increased brain size in human evolution may have had the most profound effect of all—it added enough neural connections to convert a relatively inflexible and rigidly programmed device into an organ of colored emotions and rational decisions.

Again, on the surface these trends appear to be anagenic improvements—gradual changes in brain organization and complexity produced quantitative and eventually qualitative differences in the ability to modify one's behavior through experience. On closer analysis, the steady-state model of directional change in mental evolution can also explain these grades of learning. Grades of learning ability are accounted for in this model by changes in neural organization or the generation of inductive forces that result from the action of stochastic processes operating on peripheral populations of organisms. If by chance these random selection factors continued to operate for a sufficiently long time, a new adaptive zone of neural organization would be achieved, yielding a new grade of learning. In some clades these "invasions" will cross several adaptive zones in rapid succession and the result will be that these organisms will display several grades of learning, some "higher" than others. Other less differentiated lineages (but not necessarily less recently evolved lineages) will display only the lowest grades of behavioral modification. The uninformed observer, looking at a slice of time, will see qualitative differences in the learning ability of living representatives of these clades and conclude that there has been progressive improvement in mental evolution, but with quantum leaps in the abilities of select taxa.

Steady-Statism and Constraints on Learning. In the comparative psychology of learning we also recognize qualitative differences among species under the heading of biological boundaries (Seligman & Hagar, 1972) or constraints on learning (Hinde & Stevenson-Hinde, 1973; Shettleworth, 1972). Associations made in one context may not be made in another. That is, the capacity for learning may be context specific or even stimulus specific. Thus herring gulls soon learn to distinguish their own chicks individually after a few days, but they seem unable to discriminate their own eggs even from quite differently marked eggs belonging to another species (Tinbergen, 1951). Similarly, male mute swans will repeatedly attack a harmless plastic decoy day after day so long as it has the releaser provided by the white coloring, but they immediately learn to identify their mates and their offspring (Demarest, 1981). Learning by association in both instances seems to have supplanted a more inflexible behavior pattern only when learning of this sort increased the animal's biological efficiency (i.e., control over and independence of the environment). To account for constraints on learning I would argue that during the prolonged steady-state periods of mental evolution, adaptation to local conditions permitted "adaptive specializations" to become available for use by other neural programs (Rozin, 1976). This model postulates the establishment of specialized solutions to specific problems in an organism's struggle to survive (i.e., "adaptive specializations"), and the subsequent emancipation of these programmed solutions from their original specialized context. Programs of action originally wired tightly into a particular functional system (e.g., the herring gull's species-specific reaction to its eggs, the mute swan's rigid reaction to a white, swan-shaped intruder in his territory) may be made accessible to other programs or systems in the brain (e.g., individual recognition) so that identification of one's chicks or one's mate becomes a possibility even though egg recognition and the identification of a harmless decoy are not. In Rozin's (1976) words:

> Intelligent behaviors . . . can be viewed as specific adaptations to specific problems rather than as the outputs of a superior, unitary, general intelligence sys-

> tem. . . . Since all reasonably complex animals have quite a few sophisticated circuits or programs (adaptive specializations) in their brains, I suggest that a major route to increasing flexibility and power over the environment, surely hallmarks of intelligence, would be to make these more generally available or accessible [p. 247, 256].

The steady-state model of mental evolution would argue that adaptive specializations may become accessible to already established learning abilities by means of the traditional slow, gradual process of natural selection to local ecological conditions. Viewed in this way, grades of learning and constraints on learning may well be the products of two different phases of the dynamic steady-state system: grades due to sudden, intense stochastic forces; constraints due to slow, gradual ecological selection forces. If this view has any merit we might expect to find broad phyletic similarities in the grades of learning that characterize a clade (especially for the lesser grades), but few similarities in the constraints on learning exhibited by representatives of the clade.

SCIENTISTS' ASSUMPTIONS AND THEORIES OF LEARNING

I began this chapter with a brief discussion of some implicit assumptions that underlie much of the thinking that goes on in comparative psychology. I also noted that few people are aware that their theories are founded on the adherence to one or another of these points of view and that fewer still pay any attention to this fact. For the biologically trained animal behaviorist in the 1940s and 1950s, grades of learning and constraints on learning were not considered unusual (Lorenz, 1935/1972; Tinbergen, 1951). Huxley (1953), in a popularized account of his *Evolution: The Modern Synthesis* (1943), referred several times to examples of grades and constraints on learning. Behaviorism, on the other hand, systematically avoided the study of built-in associative bias. Why has this been so? The answer, it seems to me, lies not in the obvious difference in training in evolutionary principles, for scholars in biology were and still are no less guilty of what Hodos and Campbell (1969) and Mayr (1958, 1968) call the typological approach. Mayr himself was criticized for using idealist concepts (Simpson, 1953), and the procedure of abstracting "morphotypes" of ancestral behavior suggested by Hodos (1973)—which, by the way, are more properly termed ethoclines—is a prime example of such typological thinking. I believe the real difference lies in the attitudes the different populations of scientists assumed on the issues of the tempo and direction of mental evolution.

A Scheme for Classifying the Evolutionary Positions of Learning Theories. Figure 6.4 is a classification matrix composed of four cells, with continuity and discontinuity on one axis representing two levels of the tempo of mental evolution, and anagenic and steady-state forces on the other axis repre-

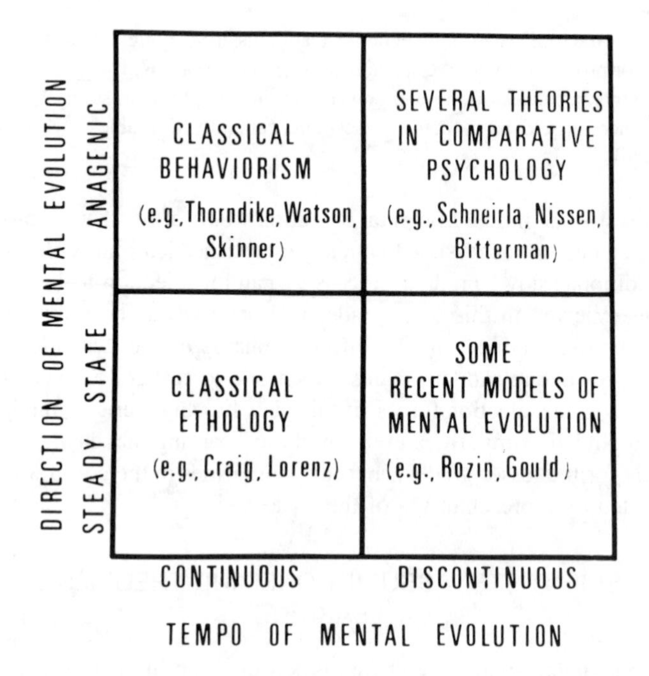

FIG. 6.4. *A Scheme for Classifying the Evolutionary Positions of Learning Theories*. Each cell in this classification matrix represents a different viewpoint regarding the direction and tempo of mental evolution. Representative theorists are included within each cell for comparison.

senting two levels of the direction of mental evolution. The assumptions of classical Behaviorism fall clearly in the anagenic-continuity cell. Recall the philosophical tradition from which behavior theory grew. The focus of John Locke and the British Empiricists was on upward progress through the steady accumulation of experience. Change had a direction, but the tempo was gradual. With the Darwinian notion of mental continuity and the romance with an appliance-centered animal psychology, Thorndike (1911) and Watson (1914) and most of all Skinner (1938) became convinced of the gradual basis of mental evolution. The laws of learning were the same, they argued. Differences, where they were found to exist, were small and quantitative. Harlow's (1958, 1959a, 1959b) comparative studies only reinforced this view, demonstrating qualitatively similar but progressively faster problem-solving ability in a sample of mammals meant to represent a phyletic sequence.

The assumptions of classical ethology, on the other hand, fall neatly into the steady-state–continuity category. Mental evolution is viewed as a process of gradual adaptation to the ecological needs of an organism. Actually, the pioneers of ethology—Heinroth, Jennings, and Whitman—were adherents to a progressionist philosophy common in the late 19th century. Like much evolutionary thinking between 1870 and 1890, ethology developed in the vogue of Herbert Spencer—a firm believer in the progressive nature of evolution (Hofstadter, 1955), It is somewhat ironic that Darwin stood almost alone in insisting that natural selection led only to increasing adaptation between organisms and their own environment and not to an abstract ideal of progress. However, the second generation of ethologists, particularly Lorenz and Craig, placed greater stress on the importance of the Darwinian idea of adaptation to local conditions, and species-specific behavior soon became the hallmark of ethology. Mental evolution was seen as a slow, gradual process of adapting to the requirements of the immediate environment and not as a sequence of stages reflecting progressive improvement in modes of behavioral adaptation. Is it any wonder ethologists and behaviorists differed so much in their respective attitudes about laboratory and field studies? Consider, for example, how an ethologist might characterize the kind of research done by the learning psychologist. With the emphasis on creating an artificial environment in order to neutralize the inherent predispositions of their animal subjects, behaviorists believed they were uncovering the general principles of learning. The result for the ethologist was the elimination of any possible chance of discovering real principles of behavior. The phenomenon of taste aversion learning (Garcia, Kimeldorf, & Hunt, 1961; Garcia & Koelling, 1966) which created such a stir among behaviorists (Revusky & Garcia, 1970) would hardly seem surprising to the ethologist.

The category typically overlooked by historians (for example, Boring, 1950) in tracing the study of mental evolution is the anagenic–discontinuity viewpoint. When Beach (1950) reprimanded comparative psychology for its failure to develop a broadly comparative program of research he prefaced his remarks with an apology: "to that tiny band of hardy souls who are herewith excepted from the general indictment that follows [p. 115]." This category consists of all the mental discontinuity theorists mentioned earlier, particularly Schneirla and Bitterman. Their point of view, that mental evolution has been a progressive but discontinuous process, is expressed in the following passage from Schneirla (1962):

> Throughout evolution, and particularly at times of crisis and other turning points, there evidently was a premium on mutations increasing the complexity and functional relationships of structures favouring wider and more versatile environmental adjustments. . . . New adaptive levels were achieved in evolution not just through added complexity of existing structures affecting function but through qualitative changes as well . . . The result presumably was not so much that mechanisms of lower levels were repeated on higher levels as that they were modified in relation to the new context. Accordingly, it is probable that in vertebrate evolution from fish to man . . . the breadth and plasticity of behaviour in its bearing on species adaptation [was widened] [p. 46].

Few animal psychologists took a position favorable to the assumption of sudden change in the tempo of mental evolution, but those who did all seem to fall in this category.

I cannot pass over Schneirla's remarks without noting a unique feature in his contribution to the comparative psychology of learning, that is, the attention he paid to grades of plasticity in ontogeny (i.e., paedomorphosis, cf. Gould, 1977c). In fact, variation in ontogenetic patterns in phylogeny was much more fundamental to his way of thinking than was change in adult abilities during phylogeny (Aronson, Tobach, Rosenblatt, & Lehrman, 1972). In our analysis of the direction and tempo of mental evolution we have overlooked the very thing an emphasis on developmental patterns provides, that is an account of the vehicle of mental evolution. Although it is beyond the scope of this chapter to trace this line of thinking, Gould (1977c) has provided an incisive review of the relationship between ontogeny and phylogeny in general, and his chapter on human evolution, in particular, delineates the significance of changes in ontogeny for mental evolution.

That leaves but one cell to the matrix, the steady-state–discontinuity attitude. It is the hypothesis concerning mental evolution that I have presented here for the first time, based on the Eldredge-Gould (1972) theory of punctuated equilibrium. I suppose it is cheating to say that this view is purely a steady-state theory because directional change must be incorporated to account for grades of learning. Unlike any other theory, however, this hypothesis proposes that directional change is sudden and dramatic, that it is a product of random, stochastic forces, and that most of the course of mental evolution is without direction.

Constraints on Learning in the Construction of Theory in Comparative Psychology. Our need to reconcile our gradualistic bias with a world full of discontinuity is a theme that appears again and again in the history of thought. We attempt to smooth the surface phenomena of saltation to an underlying continuity of process—*natura non facit saltum* (''nature does not make leaps''). Theories of change confront this dilemma at every turn. The notion of progress is another attractive idea that has dominated thinking in the western world over the last 200 years. Darwin, however, provided an alternative model of nature. Contemporary scientists, trying to find order in this diversity of opinion, have chosen to accept a limited set of these assumptions about nature and either ignored or denied the alternative attitudes. In one sense this has been similar to a biological limit on learning. As a constraint on learning will make some relationships easy to learn and others nearly impossible, our implicit assumptions make some facts easy to find while others go unnoticed (Schwartz, 1978). They affect the way we observe and interpret the world around us. According to Mendelsohn (1980): ''Continuity and discontinuity, I would argue, are not constructs of nature but constructs of the human mind used to interpret nature. It is the observer and interpreter of the physical world who posit continuities and discontinuities in the material before them. [p. 107].''

In comparative psychology our assumptions about the nature of mental evolution have shaped the theories we have devised, the questions we have asked, and the methodologies we have applied. Most animal psychologists have explicitly or implicitly incorporated the ideas of gradual, continuous, directional change in mental evolution into their theories and research programs without giving much serious attention to other points of view. These are the people who typically work at the microbehavioral level (e.g., behavioral geneticists, physiological psychologists, behaviorists, learning theorists). The Mental Discontinuity theorists differ primarily in their belief in rapid episodic changes in the tempo of evolution, and they are usually found working on macrobehavioral questions (e.g., phyletic origins, anagenesis, behavioral ecology). Few psychologists have ever worked at several levels of analysis, and this is reflected in the types of theories that have become popular, and the ease with which some criticism of comparative psychology has been accepted. Hodos and Campbell (1969), for example, were wrong when they said there is no theory in comparative psychology—there are, in fact, many of them. The theories are not particularly elegant as yet, but perhaps conceptual messiness is unavoidable in the social and behavioral sciences. We may never know whether change in mental evolution is directed or aimless, gradual or cataclysmic. Indeed there is no reason why these models cannot compliment one another. We must remember, however, that our need to find order in nature makes the search for solutions to these questions the challenge of comparative psychology—not its downfall.

ACKNOWLEDGMENTS

I would like to thank Kenneth Stunkel and Carrie Kertzman for their valuable comments, and Lorrie Demarest for help in preparing the final draft of the manuscript.

REFERENCES

Angell, J. R. The influence of Darwin on psychology. *Psychological Review*, 1909, *16*, 152–169. Reprinted in D. N. Robinson (Ed.), *Significant contributions to the history of psychology* (Series D, Vol. 3). Washington, D.C.: University Publications of America, 1978.

Aronson, L. R., Tobach, E., Rosenblatt, J. S., & Lehrman, D. S. *Selected writings of T. C. Schneirla*. San Francisco: W. H. Freeman, 1972.

Beach, F. The snark was a boojum. *American Psychologist*, 1950, *5*, 115–124.

de Beer, G. *Charles Darwin: Evolution by natural selection*. London: Thomas Nelson & Sons, 1963.

Bitterman, M. E. The evolution of intelligence. *Scientific American*, 1965, *212*, 92–100. (a)

Bitterman, M. E. Phyletic differences in learning. *American Psychologist*, 1965, *20*, 396–410. (b)

Bitterman, M. E. Thorndike and the problem of animal intelligence. *American Psychologist*, 1969, *24*, 444–453.

Bitterman, M. E., Wodinsky, J., & Candland, D. K. Some comparative psychology. *American Journal of Psychology*, 1958, *71*, 94–110.

Bonnet, C. *Contemplation de la nature* (2 vols.). Amsterdam: Marc–Michel Rey, 1764.

Boring, E. G. *A history of experimental psychology*. New York: Appleton–Century–Crofts, 1950.

Broom, R. Evolution—Is there intelligence behind it? *South African Journal of Science*, 1933, *30*, 1–19.

Bury, J. B. *The idea of progress*. London: MacMillan, 1920.

Dampier, W. C. *A history of science*. Cambridge, Eng.: Cambridge University Press, 1966.

Darwin, C. *Origin of species*. (M. Peckham, Ed.). Philadelphia: University of Pennsylvania Press, 1959. (Originally published, 1859.)

Darwin, C. *The descent of man, and selection in relation to sex*. London: Murray, 1871.

Darwin, C. *The formation of vegetable mould through the action of worms*. London: Murray, 1881.

Darwin, F. *More letters of Charles Darwin* (Vol. 1). London: Murray, 1903.

Dawson, C. *Progress and religion*. Garden City, N.Y.: Image books, 1960.

Demarest, J. Seasonal variation, sex differences and habituation of territorial behaviour in mute swans (*Cygnus olor*). In G. V. T. Matthews & M. Smart (Eds.), *Proceedings of the Second International Swan Symposium*. Slimbridge, England: International Waterfowl Research Bureau, 1981.

Dethier, V. G., & Stellar, E. *Animal behavior: Its evolutionary and neurological basis*. Englewood Cliffs, N.J.: Prentice–Hall, 1961. (also 3rd ed., 1970)

Ehrman, L., & Parsons, P. A. *The genetics of behavior*. Sunderland, Mass.: Sinauer, 1976.

Eldredge, N., & Gould, S. J. Punctuated equilibria: An alternative to phyletic gradualism. In T. J. M. Schopf (Ed.), *Models in Paleontology*. San Francisco: Freeman–Cooper, 1972.

Garcia, J., Kimeldorf, J., & Hunt, E. L. The use of ionizing radiation as a motivating stimulus. *Psychological Review*, 1961, *68*, 383–385.

Garcia, J., & Koelling, R. A. Relation of cue to consequence in avoidance learning. *Psychonomic Science*, 1966, *4*, 123–124.

Ghiselin, M. T. *The triumph of the Darwinian method*. Los Angeles: University of California Press, 1969.

Ghiselin, M. T. Darwin and evolutionary psychology. *Science*, 1973, *179*, 964–968.

Gingerich, P. D. Paleontology and phylogeny: Patterns of evolution at the species level in early tertiary mammals. *American Journal of Science* 1976, *276*, 1–28.

Gossette, R. L. Note on the calibration on inter-species successive discrimination reversal (SDR) performance differences: Qualitative vs. quantitative scaling. *Perceptual and Motor Skills*, 1970, *31*, 95–104.

Gossette, R. L. *A note on methodology in comparative behavioral analysis: Systematic variation vs. control by equation*. Unpublished manuscript, Hofstra University, Hempstead, New York, 1974.

Gossette, R. L., Gossette, M. F., & Riddel, W. Comparisons of successive discrimination reversal performances among closely and remotely related avian species. *Animal Behaviour*, 1966, *14*, 560–564.

Gossette, R. L., Kraus, G., & Speiss, J. Comparison of successive discrimination reversal (SDR) performance of seven mammalian species on a spatial task. *Psychonomic Science*, 1968, *12*, 193–194.

Gottlieb, G. The roles of experience in the development of behavior and the nervous system. G. Gottlieb (Ed.), *Neural and behavioral specificity*. New York: Academic Press, 1976.

Gould, S. J. Private thoughts of Lyell on progression and evolution. *Science*, 1970, *169*, 663–664.

Gould, S. J. Grades and clades revisited. In R. B. Masterton, W. Hodos, & H. Jerison (Eds.), *Evolution, brain and behavior: Persistent problems*. Hillsdale, N.J.: Lawrence Erlbaum Associates, 1976.

Gould, S. J. Eternal metaphors of paleontology. In A. Hallam (Ed.), *Patterns of evolution*. Amsterdam: Elsevier, 1977. (a)

Gould, S. J. *Ever since Darwin*. New York: Norton, 1977. (b)

Gould, S. J. *Ontogeny and phylogeny*. Cambridge, Mass.: Harvard University Press, 1977. (c)

Griffin, D. R. *The question of animal awareness*. New York: The Rockefeller University Press, 1976.

Hamilton, C. V. A study of trial and error reactions in mammals. *Journal of Animal Behavior*, 1911, *1*, 33–66.

Harlow, H. F. The evolution of learning. In A. Roe & G. G. Simpson (Eds.), *Behavior and evolution*. New Haven: Yale University Press, 1958.

Harlow, H. F. The development of learning in the rhesus monkey. *American Scientist*, 1959, *47*, 459–479. (a)

Harlow, H. F. Learning set and error factor theory. In S. Koch (Ed.), *Psychology: A study of a science* (Vol. 2). New York: McGraw–Hill, 1959. (b)

Hill, C. William Harvey and the idea of monarchy. In C. Webster (Ed.), *The intellectual revolution of the seventeenth century*. London: Routledge, Kegan & Paul, 1974.

Hinde, R. A., & Stevenson-Hinde, J. (Eds.). *Constraints on learning: Limitations and predispositions*. New York: Academic Press, 1973.

Hodos, W. Evolutionary interpretation of neural and behavioral studies of living vertebrates. In F. O. Schmitt (Ed.), *The neurosciences: Second study program*. New York: Rockefeller University Press, 1970.

Hodos, W. Comparative study of brain–behavior relationships. In I. Goodman & M. Schein (Eds.), *Birds: Brain and behavior*. New York: Academic Press, 1973.

Hodos, W., & Campbell, C. B. G. Scala naturae: Why there is no theory in comparative psychology. *Psychological Review*, 1969, *76*, 337–350.

Hofstadter, R. *Social Darwinism in American thought*. Boston: Beacon Press, 1955.

Holmes, S. J. *The evolution of animal intelligence*. New York: Holt, 1911.

Hull, C. L. *Principles of behavior: An introduction to behavior theory*. New York: Appleton–Century–Crofts, 1943.

Hull, D. Contemporary systematic philosophies. *Annual Review of Ecology and Systematics* (Vol. I). Palo Alto, Calif.: Annual Reviews, Inc., 1970.

Hunter, W. S. The delayed reaction in animals and children, *Behavior Monographs*, 1912, *2*, 1–85.

Huxley, J. S. *Evolution: The modern synthesis*. New York: Harper, 1943.

Huxley, J. *Evolution in action*. New York: Harper & Row, 1953.

Huxley, J. S. Cultural process and evolution. In A. Roe & G. G. Simpson (Eds.), *Behavior and evolution*. New Haven, Conn.: Yale University Press, 1958. (a)

Huxley, J. S. Evolutionary processes and taxonomy with special reference to grades. *University of Uppsala Arsskrift*, 1958, pp. 21–39. (b)

James, W. *The principles of psychology* (2 vols.). New York: Holt, 1890.

Jennings, H. S. *Behavior of lower organisms*. New York: Columbia University Press, 1906.

Jerison, H. J. *Evolution of the brain and intelligence*. New York: Academic Press, 1973.

Kohler, W. *The mentality of apes*. New York: Harcourt, Brace, 1925.

Kuhn, T. S. *The structure of scientific revolutions*. Chicago: University of Chicago Press, 1962.

Lewes, G. H. *Problems of life and mind: The study of psychology*. London: England, 1879. In D. N. Robinson (Ed.), *Significant contributions to the history of psychology* (Series A, Vol. 6). Washington, D.C.: University Publications of America, 1978.

Lewontin, R. C. Biological determinism as a social weapon. In The Ann Arbor Science for the People Collective (Eds.), *Biology as a social weapon*. Minneapolis, Minn: Burgess, 1977.

Lockard, R. B. The albino rat. A defensible choice or bad habit? *American Psychologist*, 1968, *23*, 734–742.

Lockard, R. B. Reflections on the fall of comparative psychology: Is there a message for us all? *American Psychologist*, 1971, *26*, 168–179.

Lorenz, K. Der Kumpan in der Umwelt des Vogels. *Journal of Ornithologie,* 1935, *83,* 137–213, 289–413. (Translated by R. Martin in Lorenz, K., *Studies in animal and human behavior* (Vol. 1). Cambridge, Mass.: Harvard University Press, 1972).

Lovejoy, A. O. *The great chain of being.* Cambridge, Mass.: Harvard University Press, 1936.

Lyell, C. *Principles of geology* (Vol. 2). London: Murray, 1832.

Maier, N. R. F., & Schneirla, T. C. *Principles of animal psychology.* New York: McGraw–Hill, 1935. (References in Dover edition, 1964).

Masterton, R. B., Bitterman, M. E., Campbell, C. B. G., & Hotton, N. (Eds.). *Evolution of brain and behavior in vertebrates.* Hillsdale, N.J.: Lawrence Erlbaum Associates, 1976.

Mayr, E. Behavior and systematics. In A. Roe & G. G. Simpson (Eds.), *Behavior and evolution.* New Haven: Yale University Press, 1958.

Mayr, E. The role of systematics in biology. *Science,* 1968, *159,* 595–599.

Mayr, E. *Principles of systematic zoology.* New York: McGraw–Hill, 1969.

Mayr, E. *Population, species and evolution.* Cambridge, Mass.: Harvard University Press, 1970.

Mendelsohn, E. The continuous and the discrete in the history of science. In O. G. Brim & J. Kagan (Eds.), *Constancy and change in human development.* Cambridge, Mass.: Harvard University Press, 1980.

Moore, B. R. The role of directed Pavlovian reactions in simple instrumental learning in the pigeon. In R. A. Hinde & J. Stevenson-Hinde (Eds.), *Constraints on learning.* New York: Academic Press, 1973.

Morgan, C. L. *Introduction to comparative psychology.* London: Walter Scott, 1894.

Mountjoy, P. T. An historical approach to comparative psychology. In M. R. Denny (Ed.), *Comparative psychology: An evolutionary analysis of animal behavior.* New York: Wiley, 1980.

Nelson, G. J. Origin and diversification of teleostean fishes. *Annals of the New York Academy of Sciences,* 1969, *147,* 18–30.

Nelson, G. J. Outline of a theory of comparative biology. *Systematic Zoology,* 1970, *19,* 373–385.

Nissen, H. W. Phylogenetic comparison. In S. S. Stevens (Ed.), *Handbook of experimental psychology.* New York: Wiley, 1951.

Pavlov, I. P. *Conditioned reflexes: An investigation of the physiological activity of the cerebral cortex.* Oxford: Oxford University Press, 1927. (References from Dover edition, 1960).

Popper, K. *The logic of scientific discovery.* New York: Basic Books, 1959.

Popper, K. (Ed.). *Objective knowledge: An evolutionary approach.* Oxford: Clarendon Press, 1972.

Rensch, B. Increase of learning capability with increase in brain size. *American Naturalist,* 1956, *90,* 81–95.

Rensch, B. *Evolution above the species level.* New York: Columbia University Press, 1960.

Rensch, B. The relation between the central nervous functions and the body size of animals. In J. Huxley, A. C. Hardy, & E. Ford (Eds.), *Evolution as a process.* New York: Collier, 1963.

Rensch, B. *Biophilosophy.* New York: Columbia University Press, 1971.

Revusky, S., & Garcia, J. Learned associations over long delays. In G. H. Bower (Ed.), *The psychology of learning and motivation: Advances in research and theory* (Vol. 4). New York: Academic Press, 1970.

Robinson, D. N. *An intellectual history of psychology.* New York: MacMillan, 1981.

Romanes, G. J. *Animal intelligence.* New York: Appleton, 1883.

Romanes, G. J. *Mental evolution in animals.* New York: Appleton, 1884.

Romanes, G. J. *Mental evolution in man: Origin of human faculty.* London: Kegan Paul, 1888.

Rozin, P. The evolution of intelligence and access to the cognitive unconscious. In J. A. Sprague & A. N. Epstein (Eds.), *Progress in psychobiology and physiological psychology* (Vol. 6). New York: Academic Press, 1976.

Rumbaugh, D. M. The learning and sensory capacities of the squirrel monkey in phylogenetic perspective. In L. A. Rosenblum & R. W. Cooper (Eds.), *The squirrel monkey.* New York: Academic Press, 1968.

Schneirla, T. C. Ant learning as a problem in comparative psychology. In L. R. Aronson, E. Tobach, J. S. Rosenblatt, & D. S. Lehrman (Eds.), *Selected writings of T. C. Schneirla*. San Francisco: W. H. Freeman, 1972. (Reprinted from P. L. Harriman (Ed.), *Twentieth century psychology*. New York: Philosophical Library, 1946.)

Schneirla, T. C. An evolutionary and developmental theory of biphasic process underlying approach and withdrawal. In L. R. Aronson, E. Tobach, J. S. Rosenblatt, & D. S. Lehrman (Eds.), *Selected writings of T. C. Schneirla*. San Francisco: W. H. Freeman, 1972. (Reprinted from M. R. Jones (Ed.), *Nebraska Symposium on Motivation,* 1959, *7,* 1–42.)

Schneirla, T. C. Psychology, comparative. In L. R. Aronson, E. Tobach, J. S. Rosenblatt, & D. S. Lehrman (Eds.), *Selected writings of T. C. Schneirla*. San Francisco: W. H. Freeman, 1972. (Reprinted from *Encyclopaedia Britannica,* 14th ed., 1962, *18,* 690–703).

Schwartz, B. *Psychology of learning and behavior*. New York: Norton, 1978.

Seligman, M. E. P., & Hager, J. L. *Biological boundaries of learning*. New York: Appleton–Century–Crofts, 1972.

Shettleworth, S. J. Constraints on learning. In D. S. Lehrman, R. A. Hinde, & E. Shaw (Eds.), *Advances in the study of behavior* (Vol. 4). New York: Academic Press, 1972.

Simpson, G. G. *Tempo and mode in evolution*. New York: Columbia University Press, 1944.

Simpson, G. G. *The major features of evolution*. New York: Columbia University Press, 1953.

Simpson, G. G. *Principles of animal taxonomy*. New York: Columbia University Press, 1961.

Skinner, B. F. *The behavior of organisms*. New York: Appleton–Century–Crofts, 1938.

Skinner, B. F. A case history in scientific method. *American Psychologist,* 1956, *11,* 221–233.

Spencer, H. *Principles of psychology*. London: Longmans, 1855.

Spencer, H. *Principles of psychology* (2nd ed.). London: Williams & Norgate, 1872.

Thompson, D. W. *On growth and form*. Cambridge: Cambridge University Press, 1917.

Thorndike, E. L. *Animal intelligence*. New York: MacMillan, 1911.

Tinbergen, N. *The study of instinct*. New York: Oxford University Press, 1951.

Tolman, E. C. *Purposive behavior in animals and men*. New York: Appleton–Century–Crofts, 1932.

Wagar, W. W. *The idea of progress since the Renaissance*. New York: Wiley, 1969.

Warren, J. M. The comparative psychology of learning. *Annual Review of Psychology,* 1965, *16,* 95–118. (a)

Warren, J. M. Primate learning in comparative perspective. In A. M. Schrier, H. F. Harlow, & F. Stollnitz (Eds.), *Behavior of nonhuman primates* (Vol. 1). New York: Academic Press, 1965. (b)

Warren, J. M. Learning in vertebrates. In D. A. Dewsbury & D. A. Rethlingshafer (Eds.), *Comparative psychology: A modern survey*. New York: McGraw–Hill, 1973.

Washburn, M. F. *The animal mind*. New York: MacMillan, 1908.

Watson, J. B. *Behavior: An introduction to comparative psychology*. New York: Holt, 1914.

Wilson, E. O. *Sociobiology*. Cambridge, Mass.: Harvard University Press, 1975.

Wundt, W. *Lectures on human and animal psychology,* 1894. (Translated by J. E. Creighton & E. B. Titchener, New York: MacMillan, 1907.)

Wyers, E. J., Peeke, H. V. S., & Herz, M. J. Behavioral habituation in invertebrates. In H. V. S. Peeke & M. J. Herz (Eds.), *Habituation* (Vol. 1). New York: Academic Press, 1973.

Yarczower, M., & Hazlett, L. Evolutionary scales and anagenesis. *Psychological Bulletin,* 1977, *84,* 1088–1097.

Young, R. M. The role of psychology in the nineteenth-century evolutionary debate. In M. Henle, J. Jaynes, & J. J. Sullivan (Eds.), *Historical conceptions of psychology*. New York: Springer, 1973.

Zeigler, H. P. The problem of comparison in comparative psychology. In E. Tobach, H. E. Adler, & L. L. Adler (Eds.), Comparative psychology at issue. *Annals of the New York Academy of Sciences,* 1973, *223,* p. 126–134.

7 Hybrid Models: Modifications in Models of Social Behavior That Are Borrowed Across Species and Up Evolutionary Grades

Gail Zivin
Jefferson Medical College

This chapter concerns the mental models that behavioral researchers use when doing comparative work. Mental models are those images or metaphors that clarify a phenomenon in the mind of the researcher (cf. Hempel, 1965). Comparative behavioral work attempts to understand the behavior of one organism by comparing it with similar behaviors in related species. This chapter addresses a source of possible confusion in cross-specific comparisons.

The primary goal of the traditional comparative method is to find evidence for common ancestry, and hence for clearer phylogenetic relation, in the origin of the behavior or structure in question (Alcock, 1975). The knowledge gained from noting apt comparisons, however, is used far more powerfully, if less rigorously, than just for determining relatedness in the origin of specific behaviors. When a puzzling or previously unnoticed behavior in one species is found to be related significantly to a better known behavior in another species, the researcher's understanding of the better known behavior tends to condition his or her understanding of the behavior in the other species. That is, the researcher's model (or ''notion'' or ''concept'') of the behavior in the better known examples—subsuming how the behavior works and the factors that influence it—will be the model by which the researcher initially understands the similar behavior in the less well-studied case. Regardless of how clearly the applicability of the model is tested to the new species, the model that is retained will color, shape, impel, limit further hypotheses, and affect how test results are assimilated. One example of such influence of a model borrowed from another species is found in the work that distinguishes between ''peck order'' and ''peck right'' in birds. Masure and Allee (1934) studied peck order (i.e., which birds in a hierarchy freely peck which other birds) in pigeons, using the model of peck order previously applied to chickens. The model for chickens does not provide for peck order to change across territorial boundaries within the chicken yard, despite the fact that the peck order of pigeons varies according to territorial boundaries within the larger field. Masure and Allee, unmindful of this discrepancy in their model, coined the additional construct of ''peck right'' to account for these apparent reversals of rank within the chicken model (Brown, 1975). Their comparison went beyond relatedness; they understood the behavior of one species in terms of the model borrowed from another species. Borrowing a model is dangerous when the researcher is ignorant of what he or she is borrowing and the manner in which this borrowing may limit his or her vision.

This chapter traces the history of one case of model borrowing, in which it appears that current researchers have lost sight of certain factors in their newly applied model. Their blindness is not to the elements in the model that were drawn from another set of species; that the new model is taken from other species is clearly acknowledged, and theorizing is shaped by it. Rather, the blindness seems to be to the maintenance of elements of a specific model that had been applied earlier to the species of interest. Current researchers are apparently unaware that elements from two models of dominance, one from nonhuman primates and one from a particular view of human functioning, have formed a hybrid model that is applied to humans. This leads to the conjecture presented here that much model borrowing in fact may be modified unwittingly by the knowledge a researcher already has of the species in question. It seems plausible that the context of an object of inquiry subtly influences the interpretation of new ideas that are applied to it. The unwitting modification of a model borrowed from another species may be fairly prevalent in applications of the comparative method, and comprehensible cognitive principles that act outside awareness may be responsible for such context-specific modifications.

The following history of the creation of this hybrid model of dominance in humans questions whether this modification of borrowed models is even more widespread. The data used to infer the researchers' models come as much from their operationalizations of dominance phenomena and implications drawn in discussion sections as from the conceptualizations stated explicitly in introductory sections. Lack of fit between stated conceptions and the implications of operationalizations or of claims were particularly helpful in clarifying the mixed nature of the hybrid model.

THE SOCIAL STRUCTURAL MODEL IN ANIMAL BEHAVIOR

Dominance, whether in human or nonhuman models, is associated with the priority of access to resources. In many studies of nonhuman animals, it has been demonstrated that high dominance rank yields a greater access to reproductive activities, better nutrition, and relatively less stress (Wilson, 1975). Low dominance rank yields some access to these resources, as well as opportunities to move up the hierarchy and a greater potential for safety and survival than are available outside the group (Barash, 1977). The dominance structure comprises the hierarchical set of roles attached to the priorities of access that the members of a group negotiate through competition. The potential for such a structure to characterize behavior in the group has been selected by evolution.

In the nonhuman model, dominance structure is the property not of the individual but of the group's social structure. Rank in the hierarchy is proportional to an individual's access to resources as a member of the group; one's degree of dominance can be known only by reference to one's place in the structure. This structure is not reducible to the property of individuals; rather, many researchers consider it an emergent part of the group. As demonstrated later, this is in direct contrast to the prehybrid model of human dominance, which, termed here the ''trait model,'' sees the individual as the locus of dominance.

Despite agreement on these general features, there is lack of agreement among animal behaviorists concerning the set of properties that define dominance interactions. Although most theorists (e.g., Brown, 1975) assert that the priority of access to resources is the focus of dominance interactions, some also require that interaction strategies or tone be aggressive, hostile, or agonistic, ''albeit sometimes of a subtle and indirect nature [Wilson, 1975, p. 583],'' to be considered dominance interactions. Indeed, their attempt to avoid the anthropomorphic connotations of ''aggressive'' and ''hostile'' by coining the term ''agonistic'' to refer merely to vigorous challenge seems vitiated by their frequent use of ''agonistic'' to mean ''aggressive.'' Respected researchers have disputed the necessity of using such strategies or tone to define dominance interactions, whether formalized in theory or employed in measurement (Bernstein, 1980; Hand, 1981; Rowell, 1972). Still, attempts to broaden the concept of aggressive tone to include other strategies for ordering the priority of access (e.g., submission independent of aggression by Rowell, 1972, threat of withholding social contact by Hand, 1981, or multiple control strategies by, among others, Cain and Baenninger, 1980) have not been adopted widely.

There is at least one other property whose defining relation to dominance structures is also in question, namely, whether the rankings within hierarchies vary by task, function, or context. One frequent approach to data that may suggest multiple, contextually dependent hierarchies is to consider them organizations unique to the species for which they appear and to label them a special class, such as ''relative dominance hierarchies [Wilson, 1975, p. 280],'' whose theoretical relation to the model concept of an all-context hierarchy is unspecified.

Aspects of dominance for animal behaviorists include the existence of the hierarchy as well as the individual's rank in the hierarchy. The metatheoretical role of the concept of dominance in animal behavior research usually is to explain the latter aspect rather than to explain other social phenomena by either aspect. Strayer (1980) urges a richer and more specific explanation of other social features by either aspect of dominance. When dominance is invoked as an explanation, however, it is in the first aspect and as part of a very general explanation of adaptive social order through aggression reduction and allocation of resources (Strayer & Strayer, 1976).

Many factors are studied for their possible influence on dominance rank. These include biological features "located" in an individual's group-level factors, such as parental rank, kinship lines, previous experience with other individuals, group size, and group competition. Notice, for later contrast with human models, that "aggressiveness" as a characteristic trait of an individual is only one of the many factors that explain rank. Within the animal behavior conception, "aggressiveness" is treated as a variable state of the interactants rather than as an invariable, measurable amount of a characteristic trait on which the interactants may differ. It is shown later that this latter, internal conception is prominent in models of human dominance.

The earliest model of dominance in animal behavior was an easily observed image of social structure: birds' peck order. In 1922 in German and in 1935 in English, Schjelderup-Ebbe described the social hierarchy of peck order in several species of birds. A bird's place in a peck order is indicated by the number of other birds to whom it gives and from whom it receives pecks and threats. Recent behavioral studies selected to indicate dominance status, however, reflect an increasing sophistication in the understanding of the economics of social structure and group adaptation. Such early work as that of Schelderup-Ebbe (1935) and Maslow (1936) studied harassment and conflict outcome, but later work emphasized economically relevant behaviors that reflect the priority of access to resources. For example, Brown (1963) found that the distance from nest to feeding station could be used (in interaction with maps of territories) as an indicator of rank in the jay and great titmouse. Current textbooks on animal behavior teach that any behaviors that reflect the priority of access are the heart of an adequate operationalization of rank (Brown, 1975). Some researchers include agonistic threats, initiations, and outcomes as behaviors that indicate the likely priority of access.

In the late 1960s, an additional set of behaviors was used to index rank in animal studies. These facial and postural movements, termed *dominance and submissive gestures* (Jay, 1965; Yoshiba, 1968) and *threat display* (Andrew, 1972; Smith, 1977), also may include a more general evaluation of movement style along the dimensions from defensive and hesitating to assertive and unhesitating. The discovery (Hall & DeVore, 1965) of certain facial gestures (at first thought specific to a species but eventually found to be similar across primates [van Hooff, 1971]) that ritually accompany and often replace physical attack and conflict inspired primatologists (e.g., Jay, 1965) to use these gestures as badges of rank. Dominance or threat gestures in nonhuman primates include lowered brow, stare, forward-thrust face, and elongated neck. Submissive gestures include raised brows, wide-eyed stare, lower teeth exposed in grin, and contracted spinal posture (Hall & Devore, 1965; Rowell, 1972; van Hooff, 1971). Observing these gestures in the human domain easily leads to subjective emotional expressive—rather than objective—informational interpretations, the latter of which would be consistent with the perspective of nonindividual level of group structure. Animal behaviorists, however, rarely color their interpretations of dominance and instead are likely to use ritualized-display interpretations of specific gestures. They interpret the availability of gestures to individuals of specific ranks as species-specific behaviors that are selected for their reliability in signaling rank (Brown, 1975) and are available to all group members as they move through ranks with age and situational changes (Rowell, 1972). It is shown later in this chapter that the invitation to subjective interpretation of dominance-related gestures is more consistent with tendencies in theorizing about human dominance.

The computation of a dominance hierarchy is done by placing the observed, relevant behaviors—whether pecks, permitted access to food, supplantings, outcomes of fights, or gestures—in a dominance matrix. The behaviors are observed first in laboratory pairings of all group members or in naturalistic observations of dyadic interactions in the group. (Laboratory pairings and natural observations do not necessarily yield correlated matrix results.)

As shown in Fig. 7.1, the matrix margins list all group members (*A, B, C, . . ., N*) as winners or initiators of conflicts along the top margin and in the same order as losers or recipients on the side margin. The frequencies with which *A* won and *B* lost, with which *A* won and *C* lost, and so on, are entered in the corresponding cells. The order of the names is then reorganized so that, as far as possible, the minimum number of "reversals" (e.g., a win by *A* over *B*, who has won the majority of encounters with *A*) falls below the left–right descending diagonal of the matrix. This new order constitutes a roughly linear hierarchy.

The social structure model of dominance developed in animal behavior studies thus includes a means for describing and explaining dominance at the group and not the individual level. It also includes a systemwide view to explain the phenomenon that includes but does not isolate individual aggressiveness in explaining attained rank; uses dominance as a dependent variable phenomenon to be explained by other variables; and operationalizes rank in the dominance hierarchy through observable behaviors associated with overt conflict, physical emblems of threat and submission, and instances of giving and receiving priority of access. As seen later, the trait model of dominance that originated in human personality studies contrasts with the social structure model on each of these features, and the hybrid combination of these two models includes all these features as well as new features of the old human models.

Steps in the construction of a dominance matrix.

---

(1) *Observations:* B > D, C > A, B > A, C > B, B > D, etc.*

(2) *Starting Order:* Choose an arbitrary order, *e.g.*, DEACB.

(3) *Starting Matrix:* Enter the number of wins and losses observed in the matrix:

WINNER \ LOSER	D	E	A	C	B
D		24	3	0	0
E	0		13	0	0
A	21	11		0	0
C	12	16	17		14
B	37	31	41	0	

(4) *Treatment of Reversals:* A win by one individual over another that has won the majority of encounters with the first is termed a *reversal*. Rearrange the order so that only reversals fall below the diagonal, so far as possible; that is, change the above order to CBDAE or CBEDA or CBAED.

(5) *Treatment of Nonlinearity:* An order in which an individual dominates another (wins the majority of encounters) that dominates the first is termed *nonlinear* or circular. Rearrange to minimize the inevitable ambiguity. From the circular relationship diagrammed below there are three main alternatives, as shown. In the three alternatives not shown, the departure from linearity involves two individuals rather than one.

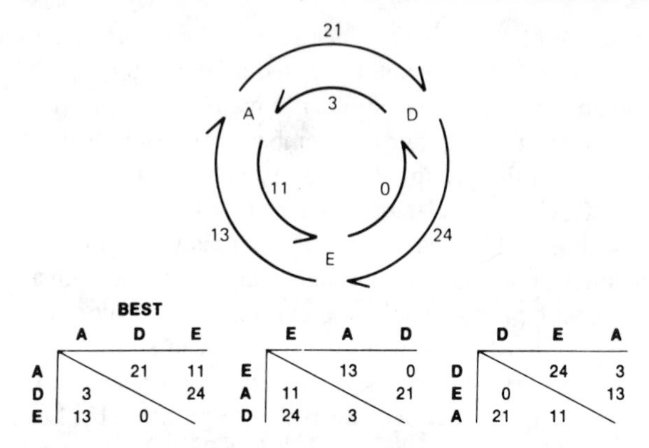

BEST

	A	D	E
A		21	11
D	3		24
E	13	0	

	E	A	D
E		13	0
A	11		21
D	24	3	

	D	E	A
D		24	3
E	0		13
A	21	11	

Place the individuals that are in the least ambiguous relationships (lowest proportion of reversals) in linear order. This procedure tends to minimize the total of encounters entered below the diagonal.

(6) *Final Matrix:* The one order that best reflects the order of dominance within the group is then CBADE. The following matrix may then be constructed:

WINNER \ LOSER	C	B	A	D	E	Wins	Losses
C		14	17	12	16	59	0
B	0		41	37	31	109	14
A	0	0		21	11	32	74
D	0	0	3		24	27	70
E	0	0	13	0		13	82

---

* B > D means B won an encounter with D. In most cases, these encounters take the form of supplanting rather than fighting.

FIG. 7.1. Dominance matrix construction, taken from J. L. Brown, 1975.

MODELS IN HUMAN BEHAVIOR

Trait Model

The earliest systematic notion of dominance in human studies was developed in the 1920s and 1930s and drew upon the social theorizing of the 19th century for some of its style (Sighele, 1892; Spencer, 1896). Although contemporaneous with the new observations in animal behavior, the original orientation in human studies was psychological, not social, and individual, not group structural. The model for human studies was trait, a quality of personality that individuals were hypothesized to ''have'' a certain amount of. Roughly speaking, a trait is a person's tendency to react to a stimulus in a particular way, independent of the nature of that stimulus. A common example is the trait of introversion; an introverted person tends to avoid interacting with others, regardless of the nature of any external situation. Psychologists as well as lay persons believe that body gestures and facial expressions can be read to reveal a person's traits. This may or may not be so, but the primary difficulty with a simple trait explanation of social interaction that does not take into account other factors such as type of previously established relationship and type of social structure is that it is only a partial explanation, with poor prediction to the style or outcome of an interaction from the individuals' general reaction tendencies.

The trait idea developed in the 1930s and 1940s, when a methodology for reliably (but not necessarily validly) measuring aspects of personality in standardized tests and special clinical practices was being formulated. Traits were measured by written questionnaires to which individuals indicated their varying degrees of agreement with certain statements. Arithmetic formulas converted sums of the degrees of agreement to estimates of relative amounts of measured traits. These developments fostered interest in the theories of personality and motivation, which proposed such stable, measurable features as traits (Allport, 1937) and needs (Murray, 1936).

Gordon Allport, who constructed a sophisticated, highly influential theory of personality traits, elevated dominance to the status of an important ''expressive common trait'' (Allport, 1937, p. 410). Although he used the words *dominance* and *ascendance* interchangeably, Allport favored ''ascendance'' as the term for this trait. By ''expressive'' Allport (1937) meant giving ''color [to] behavior that is specifically motivated to some ulterior end [p. 410]''; by ''common'' he indicated that a trait was normally distributed within a segment of the culture such that individuals could be measured against it for the relative amounts they possessed. ''Trait'' was defined as: ''a generalized and focalized neuropsychic system, peculiar to the individual, with the capacity to render many stimuli functionally equivalent and to initiate and guide consistent [equivalent] forms of adaptive and expressive behavior [p. 295]. ''They are *modi vivendi,* ultimately deriving their significance from the role they play in advancing adaptation with, and mastery of, the personal environment [p. 342].'' With these formal terms of introduction, the trait of ascendance (and its opposite trait, submission) are described thus:

> In the pursuit of almost any goal, the ascendant person will be ascendant. . . . In friendly relations as well as in unfriendly, one will dominate and be the ''victor,'' and the other will yield and become the ''vanquished.'' . . . Likewise in nonsocial situations an individual must, as a rule, either become the aggressor toward his environment or else submit to its pressure, giving up to the forces opposed to him [pp. 410–11].

Allport (1937) was aware of animal behavior approaches to dominance, as evidenced by his citation of Yerkes' (1925) and Skard's (1936) discussions of animal hierarchies. Despite these references, however, there is no mention of the influence of early animal behavior in Allport's chapter ''Ethology and the Study of Sentiments.'' (He did note that John Stuart Mill coined *ethology* in 1843 to refer to ''the science of character'' in man.) Allport proposed this still unrealized science as a wing of the new empirical psychology, because traits, as stable elements of character, now could be measured.

Allport cited several other vague and characterological notions of human dominance in early social theory (Sighele, 1892; Simmel, 1902; Spencer, 1896). These were traitlike conceptions of personal characteristics but not of social structure or relationships. Nevertheless, Allport mixed these with the hierarchical references from animal studies and used all of them to refer to what his traits of ascendance and submission labeled. By giving one term to various phenomena, Allport set an unfortunate precedent for treating various levels of analysis of dominance phenomena as equivalent. A few researchers (Anderson, 1940; Scott, 1958) have noted the mixtures and their attendant confusion, but the tendency to blur the distinctions across the several levels and manifestations of human social dominance has remained, particularly in psychological discussions of dominance.

Allport's theory and his published test for the traits of ascendance and submission stimulated research on the trait of dominance in adults and children. His *A–S Reaction Study* scale (Allport & Allport, 1928a) was used in psychological research (Allport & Allport, 1928b) and was revised for use in making personnel decisions in business. This questionnaire (Allport, 1937) reflects well the trait conception of dominance, as evidenced by a representative item:

> Someone tries to push ahead of you in line. You have been waiting for some time and can't wait much longer. Suppose the intruder is the same sex as yourself, do you usually:
>
> a. Remonstrate the intruder.
> b. ''Look daggers'' at intruder or make audible comments to neighbor.
> c. Decide not to wait and go away.
> d. Do nothing. [p. 412].

The test thus elicits reports of likely actions and reactions in various life areas, such as feelings in the presence of superiors and irresponsibility on the part of service workers or subordinates.

As interest in human dominance came to promote research during this early period, the model was based on the conception that individuals have different, measurable amounts of dominance, which is a motivational tendency toward certain behaviors. Social relations as a by-product of differential dominance relations were considered the obvious and uninteresting dyadic manifestation of these tendencies. When differentials in dominance relations were observed or tested, it was primarily as a study of the effect of the dominance trait (such as "domineeringness") on the individuals in the pair, whether in marital pairs (Lomas, 1959) or between parent and child (Biller, 1969). Although the sociologist R. F. Bales examined multiple interactions in groups, he considered dominance a trait that explained individual roles in the group (Bales, 1950; Bales & Cohen, 1979).

The term "dominance" was in active use among psychologists and reflected America's concern with authoritarianism and fascism. By the 1950s the trait of dominance was subdivided to include "domineeringness," "authoritarianism," "conservatism," and "dogmatism" in humans. At the same time that animal behaviorists were moving toward the term "agonism" to remove value judgment from their dominance characterizations, researchers of human phenomena were emphasizing the study of what were unabashedly considered societally harmful motivational tendencies related to the trait of dominance (Adorno, Frenkel-Brunswick, Levinson, & Sanford, 1950).

The metatheoretical role of the concept of dominance in human social psychological theory contrasts with its metatheoretical role in animal studies. Dominance rarely was studied as a dependent variable, although its age of onset (Jack, 1934), modifiability (Page, 1936), and causes in child rearing (Anderson & Brewer, 1946) were among the few independent variables examined in the earlier years. Much more typically, and increasingly in the 1960s, dominance was treated as an independent variable and studied for its contributions to other personality structures and to particular qualities of relationship. Theorizing remained at the individual motivational level. There was no quest as in animal behavior studies for ecological, evolutionary, or even group structural perspectives on human dominance.

Adoption of the Social Structural Model

Excitement about naturalistic observation and ethological theory in the early 1970s inspired many psychologists (primarily developmentalists), a few biologically oriented anthropologists, and some political scientists to explore—and borrow—concepts and methodology from animal behavior studies. Two developments started a new stream of social-structural empirical studies of dominance in humans: McGrew (1969) and Blurton Jones (1972) demonstrated that preschool children's classrooms lent themselves very well to the application of primatology's strongly behavioral—in contrast to psychology's previously inferential—observation techniques. Developmentalists became familiar with dominance assessments reported in the primate literature. Further, as primatologists developed interests in human social structure in the preschool classroom the blend of ethology and psychology was assured. Soon their theoretical approaches and empirical techniques were printed in the human developmental literature (e.g., Strayer & Strayer, 1976). A revolution against nonsocially oriented, nonbiologically informed, nonsystem oriented, prematurely experimental, and motivationally inferential traditions occurred in human studies. Implications of traits, as in older personality psychology, thus were avoided with the new, more behavioral techniques. It now was established that children had primate dominance hierarchies that were discernible through matrices derived from agonistic behavior and supplantings (McGrew, 1972; Missakian, 1975; Sluckin & Smith, 1977; Smith, 1974; Strayer & Strayer, 1976).

The selection of dominance- and submission-relevant behaviors at first resembled those studied in nonhuman primates: initiations of overt physical struggles for resources, supplantings, wins or losses of struggles, and initiations of aggressive (or agonistic) acts. Another behavior soon was associated with dominance: the giving and receiving of glances. The hierarchical social structure that can be formed from it is called an "attention structure." It still remains controversial, however, as to whether glances reflect an aspect of dominance, a different social structure that is correlated with dominance, or an entirely independent phenomenon (Abramovitch, 1980). According to M. R. A. Chance, who developed the concept (1967), glances are distributed within the human group in proportion to the members' importance for leadership or control functions.

In the pursuit of the behavioral and social-structural features of dominance in humans, communication gestures have come to be studied along with behaviors of access to resources. Dominance-related gestures have been sought in human studies as analogues or homologues of such gestures in nonhuman primates (Chevalier-Skolnikoff, 1973; van Hooff, 1972). Grant (1969), Blurton Jones (1967), and Montagner, Henry, Lombardot, Benedini, Restoin, Belzoni, Moyse, Humbert, Durand, Burnod, Nicolas, & Rosier (1977) noted human facial gestures that appear similar in form and context to nonhuman primate gestures. Camras (1977) found that the lowered brow and compressed lips, components of these gestures, appeared to intimidate an age peer into relinguishing a disputed object. None of these studies related these gestures to a systematic measure of dominance rank; student colleagues and this author, however, have located two facial gestures, one of which, the win-predicting Plus Face, correlates significantly with the "toughness ranking" measure of dominance hierarchy (Zivin, 1977b). Although neither face strongly resembled the threat and submission gestures of nonhuman primates, each predicted significantly that the facemaker will win or lose an encounter. These faces have been found in five age groups, from preschool and school ages (Zivin, 1977a, 1977b) through adolescents and adults (Acosta, Woodman, & Zivin, 1979; Hottenstein, 1977).

The "gesture" researchers just cited have remained firmly within the social structure model of dominance and have not treated the facial gestures as "expressions" of traits (or states). This is surprising because nonverbal gesture is one mode of behavior that is often seen to yield physical evidence of who one is or what one feels (Benthall & Polhemus, 1975; Spiegel & Machotka, 1974). The Western social scientific orientation to believe "muscular leakage" of one's true inner state harks back to Freud's famous dictum that emotions leak out at every pore and to Darwin's *Expression of Emotion in Man and Animals*. Yet, these researchers and other human ethologists studying dominance-related gestures (Masters, 1978) did not need to assume traits or states to study gestures usefully within the social structure model. Four theoretical strategies avoided simple trait assumptions for these researchers: (1) assignment of gestures to "emblem" status, which assumes a noncausal association of a gesture with a property (e.g., rank); (2) functional analyses of the contingencies upon displaying a gesture; (3) treatment of gestures as potential signals within Smith's (1977) ethological–contextual message analysis approach; and (4) experimental test between expressive and instrumental causes of gestures (Zivin, 1977b). It thereby has been possible to study the functional gesture behaviors and other variables associated with demonstrated rank within a predominantly social structure model of dominance.

Only by active self-restraint, urged by scientific logic, can the researcher avoid the assumption of traits and states behind gestures. Indeed, one group of nonethologically oriented researchers (Keating, Masur, Segall, Cysneiros, Divale, Kilbride, Komin, Leahy, Thurman, & Wirsing, 1981) is testing for the degree of perception of the trait of dominance in subjects who were shown line drawings of key facial elements. In Western culture, traits are considered natural, true assumptions. This is substantiated by the new and powerful attribution theory of social psychology (Jones & Nisbett, 1972); this theory has as one of its empirical foundations the ease and tenacity with which people attribute traits to others—but not to themselves—in interpreting the reasons for the behavior of others. Whether veridical or not, it appears that the assumption of traits comes easily to Westerners and resists being shed. This may contribute to the retention of some features of the trait model in the hybrid model described in the following section.

The indicators of the social structure model of dominance discussed so far all have been behavioral measures applicable equally across primate species. Early in the application of the social structure model to humans, however, there emerged an interesting nonbehavioral measure. Its creators used it within the social structure orientation as they searched for a structural hierarchy, yet they measured group members' perceptions of a trait of assumed relevance to dominance, "toughness." Toughness ranking is a uniquely human, cognitive, perceptual measure that was proposed to index dominance hierarchy in humans (Edelman & Omark, 1973; Omark & Edelman, 1975; Parker & Omark, 1980). It has evolved as a frequent alternative to behavioral matrices to measure human dominance hierarchy. The technique involves asking group members to use their intuition to make paired or serial ordinal judgments, adjustable in complexity for the age of the subjects and their cognitive level, of the "toughness" of all the other group members. The judgments result in an average-judged-toughness hierarchy of all group members. This form of hierarchy has been assumed tacitly to be equivalent to behavior-based hierarchies; empirical validiation is lacking, however. As Sluckin and Smith (1977) emphasize, toughness ranking is one measure of an individual's (averaged) perceptions of others filtered through their skill in ranking these perceptions along a single toughness dimension. It is not a measure of the functional dominance behaviors, and as Sluckin and Smith found, it does not correlate significantly with the hierarchy obtained from children's matrices of such dominance behaviors. Still, it remains accepted as a dominance measure, and its acceptance is informative. It indirectly measures the perception of some trait or quality on the presumption that the perception of that trait or quality allows the formation and functioning of a dominance hierarchy. This emphasis on cognitive evaluation, perceptual equivalence, and attributed or veridical trait may be fruitful in understanding human—and perhaps nonhuman—dominance structures. It brings to light one way in which the trait model of dominance has merged with the social structure model of dominance in humans toward a valuable end.

Emergence of the Hybrid Model

Over the past decade, a union has developed between the two models so far described, creating a third, hybrid model, which guides much of the current thinking on human dominance. This hybrid is applied to humans, but without the naive aspects of the trait approach that was prevalent from the 1920s to the 1950s; no researcher is reducing dominance hierarchies to differential expressions of a single motivational tendency. Yet, in this hybrid there persists from the trait model an emphasis on the motivational and subjective tendencies of the individual. These tendencies are considered important when explaining the existence of dominance structures and one's rank in the structure. A full description of the current hybrid model of dominance in humans includes three features: (1) emphasis on both group and individual levels through the orientation to explain structure and rank in terms of individual capacities and action tendencies; (2) the burgeoning ecological orientation to explain dominance structure rank in terms of group-level variables; and (3) the broadened role of dominance as a variable for explaining other group-level phenomena.

The Explanation of Structure and Rank in Terms of Individual Characteristics. Early in the work locating dominance hierarchies in human children, Omark and Edelman, as noted previously, emphasized the explanatory relevance of children's cognitive abilities to the dominance structure they may develop (Edelman & Omark, 1973; Omark & Edelman, 1975). They hypothesized that the stability and linearity in the dominance hierarchy of a group of children must influence and be influenced by the group's average capacity to perform the logical operation of "seriation" (Piaget, 1965), which involves the mental manipulations necessary to perceive, understand, and use a ranked linear grading, including the logic of the transitive rule (if $A > B$ and $B > C$, then $A > C$). Their hypothesis arose out of the correlative finding that preschoolers and kindergartners neither accurately identified the "tougher" person in paired comparisons nor seemed to comprehend the request to rank everyone in the class (Edelman & Omark, 1973). Omark and Edelman proposed that the functional relationship may work in both directions simultaneously; children cannot develop a stable dominance hierarchy until they have developed their seriating capacity, and they develop this capacity in the emotionally important, biologically scheduled arena of negotiating dominance relations as the hierarchy is being stabilized. Edelman and Omark proposed explanations of rank partially resulting from group members' perception of other members' individual "internal" characteristics. For this they asked for rankings on "niceness" and "smartness" as well as on "toughness." Judgments on the first two were found to create less clear, less reliable rankings than judgments on "toughness." Thus, the third published study of children's dominance that cited strong ethological influence (Edelman & Omark, 1973) departs from the strict primate model of the first two (Blurton Jones, 1967; McGrew, 1969) and explains the hierarchy structure and individual rank in terms of individual traits (or group members' perceptions of these traits).

Another example of emphasis on internal traits in the explanation of individual rank occurs in the work of Savin-Williams (1976, 1977, 1980). This is the only published work to have extended systematically the study of dominance hierarchies through young adolescence. In terms of the number and kinds of variables used, it is the most comprehensive study of dominance in a human group. Savin-Williams exemplifies the current hybrid by measuring hierarchy against traditional primate behaviors in a dominance matrix (appropriately adding the verbal agonism in commands and ridicule), while measuring several other variables, primarily personality, social characteristics, or traits that may explain rank in the hierarchy. Savin-Williams is careful not to claim explanatory status for these other variables, having studied them only by correlation. Among his other descriptive characterizations of the hierarchy and speculations on its effects can be discerned the attempt to explore likely traits that may be studied later for their explanation of achieved rank. Examples of such measured variables are athletic ability, physical fitness, chronological age, peer-related popularity, intelligence, physical pubertal status, the personality characteristic of locus of control, body surface area, and creativity (1976). Although significant correlations with rank were found for the first four (1976), the only variable for which Savin-Williams overtly suggests an explanatory role of rank is athletic ability (1980, p. 228).

Weisfeld (1980) also provides a clear instance of the hybrid model's double-level orientation in explaining structural rank by individual action tendencies. He integrates many traditional sociopsychological concepts concerning emotion and motivation into an explanation of social structure, showing rank as the result of a person's motivated actions. Weisfeld's relevant motivations include state-dependent motives, such as uncertainty reduction (the motive to avoid being uncertain), as well as the traditional trait concepts of self-esteem (motive to enhance one's positive view of oneself) and achievement motivation (motive to meet productive challenges). Barkow (1980), for another, employs an approach similar to Weisfeld's but focuses exclusively (in that paper) on the dominance-motivating force of the human aspirations for prestige and self-esteem.

The widely published, respected views of Robert Hinde, an animal behaviorist who now studies children as well, also reflect the use of the hybrid model. In an early treatment of dominance (1974), Hinde describes dominance both as an intervening variable at the individual level that explains the consistency of individual behaviors (which is similar to Allport's definition of trait) and as a variable located in and defining a relationship between individuals. Hinde's more recent three-level model (1976, 1980) explicitly eschews reducing the level of a relationship to the sum of the interactional qualities of events or the individuals in the relationship. Nevertheless, his emphasis on assessing the characteristics of the interacting individuals suggests a multilevel understanding of structures and relationships, not a reduction of these phenomena to variables at lower levels.

The Explanation of Structure and Rank in Terms of Group-Level Variables. The hybrid model explains dominance through explanations at both the individual and social structure levels. Although the explanation of dominance by supraindividual variables is less frequent among model users than the explanation by individual variables, its occurrence indicates the potential for further system-integrative constructions of the hybrid model.

Since the first days of ethological influence on the study of human dominance in the 1960s, natural selection explanations of the phenomenon of dominance have been invoked; a dominance hierarchy furthers fitness by distributing resources toward the more adaptive group members. Early applications of the animal model to humans also suggested that dominance structures lend stability and reduce the need for overt physical conflict in the group (Chance, 1967; Eibl-Eibesfeldt, 1970). Although disputed by current research (e.g., Barash, 1977; Rowell, 1972), one piece of empirical evidence in humans applies this older assertion to humans. Savin-Williams reported (1976) the observed reduction of physical violence among his subjects in temporal coordination with the speedy (1- to 2-week) stabilization of the new group's dominance hierarchy.

A subtler, more complex group-level explanation of dominance phenomena also is emerging. It is exemplified by Weisfeld's (1980) use of the concept "conditioning to success." This is essentially a social learning snowball that depends on the accumulation of group-level expectations and response tendencies toward individual group members and on the individual's accumulation of matching expectations of himself or herself. An individual succeeds incrementally, thus building expectations of success in himself and others, which sets the conditions for further success. This mechanism of accruing matched expectations that invite and facilitate winning or losing behavior does not reduce to the sum of individual reinforcement contingencies. It is a system-level process that orchestrates and coordinates members' actions while it rolls an initial winner or loser into consistent behaviors and expectations that fit that individual to a stable rank position. Such system-level explanatory approaches contribute an ecological sophistication to the potential uses of the hybrid model.

The Explanation of Group-Level Phenomena in Terms of Dominance. This aspect of the hybrid model, when added to the model in which dominance is a phenomenon to be explained, acquires a more powerful metatheoretical range than its predecessors, which focused on dominance either as a dependent or an independent variable. Examples of this broadened role in which social structure dominance is an explanatory variable are few but interesting. Barner—Barry (1980) uses the concept of children's dominance structures to study the development of their understanding of authority structures as the start of their integration into the highly abstract human political structure. In hypothesizing dominance analogues, Masters (1978) examines adults' reactions to photographs of political candidates in poses that look like the dominance-relevant gestures described previously. Uses of dominance as an independent variable in relation to group phenomena remain primarily the province of those human ethologists in political science who are termed, *political psychologists*. There is a new trend (Strayer, 1981a), however, to examine other group-level phenomena, such as affiliation networks, along with dominance structures in order to discover their mutual, system-structured, interactive relations.

The three features just described allow the hybrid model to reach beyond either of its progenitors in the scope and flexibility of its application. By giving attention to other system-level phenomena, whether through independent, dependent, or interactive variables' relations to dominance phenomena, the new model invites a comprehensive study of the system ramifications in which dominance is embedded ecologically. By retaining its dual-level focus, the hybrid allows what is learned of dominance at the group level to be anchored securely to concrete explanatory variables at the individual level.

HYBRID MODELS AS ADAPTATIONS

Why Hybridization?

Two hypotheses may explain the mixture of elements in the human dominance model as adaptations of researcher thinking. The first, already hinted at, is the notion that initial, widely accepted conceptions of phenomena, which remain deeply rooted in intellectual tradition, act as bridges to aid the assimilation of new conceptions. This facilitates the gradual evolution of conceptual structures, but it also implies that those who assimilate the new model may not recognize their own conceptual modifications that aid in the assimilation.

The second hypothesis suggests that there may be uniquely useful fits of an old model to the characteristics of the species for which the model was originally formed and that these characteristics may reappear tenaciously in the hybrid model. Of course, the model also will fit nonessentials or wrongly hypothesized features. Still, a more general model that fits roughly across many species may include some specific properties of a phenomenon in one species that allow a rich understanding of that phenomenon. In the present case, it may be that the individual-level interpretation and motivation tendencies add a key understanding of human social phenomena, such as dominance.

Evolutionary Grade

Poor fit of a model built on one species but applied to a second is particularly likely when the second species is of a different evolutionary grade on features salient to the modeled phenomenon. According to Wilson (1975), an evolutionary grade is: ''the evolutionary level of development in a particular structure, physiological process, or behavior occupied by a species or group of species. The evolutionary grade is to be distinguished from the phylogeny of the group'' [p. 584]. Despite new experimental technology's reduction of the apparent gap between great ape and human cognition (Chevalier-Skolnikoff & Poirer, 1978; Premack, 1977), two features on which humans still appear to be of a different evolutionary grade are spontaneous symbol use for interpreting the world and self-reflectiveness of internal states and cognitions. As these features color much human social interaction after 1 year of age and perhaps earlier (Lewis & Brooks-Gunn, 1979), it is likely they also color the phenomenon of human dominance: its circumstances of rank and structure recognition, modes of acquiring behavioral badges of rank, patterns of rank negotiation, and ramifications to other social categories. The hybrid model of dominance allows for the action of these two human cognitive features in its acceptance of the influence of a person's interpretive and motivational tendencies as partial explanations of rank and structure.

Humans' symbolic social cognition, which the hybrid model recognizes, creates considerable difficulty for the objective observation of the stimulus conditions that control human social behavior (Zivin, 1980). The complete behavior setting must somehow include the symbolic interpretations that individuals make of their environments, including other individuals. As symbolic cognition in social settings becomes more prevalent with age, it is perhaps not surprising that there are very few ethological studies of human adults (see, for example, Austin & Bates, 1974).

In attempting to accommodate ethological research to encompass influences of human social cognition, this author has employed three approaches, besides the toughness-ranking method mentioned earlier, that incorporate data from social cognitions into the full data sets for interaction observations. Brief mention of these approaches here may illustrate the nature of the methodological problems caused by humans' grade of social cognitions and may point to some helpful techniques for dealing with them. In one study (Acosta, Woodman, & Zivin, 1979), 24 cues of emotional intensity (e.g., high voice pitch, increase of gesticulation, among others) were subjected to factor analysis to identify conflict situations for adults who attempted polite and unruffled demeanors. That was a rather traditionally ethological solution to the discovery of the situation's subjective meanings. A greater departure, this time using field experiment techniques, took place in a study in which strangers were shown videotaped segments of naturally occurring conflicts that either did or did not include outcome-influencing facial gestures (Zivin, 1977b, 1982). The subjects, adults and children, were asked to report their expectations of the outcomes to test the effect on subjective social interpretation of the sight of these facial gestures. A third approach was taken from cultural anthropology and directly asked adults and children to label "what is going on" in videotapes of natural interactions of their own peers. Thus recognizing the methodological problems posed by humans' unique cognitive grade can invite solutions by which data can better capture the mixture of objective and symbolic stimuli that controls human social behavior.

Self-reflectiveness is the second feature of a distinctive evolutionary grade in humans. Its presence strengthens the argument for supplementing the study of uniquely human, symbolically transformed social phenomena by methods that elicit interactants' interpretations of situations. Of course, appropriate precautions must be used to check bias and self-deception, but it seems reasonable for the researcher to take advantage of human self-reflective capacity for getting around obstacles to simple objective behavioral observation. Seen in this light, there appears to be a good fit between problem and method by those researchers who use the hybrid model of dominance and also ask their subjects for self-reports. These include toughness ranking (Sluckin & Smith, 1977), sociometric ratings of popularity (Savin-Williams, 1976; Strayer, 1981a), informants' situational categorizing (Zivin, 1980, 1981a), and even subjects' drawings of themselves and their peers (Strayer & Strayer, 1980). Borrowing the informed methodology of animal behavior studies need not prevent the use of wisely selected, reliability-tested, and validated methods of eliciting subjects' reports. For well-chosen phenomena, self-report approaches appear to be appropriate research strategies within the limits of Gregory Bateson's (1977) pronouncement: "man can only know in the way that man can know."

CONCLUSION

One lesson from this historical view of dominance models is that when there is a prior model of a phenomenon for one species and that model is applied to a new species, the researcher must look for the indirect contributions from the old model to the new. They may be found in the total intellectual package of the model: the legitimate questions, prepotent hypotheses, and seeming implications that become the final working conception. The researcher should keep in mind, however, that there also may be some contradictions in this hybrid package, and these will require explicit logical attack. Furthermore, differences in evolutionary grade between the original and the newly modeled species may be relevant to the fit of the borrowed model. If the borrowing is up an evolutionary grade, it may be necessary to flesh out the model to take into account emergent properties of the newly modeled species. If the borrowing is down an evolutionary grade, the model may not fit at all, as the lack of particular emergent properties in the newly modeled species may not be representable by mere subtraction of features from the model. Regardless of the direction of borrowing, this history of dominance models illustrates the adaptive common sense among model borrowers, even when their formai positions are less adapted than their applied strategies.

The tenacious features of old models have been suggested to result from two processes. One is the model makers' need to retain some of their old conceptions of a phenomenon as an assimilative bridge to the new concept. The other is the possibility of a subtle fit between the newly studied properties of a species and the long-standing models of its phenomena; model makers may be wiser than they know in fitting seemingly tangential but unique properties in the model. If important properties that capture the flavor of the unique adaptions of the species may be omitted by an abstracting process, such as model borrowing, these properties may still be joined from the old model to the new to produce a hybrid that better represents the modeled phenomenon in that species.

ACKNOWLEDGMENT

Part of the work for this paper was supported by a grant from the Harry Frank Guggenheim Foundation.

REFERENCES

Abramovitch, R. Attention structures in hierarchically organized groups. In D. R. Omark, F. F. Strayer, & D. G. Freedman (Eds.), *Dominance relations*. New York: Garland, 1980.

Acosta, P. B., Woodman, L. D., & Zivin, G. *When is a fight a fight?* Paper presented at the annual meeting of the Animal Behavior Society, New Orleans, La., June 1979.

Adorno, T. W., Frenkel-Brunswick, E., Levinson, D. J., & Sanford, R. N. *The authoritarian personality*. New York: Harper, 1950.

Alcock, J. *Animal behavior*. Sunderland, Mass.: Sinauer Associates, 1975.

Allport, G. W. *Personality: A psychological interpretation*. New York: Henry Holt, 1937.

Allport, G. W., & Allport, F. H. *The A–S reaction study*. Boston: Houghton Mifflin, 1928. (a)

Allport, G. W., & Allport, F. H. The A–S reaction study. *Journal of Abnormal and Social Psychology*, 1928, *23*, 118–136. (b)

Anderson, H. H. An examination of the concepts of domination and integration in relation to dominance and ascendance. *Psychological Review*, 1940, *47*, 21–37.

Anderson, H. H., & Brewer, J. E. Studies of classroom teachers' personalities. II. *Applied Psychology Monographs*, 1946, *8*(Whole).

Andrew, R. J. The information potentially available in mammal displays. In R. A. Hinde (Ed.), *Non-verbal communication*. New York: Cambridge University Press, 1972.

Austin, W. T., & Bates, F. L. Ethological indicators of dominance and territory in a human captive population. *Social Forces*, 1974, *52*, 447–455.

Bales, R. F. *Interaction process analysis: A method for the study of small groups*. Cambridge, Mass.: Addison–Wesley, 1950.

Bales, R. F., & Cohen, S. P. *SYMLOG: A system for the multiple level observation of groups*. New York: Free Press, 1979.

Barash, D. P. *Sociobiology and behavior*. New York: Elsevier, 1977.

Barkow, J. Prestige and self-esteem: A biosocial interpretation. In D. R. Omark, F. F. Strayer, & D. G. Freedman (Eds.), *Dominance relations*. New York: Garland, 1980.

Barner–Barry, C. The structure of young childrens' authority relationships. In D. R. Omark, F. F. Strayer, & D. G. Freedman (Eds.), *Dominance relations*. New York: Garland, 1980.

Bateson, G. Personal communication, April 12, 1977.

Benthal, J., & Polhemus, T. *The body as a medium of expression*. New York: Dutton, 1975.

Bernstein, I. S. Dominance: A theoretical perspective for ethologists. In D. R. Omark, F. F. Strayer, & D. G. Freedman (Eds.), *Dominance relations*. New York: Garland, 1980.

Biller, H. B. Father dominance and sex-role development in kindergarten-age boys. *Developmental Psychology*, 1969, *1*, 87–94.

Blurton Jones, N. G. An ethological study of some aspects of social behaviour in children in nursery school. In D. Morris (Ed.), *Primate ethology*. London: Weidenfeld & Nicholson, 1967.

Blurton Jones, N. G. (Ed.). *Ethological studies of child behaviour*. London: Cambridge University Press, 1972.

Brown, J. L. Aggressiveness, dominance and social organization in the Stellar's jay. *Condor*, 1963, *65*, 460–484.

Brown, J. L. *The evolution of behavior*. New York: Norton, 1975.

Cain, N. W. & Baenninger, R. Social organization and maintenance of aggressive behavior in community-housed male Siamese fighting fish (*Betta splendens*). *Animal Learning & Behavior*, 1980, *8*(1), 171–176.

Camras, L. Facial expressions used by children in a conflict situation. *Child Development*, 1977, *48*, 1431–1435.

Chance, M. R. A. Attention structures as the basis of primate rank orders. *Man*, 1967, *2*, 503–518.

Chevalier-Skolnikoff, S. Facial expression of emotion in nonhuman primates. In P. Ekman (Ed.), *Darwin and facial expression*. New York: Academic Press, 1973.

Chevalier-Skolnikoff, S., & Poirer, F. E. (Eds.). *Primate bio-social development*. New York: Garland, 1978.

Edelman, M. S., & Omark, D. R. Dominance hierarchies in young children. *Social Science Information*, 1973, *12*, 103–110.

Eibl-Eibesfeldt, I. *Ethology: The biology of behavior*. New York: Holt, Rinehart, & Winston, 1970.

Grant, C. E. Human facial expression. *Man,* 1969, *4,* 525–536.

Hall, K. R. L., & DeVore, I. Baboon social behavior. In I. DeVore (Ed.), *Primate behavior: Field studies of monkeys and apes*. New York: Holt, Rinehart, & Winston, 1965.

Hand, J. L. *Threat of rejection: A non-aggressive mechanism for acquiring social dominance?* Paper presented at the annual meeting of the Animal Behavior Society, Knoxville, Tenn., June 1981.

Hempel, C. G. *Aspects of scientific exploration*. New York: Free Press, 1965.

Hinde, R. A. *Biological bases of human social behaviour*. New York: McGraw–Hill, 1974.

Hinde, R. A. Interactions, relationships, and social structures. *Man,* 1976, *11,* 1–17.

Hinde, R. A. *Ethology and the social sciences*. Keynote address, annual meeting of the Animal Behavior Society, Ft. Collins, Colo., June 1980.

Hottenstein, M. P. *An exploration of the relationship between age, social status, and facial gesturing*. Unpublished doctoral dissertation, University of Pennsylvania, 1977.

Jack, L. M. *University of Iowa Studies in Child Welfare,* 1934, *9*(3, Whole).

Jay, P. The common langur of North India. In I. DeVore (Ed.), *Primate behavior: Field studies of monkeys and apes*. New York: Holt, Rinehart, & Winston, 1965.

Jones, E. E., & Nisbett, R. E. *Attribution: Perceiving the causes of behavior*. Morristown, N.J.: General Learning Press, 1972.

Keating, C. F., Masur, A., Segall, M. H., Cysneiros, P. G., Divale, W. T., Kilbride, J. E., Komin, S., Leahy, P., Thurman, B., & Wirsing, R. Culture and the perception of social dominance from facial expression. *Journal of Personality and Social Psychology,* 1981, *40,* 615–626.

Lewis, M., & Brooks–Gunn, J. *Social cognition and the acquisition of self*. New York: Plenum, 1979.

Lomas, P. The husband–wife relationship in cases of puerperal breakdown. *British Journal of Medical Psychology,* 1959, *32,* 117–123.

Maslow, A. H. The role of dominance in the social and sexual behavior of infrahuman primates: IV. *Journal of Genetic Psychology,* 1936, *49,* 161–198.

Masters, R. Attention structures. In C. Barner-Barry (Chair) *Human Ethology and Political Psychology,* a symposium presented at the first annual meeting of the International Society for Political Psychology, New York, September 1978.

Masure, R. A., & Allee, W. C. The social order in flocks of the common chicken and pigeon. *Auk,* 1934, *51,* 306–327.

McGrew, W. C. An ethological study of agonistic behavior in preschool children. *Proceedings from the Symposia of the Second Congress of the International Primatological Society*. New York: Karger, 1969.

McGrew, W. C. *An ethological study of children's behavior*. New York: Academic Press, 1972.

Mill, J. S. *System of logic, book IV,* 1843.

Missakian, E. A. *Dominance relations in communally reared children in Synanon*. Paper presented at the annual meeting of the Animal Behavior Society, Boulder, Colo., June 1975.

Montagner, H., Henry, J. C., Lombardot, M., Benedini, M., Restoin, A., Belzoni, D., Moyse, A., Humbert, Y., Durand, M., Burnod, J., Nicolas, R.-M., & Rosier, M. Sur la différenciation de profils compartementaux chez les enfants de un à cinq ans à partir de l'étude des communications non-verbales. *Science Psychomotrice,* 1977, *1*(2), 53–88.

Murray, H. A. Basic concepts for a psychology of personality. *Journal of Genetic Psychology,* 1936, 15, 241–268.

Omark, D. R., & Edelman, M. S. A comparison of status hierarchies in young children: An ethological approach. *Social Science Information,* 1975, *14,* 87–107.

Page, M. L. *University of Iowa Studies, Studies in Child Welfare,* 1936, *12*(3, Whole).

Parker, R., & Omark, D. R. The social ecology of toughness. In D. R. Omark, F. F. Strayer, & D. G. Freedman (Eds.), *Dominance relations*. New York: Garland, 1980.

Piaget, J. *The child's conception of number*. New York: Norton, 1965.

Premack, D. *Intelligence in ape and man*. New York: Halsted Press, 1977.

Rowell, T. *Social behavior of monkeys*. Baltimore: Penquin, 1972.

Savin-Williams, R. C. An ethological study of dominance formation and maintenance in a group of human adolescents. *Child Development*, 1976, *47*, 972–979.

Savin-Williams, R. C. Dominance in a human adolescent group. *Animal Behaviour*, 1977, *25*, 400–405.

Savin-Williams, R. C. Dominance and submission among adolescent boys. In D. R. Omark, F. F. Strayer, & D. G. Freedman (Eds.), *Dominance relations*. New York: Garland, 1980.

Schjelderup-Ebbe, T. Beitrage zur sozialpsychologie des haushuhns. *Zeitschrift fuer Psychologie*, 1922, *88*, 225–252.

Schjelderup-Ebbe, T. Social behavior of birds. In C. Murchison (Ed.), *Handbook of social psychology*. Worcester, Mass.: Clark University Press, 1935.

Scott, J. P. *Aggression*. Chicago: University of Chicago Press, 1958.

Sighele, S. *Le crine á deux* Paris, 1892.

Simmel, G. The number of members as determining the sociological form of the group II. *American Journal of Sociology*, 1902, *8*, 1–46.

Skard, A. G. Studies of the psychology of needs: Observations and experiments on the sexual needs in hens. *Acta Psychologia*, 1936, *2*, 175–232.

Sluckin, A. M., & Smith, P. K. Two approaches to the concept of dominance in children. *Child Development*, 1977, *48*, 917–923.

Smith, P. K. Aggression in a preschool playgroup. In J. de Wit & W. W. Hartup (Eds.), *Determinants and origins of aggressive behavior*. The Hague: Mouton, 1974.

Smith, W. J. *The behavior of communication: An ethological approach*. Cambridge, Mass.: Harvard University Press, 1977.

Spencer, H. *Principles of ethics*. New York: Appleton, 1896.

Spiegel, J. P., & Machotka, P. *Messages of the body*. New York: Free Press, 1974.

Strayer, F. F. Current problems in the study of human dominance. In D. R. Omark, F. F. Strayer, & D. G. Freedman (Eds.), *Dominance relations*. New York: Garland, 1980.

Strayer, F. F. *Attention, popularity and dominance*. Paper presented at the Biennial Meeting of the Society for Research in Child Development, Boston, April, 1981. (a)

Strayer, F. F. Investigating the nature of social learning. In W. Charlesworth (Chair) *Back to nature, changing paradigms and child ethology ten years later*. A symposium presented at the biennial meeting of the Society for Research in Child Development, Boston, April 1981. (b)

Strayer, F. F., & Strayer, J. An ethological analysis of social agonism and dominance relations among preschool children. Child Development, 1976, *47*, 980–989.

Strayer, J., & Strayer, F. F. The representation of social dominance in children's drawings. In D. R. Omark, F. F. Strayer, & D. G. Freedman (Eds.), *Dominance relations*. New York: Garland, 1980.

van Hooff, J. A. R. A. M. *Aspects of the social behaviour and communication in human and higher nonhuman primates*. Rotterdam: Bronder-Offset, 1971.

van Hooff, J. A. R. A. M. A comparative approach to the phylogeny of laughter and smiling. In R. A. Hinde (Ed.), *Non-verbal communication*. Cambridge, Eng.: Cambridge University Press, 1972.

Weisfeld, G. E. Social dominance and human motivation. In D. R. Omark, F. F. Strayer, & D. G. Freedman (Eds.), *Dominance relations*. New York: Garland, 1980.

Wilson, E. O. *Sociobiology: The new synthesis*. Cambridge, Mass.: Harvard University Press, 1975.

Yerkes, R. M. *Almost human*. New York: Century, 1925.

Yoshiba, K. Local and intertroop variability in ecology and social behavior of common Indian langurs. In P. C. Jay (Ed.), *Primates*. New York: Holt, Rinehart, & Winston, 1968.

Zivin, G. On becoming subtle: Age and social rank changes in the use of a facial gesture. *Child Development*, 1977, *48*, 1314–1321. (a)

Zivin, G. Preschool children's facial gestures predict conflict outcomes. *Social Science Information*, 1977, *16*, 715–730. (b)

Zivin, G. *Making sense of multi-level behavioral analysis*. Paper presented at the annual meeting of the Animal Behavior Society, Ft. Collins, Colo., June 1980.

Zivin, G. Making a multi-level behavioral record for several interactants. In G. Zivin & L. Stettner (Chairs), *Capturing the quality of a relationship*, a symposium presented at the biennial meeting of the Society for Research in Child Development, Boston, April 1981. (a)

Zivin, G. *Why so many children: Comparative issues illustrated by the application of primate models to human social behavior*. Paper presented at the annual meeting of the Animal Behavior Society, Knoxville, Tenn., June 1981. (b)

Zivin, G. Watching the sands shift: Conceptualizing development of nonverbal mastery. In R. S. Feldman (Ed.), *The development of nonverbal communication in children*. New York: Springer-Verlag, 1982.

8 On the Process and Product of Cross-Species Generalization

Stephen J. Suomi
University of Wisconsin at Madison

Klaus Immelmann
University of Bielefeld

INTRODUCTION

There is virtually nothing of greater fundamental importance for the progress of science than the process of generalization. It is through generalization that specific observations are transformed and translated into scientific laws and principles. Scientific observations gathered in the most rigorous and objective fashion are meaningless unless they can be organized into a more general set of principles. "Scientific" laws that have not been derived from sets of actual data are in reality only speculative conjectures. The process of generalization provides the basis for converting specific data sets into larger explanatory principles. Without generalization there could be no science as we know it.

There are numerous different forms of generalization that can be found throughout the many fields of scientific inquiry. One such form common in the biological and behavioral sciences is that of *cross-species generalization,* in which general laws and principles are derived from data representing more than one species of subjects. Cross-species generalization provides the fundamental basis for all areas of comparative research in disciplines from anatomy to zoology. In each field a basic goal of research is to discover in which ways the species being compared are similar—and in what ways they differ from one another. When equivalent characteristics and/or phenomena are found in different species (i.e., when the characteristics and/or phenomena *generalize* between these species), we can begin to make assumptions about factors that cause and factors that control the phenomena in one species, based on what we know about them in the other(s). Conversely, we learn the limits of any explanatory principles when we fail to find generalization between the species under comparison. It is in the emerging patterns of cross-species similarities and differences that comparative sciences are developed and expanded.

In this chapter we present our views and interpretations of the process of generalization among different species of animals, including *homo sapiens*. We begin by describing what to date has been the classical biological approach to making cross-species comparisons. We then consider some factors that suggest an alternative basis for establishing generalizations among species—the ultimate factors that cause and the proximate factors that control the characteristic or phenomenon in question. Next we examine the different degrees of generalization that may exist between species in terms of these factors. The degree of generalization that one discovers in any particular case often depends on the level(s) of analysis that one is employing, as well as the ontogenic status of the organisms under study; relevant examples are provided. We then present some additional thoughts about the relationship between cross-species generalization and the phylogenetic similarity of the species being compared. Finally, we discuss some practical applications of knowledge about cross-species generalities, with a special emphasis on comparisons between human and nonhuman subjects.

CLASSICAL BIOLOGICAL APPROACHES TO CROSS-SPECIES GENERALIZATIONS

Cross-species comparisons have always been considered important in biology, and over the years, they have been formalized largely in those areas of the discipline concerned with taxonomy. Classification is unquestionably a basic part of biological study; one of the very first things beginning students in high school biology courses are taught is the general taxonomic organization of the animal kingdom. This organization is based largely on phylogenetic considerations. Today's taxonomists may still argue among themselves as to the most effective strategy for assigning individual specimens to particular taxa (and vice versa), but virtually all their different strategies involve phylogenetic comparison at one stage or another (Mayr, 1981).

The notion of phylogenetic relatedness has traditionally played an important role for biologists when evaluating apparent similarities of characteristics or phenomena between different species. Cross-species generalizations can be based on comparisons not only between species that show close phylogenetic relationships but also between others that belong to very different taxonomic categories. In both cases important conclusions about the cross-species generality of the characteristic or phenomenon can be duly derived—at least in theory. However, to a traditional biologist these two cases would not carry the same relative weight in terms of overall significance or generality of principle. In the case of similarity between two species that are phylogenetically close, the basis for the similarity is usually assumed to lie in the common ancestry of both species. The characteristic or phenomenon in question appears in like form in these species presumably because it has been "preserved" in the genomes of successive generations of the common ancestor's lines of descendants currently represented by the two species being compared. In contrast, most similarities that appear between species representing very different taxa are typically thought to be the result of parallel adaptation to the same or equivalent environmental conditions and demands. Here, the source of information is different for the respective species, whereas it is identical for closely related species.

Characteristics or phenomena that occur in like form in closely related species and therefore are thought to be of common phylogentic origin are termed *homologies*. One example of a relatively clear-cut homology between species can be found in the visual system of human beings and chimpanzees (*Pan*). Here, the basic sensory components of the system (e.g., cornea, iris, lens, rods, cones, bipolar cells, optic nerve) are essentially identical in these different species. Moreover, the mode of operation of the visual system, its neural connections to the brain, its ontogenic development, and even the forms of visual disorders (e.g., myopia, astigmatism, and glaucoma) that can occur in some individuals are equivalent in humans and chimps. Presumably, both sets of ancestors retained the same basic visual systems from their common progenitor, whatever it might have been, and these systems have not changed much since then.

Characteristics or phenomena that share similar features, but occur in phylogenetically distant species and therefore are thought to represent parallel adaptations to comparable environmental demands, are called *analogies* in traditional biological terminology. A classic example of analogous structures involves the wings of birds and those of insects. Both types of wings serve the same general purpose (propulsion of the organism through the air) and both clearly have some similarities in overall appearance. However, there are important fundamental differences between the wings of birds and those of insects, not the least of which concern their separate phylogenetic origins. Ontogenic differences are also clear: a bird's wing develops out of forelimb skeletal tissue, whereas an insect's wing grows out of dorsal skin. Other basic differences include mode of operation (flapping and fluttering versus buzzing), external covering (feathers versus scales) and number per organism (typically, one versus two bilateral pairs). Nevertheless, flying birds and flying insects share the ability to locomote through the air, and in that respect they have more in common with each other than with virtually any other member of the animal kingdom, phylogenetic origins notwithstanding.

There is a third type or class of cross-species similarity classically recognized by biologists; it is termed *convergence*.[1] Cases of convergence are interesting in

---

[1]Editor's note: Cases wherein structural or behavioral similarities are based on a combination of homologous and analogous factors have also been highlighted by Eibl-Eibesfeldt (this volume), who applies the label *homoiologies* to these sorts of phenomena. A decision to call these cases convergences or homoiologies seems less important at the moment than the recognition of their importance in comparative matters.

that they incorporate elements of both homologies and analogies—they involve features of clearly homologous origin that have adapted to similar environmental demands, although their respective modes of adaptation are unquestionably different from each other. One well-known example of convergent phenomena concerns the forelimbs of some diverse species whose ancestors were terrestrial but who later became basically aquatic and whose forelimbs in the process changed back into fins or flippers. Such evolutionary changes have occurred at least four different times and ways among vertebrates: in reptiles (ichthyosauruses), in birds (penguins), and independently in two mammalian groups, seals and whales. Technically, the forelimbs of all these vertebrate species are homologous, in that the ancestry of each can be traced to a primitive fish fin several hundred million years ago. In the interim each evolved into a leg or wing more suited for terrestrial life. However, each of these species lines subsequently readapted to aquatic conditions, albeit in quite different fashions, and in this sense they are clearly analogous. In other words, although these four types of forelimbs initially evolved separately from a common ancestrial form, more recently they have become more similar due to convergent evolution. Convergence, therefore, represents a *process* of evolution, whereas homologies and analogies are the *results* of this and other evolutionary processes. Cases of convergence represent alternative solutions to the same basic problem of adaptation, given some common material with which to work.

Homologies and analogies can be viewed as points on a continuum of generality, with generalizations based on homologies seen as more "appropriate" or "truer" than those based on mere analogies. This has been the basic position of most taxonomists, and it has proved to be reasonably useful, especially for cross-species comparisons involving anatomy. But similarity of phylogenetic origin is not the only basis on which any two species can be compared, and in many areas outside of anatomy it may not always be an entirely satisfactory or appropriate basis for comparison.

For example, consider a comparison of vocal communication among different species of primates, including humans. From the standpoint of phylogeny, one would most likely expect vocal communication in humans to resemble more closely that in gorillas and other great apes than in any other primate species because these are the species that are phylogenetically most similar to *Homo sapiens*. And indeed, human ears, being homologous with the ears of the great apes, do in fact more closely resemble those of gorillas than they do the ears of more "primitive" New World monkeys in terms of physical structure, effective frequency range, and physiological mode of operation. But gorillas are relatively silent creatures, communicating more with facial expressions, posture, and gestures and relying less on vocal exchanges than many other primate species; in contrast, marmosets, tamarins, and many other New World monkeys rely heavily on vocal exchanges in their communication systems. As a result, Snowdon (in press), among others, has argued that the vocal communication systems of marmosets and tamarins make much more appropriate models for human communication than do those of gorillas, even though the "receiving end" of the communication systems are much more homologous between humans and gorillas than they are between humans and marmosets.

Indeed, similarity of anatomical structure is not always guaranteed by relative closeness of phylogenetic relatedness. Return for a moment to the previous example of convergence—the fins and flippers of penguins, whales, and seals. Phylogenetically, whales are much closer to seals than they are to penguins; being mammals, whales share a more "recent" common ancestor with the mammalian seals than they do with the avian penguins. But in terms of mode of operation and range of uses for these forelimb structures, the fins of whales seem more like those of penguins than those of seals, and indeed (except for their external covering) whales' front fins look more like the forelimbs of penguins than the flippers of seals.

Thus, consideration of phylogenetic similarity provides a basis for cross-species comparisons—as embodied in the terms homology, analogy, and convergence—that has been generally quite useful for biologists, especially those concerned with taxonomy. But phylogenetic similarity is not the only basis for comparing different species of animals, and in some cases (especially those not directly involving anatomical generalizations) it does not always provide an entirely appropriate or useful standard for comparison. There are other factors that can be considered when making generalizations between different species, and two important sets of these factors are examined in the next section.

ULTIMATE AND PROXIMATE FACTORS

What sort of criteria besides similarity of physical appearance and extent of common phylogenetic ancestry might be appropriate for establishing generality between any two species? One important type of such criteria involves the degree to which factors that clearly influence the characteristic or phenomenon in question in one species do likewise in the other. In other words, one can judge the relative similarity of the two species by the extent to which they are affected in like manner by the same set of influences.

Psychologists often refer to these various influences as "independent variables," and they cover a large number of diverse factors, ranging from aspects of the species' respective physical and social environments to nutritional and pharmacological manipulations administered by experimenters. In contrast, most biologists have employed different terminology when discussing these sets of variables, and they have traditionally viewed them in a different conceptual light than typically has been used by experimental psychologists. Biologists tend to distinguish between those factors whose influence is "proximal" and those whose influence is "ultimate," at least as seen from an evolutionary perspective, whereas the terms "ultimate factors" and "proximate factors" are foreign to most psychologists (Immelmann & Suomi, 1981).

The "ultimate–proximate" dichotomy arises from the fact that those factors responsible for the appearance or cause of a given characteristic or phenomenon in a species may be quite different from those factors that actually control its appearance in individual members of the species. This dichotomy was first recognized from studies of reproductive behavior in diverse species of birds. When ornithologists started to look at breeding seasons of birds and compared different species and different geographic areas they soon became aware that every species produced its young at exactly those times of the year during which young could most successfully be raised. It appeared obvious in these species that reproductive activities were being restricted to the most "profitable" periods, and that such adaptation most likely evolved by the gradual elimination of the genes of those pairs who bred too early or too late in the season.

The main selective factor in this respect appears to be an adequate food supply for parents and young. Those pairs who raise their offspring during periods when the particular kind of food the species needs is most abundant and most easily accessible are most likely to be successful. The problem is that in most climatic zones of the earth, the time period of the year during which optimal food conditions prevail is rather short. On the other hand, the reproductive cycle of a bird, involving gonad growth, nest building, courtship, egg laying and incubation, may take weeks or—in larger species—even months. In order for these birds to be able to raise their young at the best time of the year, therefore, the early stages of the breeding cycle must take place *prior* to the period of optimum food availability, and this in turn requires a reaction to some kind of environmental information that "forecasts" the time of favorable food conditions to come.

For the vast majority of birds, it is the increasing day length after the winter solstice that—through a complicated control system that involves photoreceptors, parts of the brain, the hypophysis, as well as the gonads and the hormones produced by them—brings the individual into breeding condition and controls the occurrence of the early stages of the reproductive cycle at the most appropriate time. The reaction to increases in day length are so strong that, under laboratory conditions, birds can be brought into full breeding condition merely by offering them a few extra hours of light per day, even in autumn or winter and even with a rather poor food supply.

In describing these differences, Baker (1938) first introduced the terms *ultimate* and *proximate causes,* and Thomson (1950) subsequently proposed a change to *ultimate* and *proximate factors,* two terms that have since been used in the ethological literature. In this terminology, ultimate factors are those environmental variables that, through natural selection, have led to the most appropriate timing of avian reproduction, whereas proximate factors are those ambient factors that actually regulate the physiological processes involved in the different stages of the reproductive cycle. Once the differences between ultimate and proximate control of the annual cycles of birds were recognized, it soon became apparent that a similar distinction between factors that cause and factors that control a particular characteristic or phenomenon could also be found in many other biological processes, including behavior.

In the behavioral area, habitat selection provides a frequently cited example of the difference between ultimate and proximate factors (Immelmann & Balda, Note 1). Most species of animals live in a particular type of environment. Ultimate factors that during the course of evolution have led to a higher reproductive success for those individuals living in the most suitable types of environment and thus leaving more offspring than those that settle in suboptimal habitats, are, for example, an adequate food supply, camouflage protection from predators, favorable conditions for reproduction, and many others. Proximate factors, on the other hand, are the specific characteristics of this type of environment by which the single individual is able to recognize the most suitable habitat.

Other examples are provided by the social systems of animals. Within limits, the social organization of animals is species typical. Some live a solitary life, joining another individual only for a brief period of mating, whereas others live in pairs, in small family groups, in harems, or in big flocks, or (like some insects) in societies containing millions of individuals. A wealth of information has shown that such systems tend to be adapted to the specific environment of the species and have been ultimately selected for by factors like the spatial and temporal distribution of food, the type of vegetation, and the kind and amount of protection against predators it offers, as well as other factors (Hladik, 1975). The actual regulating factors that control the formation and maintenance of the species-typical social system, on the other hand, represent physiological and behavioral adaptations (e.g., the occurrence of appeasement gestures, the establishment of a rank order, or the existence of some kind of a ''social attraction'' that keeps members of a pair or a group together [Hinde, 1974]).

As a final example, in many species of animals the learning of species-specific characteristics by which members of the same species recognize each other, as well as the learning of distinguishing features of the immediate habitat, is restricted to particular stages in the development of the organism. This learning process is called *imprinting*, and it occurs only during so-called sensitive phases (periods) in development (cf. Immelmann, 1975). The onset and duration of such sensitive phases, which are different in different species, have ultimately been selected to occur at those times at which the opportunity to acquire biologically relevant information is also greatest (e.g., because the offspring is still with its parents and is living in a habitat that has already proved to be suitable for reproduction for this particular species). Proximate factors, on the other hand, usually involve physiological processes (e.g., a specific hormonal state and/or developmental processes in the brain) that actually control the onset and the end of sensitive phases.

The list of examples for the involvement of two different sets of factors could be continued considerably. The use of the terms ''ultimate'' and ''proximate'' factors, therefore, has become more and more widespread in the biological literature, especially since Lack (1954) broadened the definition of the terms: "Ultimate factors are concerned with survival value, proximate factors with adaptations in physiology and behavior." In contrast, most psychologists have paid little attention to ultimate factors and instead have tended to concentrate on study of proximate factors, which they typically call "mechanisms" (Immelmann & Suomi, 1981).

It should be pointed out that in some cases, ultimate and proximate factors may not be so distinct from one another and instead may involve the same basic set of variables (cf. Immelmann & Balda, Note 1). Nevertheless, each term represents a different focus or emphasis on factors that can influence a given organism one way or another. And, from the standpoint of cross-species generalization, ultimate and proximate factors represent different dimensions on which any two species can be compared in terms of possible similarities and differences.

DEGREES OF GENERALIZATION

The preceding discussion should make it apparent that there are a number of different criteria on which comparisons between species can be based. Cross-species generalizations can focus on similarity of physical appearance, extent of common ancestry, relative sensitivity to various ultimate factors, or degree of control exerted by the same set of proximate factors. These different criteria are to some degree orthogonal to one another; just because a certain characteristic is homologous in two species does not imply that it has been shaped by the same ultimate factors in both species. For example, the wings of eagles and the wings of sparrows are homologous, but they clearly have been subjected to quite different selection pressures. Along the same lines, the fact that two species are influenced by the same proximate factors does not insure that resulting behavioral changes will be identical or even grossly similar. For example, administration of low doses of amphetamine to rats and rhesus monkeys results in the same kind of chemical changes in brain neurotransmitters in both species, but whereas rats speed up their behavior following amphetamine treatment (Randrup & Munkvau, 1967), rhesus monkeys appear to become physically less active (Rush, Note 2; Segal, 1975).

Thus, similarity of physical appearance, extent of common ancestry, sensitivity to the same ultimate factors, and comparable influence by common proximate factors represent basically different domains for considerations of cross-species generalities. Is any one of these areas (or others not mentioned thus far) any more important or valid for generalizing than the others? It all depends on the particular purpose of the generalization at hand, as we detail later.

On the other hand, it is clear that some generalizations are more "complete" than are others. Although the aforementioned criteria have been described as basically orthogonal to each other, this does not mean that more than one cannot be satisfied in any given cross-species comparison. There *are* cases in which a certain characteristic or phenomenon is not only homologous between two species but also shaped by the same ultimate factors in like fashion as well. Similarly, two species that are influenced in like fashion by equivalent proximate factors might be virtually identical in physical appearance (especially after the same proximate influences have been evoked). Of course, the best of all worlds occurs when both species closely resemble each other with respect to each of the different criteria. When that happens, the generalization becomes considerably more than a matter of mere faith (cf. Harlow, Gluck, & Suomi, 1972). Generally speaking, the more criteria on which the species under comparison are judged to be similar, the more compelling and valid will be the generalization.

For example, let us return to the previous comparison of the visual systems of humans and chimpanzees. The comparison involves a homology, in that humans and chimpanzees are quite closely related phylogenetically (Lovejoy, 1981, among others, has estimated that any one human shares 98% of his or her nonreplicated DNA with every normal chimpanzee!). Also, there is no question that the visual systems of humans and chimpanzees are virtually identical in terms of physical structure. As mentioned earlier, the two visual systems share the same basic components (e.g., cornea, iris, lens, rods, cones, fovea, bipolar cells) arranged in the same basic structural order. Although it is exceedingly difficult, if not impossible, to reconstruct evolutionary history with perfect precision, the fact remains that numerous operating characteristics of one system are matched by the other (e.g., relatively poor night vision, comparable visual acuity, similar changes with increasing age), suggesting parallel selection pressures. Furthermore, the same set of proximate factors seems to influence the chimpanzee and human visual system in like fashion.

Thus, there is virtually complete generalization between the visual systems of chimpanzees and humans. Indeed, it seems that one could *substitute* one system for the other with little apparent change (loss or gain) in the host organism, assuming that any technical and ethical/moral difficulties in such a transplant operation could be worked out in actual practice. In theory, one could have a chimpanzee "eye bank" for humans—or a human "eye bank" for chimpanzees. The point is that under conditions of complete cross-species generalization, the characteristics or phenomena in question are so equivalent in the comparison species that substitution of one for the other should produce few, if any, noticeable differences in appearance or performance.

Such complete cases of generalization are usually not the rule in cross-species comparisons. Instead, cases of incomplete generalization are far more common. Here, the characteristic or phenomenon in question may be highly similar in appearance, homologous in origin, and shaped by the same set of ultimate factors, but controlled by different proximate factors. Or it could be influenced by the same proximate factors but be of somewhat different origin in the comparison species. In each case, there is real generality between the species according to some criteria but not others.

For example, consider the coloring of Arctic hares, least weasels, Arctic foxes and ptarmigans. During the spring, summer, and fall the external coloring of each is grey or brown or some combination thereof. But during the winter the coat or plumage of each species turns to white. In each case, the change in color clearly serves the function of camouflage, as each species resides in a habitat that is certain to be snow covered during winter months; the same set of ultimate factors seems to be operating for each species. However, the underlying physiological mechanisms responsible for the annual color change are not identical in these species when feathers rather than hairs are changing color. Thus, there are differences in the proximate factors controlling the phenomena in the different species, and in that sense the generalization is incomplete.

One might choose to characterize these examples of incomplete generalizations as cases of cross-species *simulations*. Here, complete substitution of the characteristic or phenomenon in question from one species to another would be inappropriate, for it is not identical in all respects in the two species. Nevertheless, it is possible in many cases to *simulate* certain aspects of the first species' responses to a given stimulus by following the responses to the same stimulus conditions in the second species. When such simulations are successful, the essential features of the first species' prototypical reactions are captured in those of the second species, even though the two reactions may not be identical in every respect. It is not substitution per se.

For example, consider the case of chronic administration of amphetamine to primate subjects. In human primates, chronic amphetamine (ab)use invariably results in excessive stereotypic activities (e.g., counting the number of "flakes" in a box of dried breakfast cereal), visual hallucinations, and displays of obvious paranoia; this constellation of behaviors has been termed "amphetamine psychosis" by a number of psychiatric investigators (Argrist & Gershon, 1970; Ellinwood, 1971; Griffith, 1970). Is it possible to simulate amphetamine psychosis in nonhuman primate subjects via chronic administration of amphetamine? Haber, Barchas, and Barchas (1977) gave socially living rhesus monkeys moderate daily doses of amphetamine over a 10-week period. They found that relative to untreated control monkeys, subjects administered amphetamine developed obvious stereotypic behaviors, displayed inappropriate social gestures in the absence of any social stimulation, and spent an inordinate amount of their waking hours monitoring the activities of cage mates they previously had largely ignored. It seems reasonable to conclude that many of the basic symptoms characteristic of human amphetamine psychosis were simulated in these rhesus monkeys chronically treated with the drug. However, there were also obvious differences between Haber et al.'s (1977) rhesus monkeys and human "speed freaks," verbal expressions of paranoia by the latter providing one example. The phenomenon in monkeys clearly simulates amphetamine psychosis in humans, but the simulation only provides a model of the human disorder; it is not the disorder itself (McKinney, 1974).

Such is the case for cross-species simulations in general: there may be clear-cut parallels between the species being compared, but the generalization is at best incomplete, and total substitution from one species to the other is not possible. The more complete the generalization (e.g., the greater the similarity of physical appearance or the larger the number of ultimate and/or proximate factors that are shared between the two species), the more accurate (if not compelling) will be the simulation. As is discussed subsequently, compelling simulations can be quite useful, even if they represent somewhat incomplete generalizations.

A final type of generalization occurs when the characteristic or phenomenon under consideration is only superficially similar in the species being compared, but some "larger" common principle seems to be involved. In this type of generalization, not only are there obvious species differences in overall appearance, but also the comparisons may involve analogies rather than homologies, different selection pressures (as expressed in the respective ultimate factors involved) have been in operation, and the characteristics or phenomena are controlled by different sets of proximate factors. Given the lack of precise parallels between the comparison species with respect to these criteria, any conclusions derived from such a generalization are limited to gross similarities that usually vanish upon close scrutiny.

For example, numerous authors (e.g., Bowlby, 1969; Hoffman & Ratner, 1973) have pointed out similarities between the phenomenon of filial imprinting in newly hatched ducklings and the development of mother-directed attachment by human infants. In both cases, the ducklings and the human infants must be exposed to their mothers during sensitive phases early in life in order for species-typical social relationships to emerge and be maintained later in life (cf. Bowlby, 1969; Immelmann & Suomi, 1981; Lorenz, 1935). However, these phenomena are clearly not homologous, the parameters of the respective sensitive phases are different in keeping with the ducklings' and human infants' different specific needs in the timing of their optimal sensitivity to social signals, and the actual mechanisms involved in forming and maintaining the social bonds with mother are obviously different in ducklings and in human infants (Harlow et al., 1972). Yet, there are clearly some common features in these phenomena; both involve sensitive phases that may have lifelong consequences for the behavioral repertoires of the infants involved. Still, it is difficult to simulate realistically most aspects of human mother–infant attachment within the framework of filial imprinting, at least as it occurs in ducklings.

Despite the fact that cross-species similarities of the type represented by the aforementioned imprinting example permit neither substitution nor simulation of characteristics or phenomena from one species to another, generalizations based on such similarities can serve some other useful purposes. One valuable contribution they can make concerns what might be called "parsimony of explanatory principles.'' Any clever investigator can easily come up with scores of reasons why it makes sense for ducklings to become imprinted on their mothers or why human infants should become attached to their mothers. Focusing on the similarities inherent in these two sets of phenomena, however, yields more general principles such as the notion of sensitive phases and the idea that socialization and species recognition begins with the early formation of a one-to-one bond. The heuristic value or insight provided by cross-species similarities of this type can be considerable, as has been eloquently pointed out by McKinney and Bunney (1969) and as is discussed later in this chapter.

LEVELS OF ANALYSIS AND ONTOGENIC PARALLELS

That there clearly are different degrees of cross-species generalization should be apparent from the preceding paragraphs. Such differences of degree are based primarily on the number of different criteria (regarding similarities of appearance, ancestry, and ultimate and proximate factors) that are satisfied in the cross-species comparison at hand. Yet, these criteria are far from absolute in their own right. Instead, judgments regarding the degree to which they have been satisfied in any cross-species comparison depend on the perspective of the judge—and that, in turn, depends on the *level(s) of analysis* under scrutiny and the *ontogenic stage* under consideration.

When comparing two species, it is often the case that different degrees of generalization exist for a particular characteristic or phenomenon, depending on the level of analysis utilized by the researcher. Return for a moment to the previous example of amphetamine effects in laboratory rats and in rhesus monkeys. If the level of analysis is restricted to examination of pharmacological effects in the two species, then one finds almost total generality between them, virtually a case of substitution. There is similarity of phenomena: amphetamine acts on the same dopamine transmitter system in the brain of rats as in the brain of monkeys. There is clearly a homology involved: dopaminergic systems are found in the brains of all mammals, derived from some distant common ancestor. In addition, it seems likely that the same set of ultimate factors influenced the selection of these common dopaminergic systems, at least in those parts of the brains of rats and monkeys that contain these systems (Panksepp, 1982). This is because the dopaminergic system is widely represented in mammalian brains, consistent with the previously mentioned notion of a common ancestor possessing such a system. Finally, the same set of proximate factors appears to influence pharmacological effects of amphetamine in both rats and monkeys alike. For example, amphetamine antagonists have parallel pharmacological consequences for amphetamine treatment in rats and monkeys.

Thus, from a purely pharmacological level of analysis, amphetamine effects in rats can essentially be substituted for amphetamine effects in monkeys (and vice versa) with no major changes in outcomes or conclusions. Viewed as living *in vitro* preparations, the two species are difficult to distinguish from one another—as long as one is concerned with pharmacological effects alone. However, a very different picture emerges if one begins to look at behavioral consequences of amphetamine administration in each species.

As mentioned previously, amphetamine treatment "speeds up" rats but "slows down" monkeys in terms of overt behavioral activity. It is difficult to describe these two respective behavioral reactions as "similar" in any manner that fails to strain credulity. Moreover, the behavioral reactions are not homologous and only perhaps analogous (for pertinent arguments see Rush, Note 2). Furthermore, the various ultimate factors shaping the respective behavioral reactions to amphetamine are obviously different. Thus, at a behavioral level of analysis there is little generality between rats and rhesus monkeys, even though at a pharmacological level the generality is nearly perfect.

The amphetamine example seems to illustrate that generality of characteristic or phenomenon between species does not "generalize" a priori across all levels of analysis possible. It may generalize at more "reductionistic" levels but not at "higher" ones, as in the amphetamine example. The reverse tendency could be true for other cases, as in the winter color change of coat/plumage example described earlier. Alternatively, the generality could be compelling at virtually every level of analysis, as in the human–chimpanzee visual system example or, conversely, between-species similarities could be evident only at the grossest levels of analysis, as in the filial imprinting–human mother–infant attachment comparison. Thus, cross-species generalizations may cut across several levels of analysis or they may be restricted to a single level. Moreover, substitutions at one level of analysis may only represent simulations or similarities at other levels for any given phenomenon. It seems reasonable to argue that the more levels of analysis for which a generalization "holds true," especially in terms of substitution, the more compelling and ultimately useful will be the generalization.

Another dimension on which cross-species generalizations can be considered and evaluated concerns the *ontogenic status* of the characteristic or phenomenon being compared between different species. Most organisms undergo fundamental changes in their respective anatomy, physiology, cognitive capabilities, and behavioral repertoire as they pass from birth to adulthood, and strong cross-species parallels that exist at one stage of development may not be so apparent at other stages. On the one hand, species that differ considerably from one another early in life may tend to converge with respect to particular characteristics and phenomena when they reach adulthood. For example, most reptiles are much more similar to most amphibians as adults than they are shortly after hatching. At hatching, amphibians are aquatic, whereas reptiles are terrestrial.

On the other hand, species that appear highly similar as infants may diverge considerably as they grow older. Indeed, this is a common occurrence among closely related species, at least within various mammalian orders. As Gould

(1977) has eloquently pointed out, although ontogeny does not always recapitulate phylogeny (or does so incompletely), different selection pressures may be operating during periods of development than during adulthood—and those "developmental" pressures will have chronological precedent over those operating on adult organisms. The result is that phylogenetically similar organisms may share more aspects of their fetal, neonatal, and childhood lives than they will share as adults.

A case in point can be seen in the cognitive development of macaques, apes, and humans. In each of these groups of primates, most noncortical regions of the brain are developed prenatally, whereas most cortical growth takes place postnatally, continuing until the beginning of adolescence (Gibson, 1977). At and shortly after birth, macaques, apes, and human babies alike are capable of classical (Pavlovian) and instrumental (operant) conditioning and simple discrimination tasks (Haith & Campos, 1977; Harlow, 1959; Rumbaugh & Gill, 1973); each also possesses a comparable repertoire of neonatal reflexes. By 1 month of age in macaques, 2–2½ months in apes, and 3–4 months in human infants, neonatal reflexes have largely disappeared or are under voluntary control, more complicated discriminations are possible, and short-term memory is clearly evident. By 3 months of age in macaques (7–8 months in apes and 8–12 months in human infants), fear of strangers has been developed by these infants (Eibl-Eibesfeldt, in press; Sackett, 1966). The age of 6 months in macaques (1½ years in apes and 2 years in humans) marks another developmental milestone for cognitive capabilities in each of these primate species. At this point, the Piagetian sensorimotor period has been completed, the ability to form learning sets has emerged (although it will expand greatly in succeeding months and years), and play patterns start to become increasingly complex. Thus, if one allows for the fact that macaques mature at roughly 2½ times the rate of apes and 4 times that of humans, there is remarkable generality in the pattern and extent of cognitive development among these primates, human and nonhuman (Kagan & Suomi, Note 3).

However (as we all know), the cognitive capabilities of adult humans far outstrip those of adult apes, just as mature apes are clearly superior to mature macaques in terms of brainpower. The species begin to separate ontogenically in this respect soon after the end of their respective sensorimotor stages. For example, rhesus monkeys and other macaques apparently fail to develop the clear-cut concept of self-image that apes begin to exhibit between 1½ and 2 years and human toddlers display by 2½–3 years (Gallup, 1977; Lewis & Brooks-Gunn, 1979). Similarly, human toddlers rapidly leave infant apes behind when they begin to make extensive use of spoken language during their third year of life.

Thus, the degree of generality among the cognitive capabilities of macaques, apes, and humans is considerable early in life but becomes less so as the individuals representing each species grow older. This example serves to illustrate again the point that cross-species generality expressed at one developmental stage does not guarantee comparable generality during other stages. As was the case with different levels of analysis, generalizations that hold over different stages of development cannot be assumed a priori. Instead, they must each be demonstrated empirically.

CROSS-SPECIES GENERALITIES AND PHYLOGENETIC SIMILARITIES

By now it should be apparent that cross-species generalities can differ in degree and in the levels of analysis and stages of development for which they are appropriate. It has been argued that cross-species generalities that involve substitutions are basically more valid than those involving simulations or mere similarities, whereas generalities that cross multiple levels of analysis or stages of development are of broader scope and are usually more compelling than those limited to a single analytic level or ontogenic stage. Given these considerations, let us return to a comparison of different generalities mentioned at the beginning of the chapter—those involving species that are phylogenetically "close" versus those that involve species that are phylogenetically more "distant." Is it really the case, as many biologists would have it, that the phylogenetically close generalizations will always be superior (or, at least, never inferior) to phylogenetically more distant ones? The answer turns out to be not so simple. It is generally true that most cross-species comparisons that involve true homologies are far more likely to contain other common features than are most cross-species comparisons that only involve analogies. However, as mentioned earlier, this is not always a hard-and-fast rule and exceptions do exist. Moreover, for cross-species comparisons that involve actual homologies, the "phylogenetic similarity" rule all but goes out the window. Here, species that are "closest" phylogenetically do not always provide the most compelling cross-species generalities—especially when they reside in vastly different physical and social environments.

A typical example of this latter point can be found in female reproductive behavior and physiology among higher primates, specifically rhesus monkeys, gorillas, and humans. The general physiology of the female reproductive system is clearly homologous in these three primate species, the same organs, glands, and hormones comprise the system in all three species, and each is characterized by a menstrual cycle that averages 28 days in duration (Goldfoot, Swanson, Neff, & Leavitt, in press). Yet, the timing of female reproductive activity is not the same in the three species, and the pattern among them does not match the pattern of their phylogenetic relatedness.

Specifically, rhesus monkey females closely resemble human females in terms of sexual receptivity throughout the whole of the menstrual cycle. Although there are overall "peaks" and "valleys" during the 28-day cycle, actual copulation is not restricted to particular days during this period and, in fact, pregnancies can be initiated on any one of the 28 days. In contrast, gorilla females are much more similar in this respect to rat and guinea pig females than they are to rhesus monkey and human adult females. Their copulating behavior is limited to a 2- to 3-day period immediately following ovulation, and for the rest of each lunar month gorilla females are sexually abstinent (Nadler, 1975). Because gorillas as a species are phylogenetically much closer to humans than they are to rhesus monkeys (or rats or guinea pigs), these phenomena are contrary to the notion that greater phylogenetic closeness always dictates greater cross-species generality.

Another pertinent example can be found in comparisons of reactions to early social isolation among phylogenetically close species of primates. Classic experiments carried out in the early 1960s revealed that rhesus monkey infants separated from their mothers in their first postnatal week and subsequently reared for at least the first 6 months of life in tactile and visual isolation from all conspecifics failed to develop species-appropriate social behavioral repertoires and instead exhibited a wide range of bizarre self-directed behavior never displayed by socially reared rhesus monkeys. These abnormal behaviors did not disappear when the isolated subjects were finally introduced to other monkeys, and their gross social deficits persisted into adulthood (Harlow, Dodsworth, & Harlow, 1965; Mason, 1960; Sackett, 1968). Other researchers performed similar isolation experiments with chimpanzees and reported basically the same results: early social isolation had devastating, essentially permanent consequences for species-normative behavioral development (Menzel, Davenport, & Rogers, 1963). These findings seemed consistent with anecdotal reports of extreme social deficits shown by "feral" human children who had grown up isolated from all other humans (Itard, 1932). Thus, evidence accumulated from rhesus monkeys, chimpanzees, and humans strongly suggested that the severely debilitating and long-lasting consequences of isolation rearing generalize across the higher primates.

This apparently clear picture of cross-species consistency of isolation effects among primates has been clouded considerably by recent findings reported by Sackett and his colleagues (Sackett, Holm, & Ruppenthal, 1976; Sackett, Ruppenthal, Fahrenbach, Holm, & Greenough, 1981). These researchers found that members of some species of macaques belonging to the same genus as rhesus monkeys (*Macaca*) were only moderately and transiently affected by isolation rearing. Specifically, pigtail macaque (*M. nemestrina*) infants reared according to the very same conditions and duration of social isolation as rhesus monkey infants essentially displayed spontaneous recovery of social behavior when introduced to socially normal peers, although they continued to exhibit abnormal, self-directed behaviors not shown by the normal controls. Crab-eating macaques (*M. fascicularis*) reared under identical conditions of social isolation from birth failed to show even the self-directed behavioral abnormalities displayed by the pigtails. Social isolation thus had minimal consequences for crab-eating macaques, in sharp contrast to rhesus monkey subjects.

These examples serve to illustrate the basic point that phylogenetic proximity is not always perfectly correlated with degree of cross-species generality for any given characteristic or phenomenon. Under some circumstances, generalizations among species belonging to the same genus may be less valid or complete than comparisons involving species outside the genus. Such occurrences are especially likely when both the within- and between-genus comparisons involve homologies, and the species in the between-genus comparisons share the same basic physical and/or social environments, whereas those in the within-genus comparisons do not. However, as was demonstrated in the foregoing example involving isolation effects, the underlying reasons why closely related species are less similar than more distantly related ones for a given characteristic or phenomenon may not always be readily obvious.

Of course, this is not to say that generalizations involving closely related species are always inferior to those involving more distantly related ones; for most phenomena the opposite is more often the case. Instead, the point to be made is that the degree to which any particular cross-species generalization parallels the relative phylogenetic relationships between the species being compared is ultimately an empirical question. A priori assumptions about a given generality that are based solely on knowledge about phylogenetic relatedness can be risky indeed.

PRACTICAL APPLICATIONS OF KNOWLEDGE CONCERNING CROSS-SPECIES GENERALIZATIONS

The issues considered in previous sections should make it apparent that establishing and evaluating cross-species generalizations can be a complicated and indeed almost endless business. Comparisons between different species can be based on several distinct sets of criteria, and they can be made at any number of different levels of analysis. Given these considerations, it should be apparent that all cross-species generalities do not share equivalent features, and some are more complete and/or appropriate than are others. Furthermore, the limits of any particular cross-species generalization must be determined on an empirical basis, rather than be assumed to exist a priori, and such a determination often turns out to be far from simple or straightforward.

Yet, once a cross-species generalization has been established and its limits understood, it can be of substantial practical utility. Consider the case of a cross-species generalization that is clearly substitutive in nature, at least at some levels of analysis. At those levels of analysis the substitution will be complete; empirical knowledge obtained from one species can be applied directly to the other. Delineation of the *limits* of such a substitutive generalization can also be quite useful in a variety of respects. For example, knowledge of the point(s) in ontogeny at which a characteristic that earlier was substitutive between two species ceases to be so can serve to identify developmental periods when different sets of adaptive pressures have come into play for the two species. Such knowledge can be exceedingly useful in generating plausible hypotheses about the natural history of the respective species. In other cases, knowledge of the level(s) of analysis at which an otherwise substitutive generalization no longer is sustained can be used to facilitate study of proximate factors or mechanisms that control the phenomenon in the two species. Finally, knowledge of the number and distribution of species for which the characteristic or phenomenon is substitutive can provide a basis for resolving questions about the taxonomy of the species involved.

The immediate practical benefits of establishing a cross-species generalization that involves simulation tend to be more limited in scope than those deriving from substitutive generalizations. Here, of course, one cannot simply "replace" the phenomenon or characteristic in one species with that in the other, as in a substitutive generalization. However, if one is aware of those aspects of the simulative generalization that are indeed parallel in the two species, then one can use parametric information from one species to generate comparable information (at least of an ordinal form) about the other. Moreover, the other practical benefits of substitutive generalizations that can be gained from knowledge of their respective limits are generally shared by generalizations that involve "mere" simulations. Thus, important hypothesis-generating information about ultimate factors, proximal mechanisms, and taxonomic relationships can be gleaned from knowledge about the levels of analysis or ontogenic stages at which a simulative generality breaks down.

Finally, cross-species generalizations that involve only gross similarities rather than substitutions or simulations can still be useful in a number of respects. As mentioned earlier, these "loose" generalizations can be of considerable heuristic value, often providing insights as to the organization of phenomena in one species based on what is known in another. Moreover, generalizations of this sort may enable investigators to identify the most parsimonious of a range of explanations concerning a given characteristic or phenomenon in one species when one can detect even gross parallels in another species. Of course, these gross generalizations are more limited in their practical applicability than are those based on cross-species substitutions or even simulations.

Considerations concerning practical utility are especially relevant for one particular type of cross-species generalization—those involving *homo sapiens* as one of the species under comparison. Such generalizations are often referred to as *animal models*. Animal models form the cornerstone of an enormous amount of current biomedical and behavioral research. They generally involve attempts to mimic a human disease or disorder in nonhuman animal species; knowledge about the modeled phenomena in animals is then applied to the human cases

(Suomi, 1982). Of course, such models are only as sound as the cross-species generalization on which they have been based—and some generalizations, as we have seen, are obviously more fundamentally sound than are others.

The most valid and practically useful animal models are those that are based on substitutive generalizations between humans and animals. For these models there are few, if any, limitations. Nonhuman subjects from the generalizing species can essentially be used as full-blown substitutes for human patients or clients in developing, practicing, and assessing diagnostic and treatment procedures. The practical implications and potential applications of such models are enormous—they permit far more carefully controlled and rigorous experimentation than is almost ever possible with human subjects, and if the animal species that models the human phenomenon or characteristic also grows up more rapidly than its human counterpart (e.g., as is the case for all nonhuman primate species), longitudinal studies extending to the life-span often are far more practical to carry out than comparable longitudinal studies with human subjects (Suomi, 1982). Of course, the other advantages characteristic of substitutive cross-species generalizations discussed previously also hold for these substitutive animal models.

Animal models based on simulations rather than substitutions are more limited in their immediate practical applications. These models involve cross-species comparisons that do not yield precisely parallel phenomena or characteristics between human and nonhuman subjects. As a consequence, direct application of animal data of this type to human cases or situations carries with it some risk—and the risk increases the less complete the animal simulation becomes. Thus, to use an earlier example, direct applications of the results of studies of behavioral effects of chronic amphetamine treatment in rats to cases of human amphetamine abuse is riskier than if the nonhuman subjects were rhesus monkeys. Similarly, one might have more faith in female contraceptive devices that had been tested successfully on female rhesus monkeys than those tested successfully on female gorillas, given what we know about the generality between their respective reproductive systems and those of human females (Nadler, 1975). It is wise to remember the exact limits of any simulative generalization between human and nonhuman species when using the generalization in an animal model—and it is usually even wiser to avoid direct practical applications in areas where such limits are not known. Nevertheless, simulative animal models have enjoyed widespread usage and relative success over the years. Most biomedical and toxicology animal models are simulative in nature.

Finally, the possible practical applications of animal models based on cross-species comparisons that involve neither substitutive nor simulative generalities are quite limited indeed. In some cases they can be used as diagnostic tools. Animal models are sometimes employed as screening devices for new pharmacological compounds designed to treat various human disorders, even though the drug effects in the animal subjects may bear only superficial resemblance to the human disorder itself (Kornetsky, 1977). Of course, these kinds of animal models can provide important heuristic insights or help integrate disparate sets of cross-species data, even if their immediate practical applications are usually more restricted than are those of animal models based on more compelling generalizations.

SUMMARY AND CONCLUSIONS

In this chapter we have examined the processes and products of cross-species generalizations. We have pointed out that the traditional biological approach to cross-species comparisons—involving homologies, analogies, and cases of convergence—is not the only way to formulate and evaluate these generalizations. Other means of comparison exist, and two important considerations in making such comparisons center around examination of ultimate and proximate factors associated with the characteristics or phenomena under scrutiny. We have argued that different degrees of cross-species generality are clearly possible and that they are not necessarily consistent across different levels of analysis or stages of ontogeny. Thus, the actual limits of a given cross-species generalization are not always apparent a priori; instead, they must be determined empirically.

Determination of the limits of any cross-species generalization is not only useful from a conceptual and/or theoretical perspective, but also it can be of considerable practical relevance if the generalization involves humans as well as nonhuman species. Animal models based on incomplete or inappropriate cross-species generalizations may turn out to be practically useless or even dangerous. But animal models based on sound generalizations, in which both the cross-species parallels and their respective limits are recognized and understood, can be of enormous value. Not only can they extend our theories and knowledge of animal (and human) behavior and physiology in the most fundamental of fashions, but also they can contribute to practical improvements in the health and well-being of our fellow humans as well.

ACKNOWLEDGMENTS

Part of this chapter was written while Professor Immelmann was on sabbatical leave from the University of Bielefeld and in residence at the University of Wisconsin–Madison; support for Professor Immelmann was provided by the University of Wisconsin Graduate School. We extend our gratitude to Ms. Helen A. LeRoy, who helped edit the chapter, and Ms. Judy Markgraf, who prepared the manuscript.

REFERENCE NOTES

Note 1. Immelmann, K., & Balda, R. Ultimate and proximate control of behavior. Manuscript in preparation.

Note 2. Rush, D. Effects of amphetamine administration on the cognitive and social behavior of rhesus monkeys. Unpublished doctoral dissertation, University of Madison—Wisconsin, 1980.

Note 3. Kagan, J., & Suomi, S. J. Manuscript in preparation.

REFERENCES

Argrist, B., & Gershon, S. The phenomenology of experimentally induced amphetamine psychosis—preliminary observations. *Biological Psychiatry,* 1970, *2,* 95–107.

Baker, J. R. The evolution of breeding seasons. In G. R. deBeer (Ed.), *Essays* on *aspects of evolutionary biology*. New York: Oxford University Press, 1938.

Bowlby, J. *Attachment and loss*. New York: Basic Books, 1969.

Eibl-Eibesfeldt, I. Fear of strangers. In A. Oliverio & M. Zappella (Eds.), *The behaviour of human infants*. New York: Plenum Press, in press.

Ellinwood, E. H. Effect of chronic methamphetamine intoxication in rhesus monkeys. *Biological Psychiatry,* 1971, *3,* 25–32.

Gallup, G. G., Jr. Self-recognition in primates: A comparative approach to the bidirectional properties of consciousness. *American Psychologist,* 1977, *32,* 329–338.

Gibson, K. Brain structure and intelligence. In S. Chevalier-Skolnikoff & F. Poirier (Eds.), *Primate bio-social development*. New York: Garland Press, 1977.

Goldfoot, D. A., Swanson, L. J., Neff, D. A., & Leavitt, L. Hormonal and stimulus mediation of maternal behavior of primates and rodents: Review and hypotheses. *American Journal of Primatology,* in press.

Gould, S. J. Ontogeny and phylogeny. Cambridge, Mass.: Belknap Press, 1977.

Griffith, J. D. Experimental psychosis induced by the administration of D-amphetamine. In E. Costa & S. Garantinni (Eds.), *Amphetamines and related compounds*. New York: Raven Press, 1970.

Haber, S., Barchas, P. R., & Barchas, J. D. Effects of amphetamine on social behaviors of rhesus macaques: An animal model of paranoia. In I. Hanin & E. Usdin (Eds.), *Animal models in psychiatry and neurology*. New York: Pergamon Press, 1977.

Haith, M. M., & Campos, J. J. Human infancy. *Annual Review of Psychology,* 1977, *28,* 251–293.

Harlow, H. F. The development of learning in the rhesus monkey. *American Scientist,* 1959, *47,* 459–479.

Harlow, H. F., Dodsworth, R. O., & Harlow, M. K. Social isolation in monkeys. *Proceedings of the National Academy of Sciences,* 1965, *54,* 90–96.

Harlow, H. F., Gluck, J. P., & Suomi, S. J. Generalization of behavioral data between nonhuman and human animals. *American Psychologist,* 1972, *27,* 709–716.

Hinde, R. A. *Biological basis of social behavior*. New York: McGraw–Hill, 1974.

Hladik, C. M. Ecology, diet, and social patterning in old and new world primates. In R. H. Tuttle (Ed.), *Socioecology and psychology of primates*. The Hague: Mouton, 1975.

Hoffman, H. S., & Ratner, A. M. A reinforcement model of imprinting: Implications for socialization in monkeys and men. *Psychological Review,* 1973, *80,* 527–544.

Immelmann, K. Ecological significance of imprinting and early learning. *Annual Review of Ecology and Systematics,* 1975, *6,* 15–37.

Immelmann, K., & Suomi, S. J. Sensitive phases in development. In K. Immelmann, G. Barlow, L. Petrinovich, & M. Main (Eds.), *Behavioral development*. New York: Cambridge University Press, 1981.

Itard, J.-M.-G. *The wild boy of Aveyron*. New York: Century, 1932.

Kornetsky, C. Animal models: Promises and problems. In I. Hanin & E. Usdin (Eds.), *Animal models in psychiatry and neurology*. New York: Pergamon Press, 1977.

Lack, D. *The natural regulation of animal numbers*. Oxford: Oxford University Press, 1954.

Lewis, M., & Brooks-Gunn, J. *Social cognition and the acquisition of self*. New York: Plenum Press, 1979.

Lorenz, K. Der Kumpan in der Umwelt des Vogels. *Journal für Ornithologie,* 1935, *83,* 137–213.
Lovejoy, C. O. The origin of man. *Science,* 1981, *211,* 341–350.
Mayr, E. Biological classification: Toward a synthesis of opposing methodologies. *Science,* 1981, *214,* 510–516.
Mason, W. A. The effects of social restriction on the behavior of rhesus monkeys. I. Free social behavior. *Journal of Comparative and Physiological Psychology,* 1960, *53,* 583–589.
McKinney, W. T. Animal models in psychiatry. *Perspectives in Biology and Medicine,* 1974, *17,* 529–541.
McKinney, W. T., & Bunney, W. E. Animal model of depression. *Archives of General Psychiatry,* 1969, *21,* 240–248.
Menzel, E. W., Jr., Davenport, R. K., & Rogers, C. M. Effects of environmental restriction upon the chimpanzee's responsiveness in novel situations. *Journal of Comparative and Physiological Psychology,* 1963, *56,* 329–338.
Nadler, R. D. Sexual cyclicity in captive lowland gorillas. *Science,* 1975, *189,* 813–814.
Panksepp, J. *The role of endogenous brain opiate systems in the establishment and modulation of social behaviors.* Paper presented at the Society for Research in Child Development Workshop on the Development of Aggression and Altruism, Bethesda, Md., March 1982.
Randrup, A., & Munkvau, I. Stereotyped activities produced by D-amphetamine in several animal species and man. *Psychopharmacologia,* 1967, *11,* 300–310.
Rumbaugh, D. M., & Gill, T. V. The learning skills of Great Apes. *Journal of Human Evolution,* 1973, *2,* 171–179.
Sackett, G. P. Monkeys reared in visual isolation with pictures as visual input: Evidence for an innate releasing mechanism. *Science,* 1966, *154,* 1468–1472.
Sackett, G. P. Abnormal behavior in laboratory-reared rhesus monkeys. In M. W. Fox (Ed.), *Abnormal behavior in animals.* Philadelphia: Saunders, 1968.
Sackett, G. P., Holm, R., & Ruppenthal, G. C. Social isolation rearing: Species differences in behavior of macaque monkeys. *Developmental Psychology,* 1976, *10,* 283–288.
Sackett, G. P., Ruppenthal, G. C., Fahrenbruch, C., Holm, R. A., & Greenough, W. A. Social isolation effects in monkeys vary with genotype. *Developmental Psychology,* 1981, *17,* 313–318.
Segal, D. S. Behavioral characterization of D- and L-amphetamine: Neurochemical implications. *Science,* 1975, *190,* 475–477.
Snowdon, C. T. Primate communication. *Annual Review of Psychology,* in press.
Suomi, S. J. Relevance of animal models for clinical psychology. In P. C. Kendall & J. M. Butcher (Eds.), *Handbook of research methods in clinical psychology.* New York: John Wiley & Sons, 1982.
Thomson, A. L. Factors determining the breeding season of birds: An introductory review. *Ibis,* 1950, *92,* 173–184.

Reader's guide to Section 3
BEHAVIOR AND BIOLOGY: GENETIC CONSIDERATIONS IN A COMPARATIVE APPROACH

In Chapter 9, Colwell and King discuss three major points:

1. How similar are we genetically to our nearest living animal relatives? What implications does this degree of similarity have for the comparative study of behavior?
2. To what extent can Darwinian principles be applied to the study of human social behavior?
3. What tools are available for the genetic analysis of human behavioral traits?

Through answers to these questions the authors argue that, although there are fundamental biochemical similarities between humans and our closest primate relatives, a formal, direct genetic investigation of human behavior has not necessarily provided us with definitive conclusions concerning our social-psychological human nature. More importantly, perhaps, the chapter makes a clear case that any interpretation of human behavioral traits that merely *infers* the action of genetic mechanisms as the basis for those traits—or their transmission across generations—is open to serious error without proper safeguards. They show how cultural transmission, rather than genetic transmission, may fully account for variations in most human behaviors, even those that have reproductive or evolutionary implications, or those that have been labeled "universal." The Colwell-King paper should therefore serve as a brake on improper or careless appeals to principles or theories of genetics as a justification for arguments about comparable (or not) human and nonhuman behavior.

An important theme of Zawistowski and Hirsch (Chapter 10) is that if the concept of "homology" is to be employed in any analysis of behavior, its proper use is dictated by the fundamentally genetic nature of biological homology. Homologous genes are similar in *form;* resemblances in sequence, composition, and structure are what define biochemical homologies. What, then, is the relationship of genes to behavior? The authors state that "as behavior-geneticists we wish to analyze the genetic correlates of behaviors. To perform a genetic analysis it is necessary to be able, reliably and accurately, to categorize and/or measure individual differences in phenotypic expression. It is through phenotypic variation that we gain access to the genetic system." Concerning two dimensions of behavior—function and form—they propose that homologies be studied as form, not function. This choice avoids the difficulties that function is often dependent on context, that functional analogies can be confused with homologies, and that the same form can have different functions (ritualizations) in related species. Given these considerations, Zawistowski and Hirsch provide a description of their current behavior-genetic analyses of a species: "By studying the feeding behavior of the blow fly we are learning how to analyze the genetic correlates of behavior in any species. This is consistent with the pattern of history in genetics. Even though Mendel worked with peas, the classic crosses he developed have been used time and time again to analyze new and different diploid species."

9 Disentangling Genetic and Cultural Influences on Human Behavior: Problems and Prospects

Robert K. Colwell

Mary-Claire King
University of California at Berkeley

How different are we from other animals? How much can we learn about the determinants of human behavior by studying our phylogenetically close—and not so close—animal relatives? How much can we learn about the determinants of human behavior by studying ourselves? Each of the chapters in this volume addresses these issues. Here, we approach the problem from the perspective of evolutionary genetics, discussing three specific questions:

1. How similar are we genetically to our nearest living animal relatives? What implications does this degree of similarity have for the comparative study of behavior?
2. To what extent can Darwinian principles be applied to the study of human social behavior?
3. What tools are available for the genetic analysis of human behavioral traits?

Because each of these topics is the focus of a large and rapidly growing body of technical literature, we do not pretend to provide a complete review. The references we cite are intended only to provide an entree into the literature, or to provide illustrative examples.

GENETIC SIMILARITIES AND MORPHOLOGICAL DIFFERENCES BETWEEN HUMANS AND CHIMPANZEES

Chimpanzees and gorillas are our closest living relatives, and—perhaps for that reason—we and chimpanzees appear fascinated by each other (Goodall, 1971). Recent evidence indicates that at the genetic level, humans and chimpanzees are very similar indeed (King & A. C. Wilson, 1975; A. C. Wilson, 1975). Several biochemical techniques have been used to estimate the degree of similarity of human and chimpanzee proteins, and of human and chimpanzee DNA. From each of these comparisons, we can obtain an estimate of the degree of genetic difference between humans and chimpanzees. Because some of the same approaches have been used to estimate the degree of genetic difference between other taxa, the human–chimpanzee genetic difference may be compared to other genetic differences.

Human and chimpanzee proteins have been compared by directly identifying the sequences of amino acids that comprise homologous proteins in the two species, by comparing the immunological properties of homologous proteins, and by comparing homologous proteins' size and electric charge. These approaches all indicate that the amino acid sequence of the average human protein is more than 99% identical to its chimpanzee homologue. The degree of genetic difference between humans and chimpanzees, based on these protein comparisons, is very small, corresponding to genetic differences between very close, sibling species of other mammals or of *Drosophila*. At the genetic level, nonsibling species in the same genus generally differ more from one another than do humans and chimpanzees. The genetic differences among species in different genera are considerably larger than those between humans and chimpanzees.

But humans and chimpanzees are not just in different genera, but in different families. Does this reflect simply the anthropocentrism of biologists (Schopf, Raup, Gould, & Simberloff, 1975), or are the two species in fact very different at higher levels of organization than genes and proteins? The latter appears to be the case. The molecular similarity between chimpanzees and humans is extraordinary because they differ far more than sibling species in morphology (Cherry, Case, & A. C. Wilson, 1978). Humans and chimpanzees are rather similar in thorax and arm structure. However, they differ substantially not only in brain size, but also in the anatomy of the pelvis, foot, and jaws, as well as in the relative lengths of limbs and digits (Bourne, 1970). Humans and chimpanzees are sufficiently different anatomically that nearly every bone in the body of a chimpanzee can be readily distinguished by shape or size from its human counterpart. Associated with these anatomical differences are major differences in posture, mode of locomotion, methods of procuring food, and means of communication. It is these differences in anatomy and way of life that have led to the classification of humans and chimpanzees in separate families (Simons, 1972; Simpson, 1961; Washburn, 1963). Molecular and organismal methods of evaluating the human–chimpanzee difference apparently yield quite different conclusions.

Evolutionists recognize that the anatomy and way of life of chimpanzees has changed far more slowly than human anatomy and adaptive strategy since the two species diverged (Simpson, 1963). However, at the level of genes and proteins, humans and chimpanzees have evolved at roughly similar rates. This contrast is illustrated in Fig. 9.1.

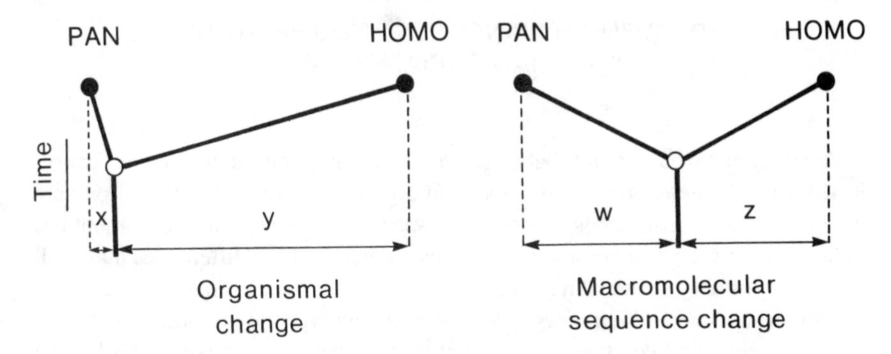

FIG. 9.1. Contrast between biological evolution and molecular evolution since the divergence of the human and chimpanzee lineages from a common ancestor. As shown on the left, zoological (organismal) evidence indicates that far more biological change has taken place in the human lineage (y) than in the chimpanzee lineage ($y > x$). As shown on the right, both protein and nucleic acid (macromolecular) evidence indicate that as much change has occurred in chimpanzee genes (w) as in human genes ($w = z$). Reprinted from Mary-Claire King and A. C. Wilson, "Evolution at two levels in humans and chimpanzees," *Science,* 11 April 1975, Volume 188, pp. 107–116. Copyright 1975 by the American Association for the Advancement of Science.

How can this paradox be resolved, and what are the consequences for the comparative study of human behavior and the behavior of other species? It may be that the anatomical changes leading to behavioral adaptations are not caused by changes in amino acid sequences of proteins. Instead, major anatomical differences between species may be due to differences in the timing of gene expression. In other words, homologous proteins of chimpanzees and humans have nearly identical sequences. What may differ are the times during embryonic development and later life that these proteins appear. In this way, a relatively small number of genetic changes regulating the expression of genes may account for the major organismal differences between humans and chimpanzees (A. C. Wilson, Carlson, & White, 1977).

The genetic similarity of humans and chimpanzees might lead us to believe we could confidently apply results of behavioral studies in chimpanzees to human behaviors. However, three factors intervene to make this appealing proposition a risky one. First, the uncoupling of molecular and morphological evolution suggests the operation of novel selective forces in the history of our rapid divergence from chimpanzees. Second, although behavior is ultimately dependent on morphology, behavioral evolution quite often proceeds even more rapidly than morphological evolution (Ehrman & Parsons, 1976; Futuyma, 1979; E. O. Wilson, 1975). To these two levels of uncoupling, we must of course add the complex effects of human culture, which, though it rests ultimately on evolved morphological and behavioral capacities, can generate rapid behavioral change at a rate effectively independent of genetic evolution.

DARWINIAN PRINCIPLES AND THE STUDY OF HUMAN SOCIAL BEHAVIOR

No evolutionist would doubt that differences among species in the general capacity for social interaction, or in modes and capacities for communication, are based on genetic differences among those species. As the previous argument has indicated, between humans and their closest relatives, these differences may well lie in evolutionary changes in gene expression, rather than in qualitative changes in protein sequence. Controversy over the comparative study of human behavior focuses instead on two other issues. First, to what extent have *particular* human social behaviors been *produced by* natural selection of alternative genotypes? And second, to what extent are *differences* in behavior among individuals or groups of individuals *caused by* genetic differences among them? These questions are most dramatically posed in the comparative study of behavior. If a biologist can demonstrate that behavior in another species is the result of natural selection, it might be concluded that an apparently similar human behavior is also the result of natural selection, especially if the behavior may appear to be related to reproductive success. Furthermore, if differences in behavior among individuals of another species are demonstrably caused by genetic differences among them, the biologist might conclude that seemingly parallel (behavioral) variation among people is also caused by genetic differences among them. The more closely related the other species is to humans, the more tempting these inferences become.

Beginning with Darwin, it has frequently been proposed that the evolutionary significance of a wide array of human social behaviors can be illuminated by the comparative study of other species. Although there are other important contemporary currents in the study of comparative behavior, the most visible and controversial approach in recent years has been sociobiology, which indeed lays claim to being the legitimate heir of Darwin (Alexander, 1979; Barash, 1977, 1978a, 1978b, 1980; Dawkins, 1976, 1980; Hamilton, 1975; Lumsden & Wilson, 1981; Trivers, 1971; and E. O. Wilson, 1975, 1978, 1980). In their efforts to seek evolutionary explanations for recurring patterns of social behavior among all animal species, sociobiologists have encountered a diverse array of criticism for their analysis of human social behavior. The emotional and rhetorical excesses of the initial response to (and defense of) sociobiology, attendant on the publication of Wilson's (1975) book, have been followed by carefully reasoned critiques, by biologists (Beach, 1978; Cavalli-Sforza & Feldman, 1981; Futuyma, 1979; Gould, 1980a, 1980b; J. C. King, 1980; Lewontin, 1978, 1979; Washburn, 1978), social scientists (Adkins, 1980; Leacock, 1980; Freeman, 1980; Layzer, 1980; Grene, 1978; Harris, 1980; Livingstone, 1980; Shields, 1980; Silverberg, 1980; Williams, 1980), and philosophers (Caplan, 1980; Hull, 1980; Midgley, 1980a, 1980b). It now appears that human sociobiology, as envisioned by its progenitors, suffers from serious and challenging biological, methodological, and philosophical difficulties. Here, we attempt to present a fresh way of looking at the subset of these difficulties that have to do specifically with genetic evolution in humans.

Risky Inference: The Sower's Fallacy

A purely inferential approach to the investigation of evolutionary adaptation can lead to serious error, especially in the analysis of human social behavior, but in principle for any trait in any species. In its purest form, this approach consists of three steps. The first step is the identification of a morphological, developmental, physiological, or behavioral pattern (we will call it the "focal trait") in an animal or plant species. The second requirement is the experimental, mathematical, or rhetorical demonstration that those individual organisms (or lineages) that manifest the focal trait have a survival or reproductive advantage over those individuals (or lineages) that manifest alternative traits. In other words the focal trait must be shown to be adaptive (in the sense of Clutton-Brock & Harvey 1979). The alternative patterns may themselves be identified in living members of the species, if variability exists. On the other hand, if the focal trait is universal in the species in question, the alternative patterns are postulated to have existed among ancestors of the species. In this case the alternatives are either modeled after existing patterns in other species, preferably closely related ones, or are simply constructed by plausible argument.

It is the next (third) step that is problematic. Assuming that steps one and two have been properly carried out, it has so far been shown that the focal trait increases the Darwinian fitness of those individuals (or lineages) that manifest it. Thus it is tempting to infer that natural selection of alternative genotypes must be responsible for the persistence or prevalence of the focal trait.

In fact, the inference that natural selection is or has been at work requires an assumption that is very often unstated, almost never tested (even with nonhuman organisms), but absolutely critical; the assumption is that the focal trait is transmitted genetically from one generation to the next. (In this Chapter we use the term *genetically transmitted* to designate any trait that depends causally on the genotype of its bearer and is or was expressed differently in individuals of alternative genotypes.)

In practice, the assumption of genetic transmission is often treated (fallaciously) as a conclusion, made plausible by the existence of the focal trait and the demonstration of reproductive advantage. Put more formally, reproductive advantage of one phenotype over another is a necessary condition for natural selection, but is never a sufficient condition. Genetic transmission of phenotypic differences is an equally necessary condition for natural selection (and likewise not a sufficient one, because phenotypes may not differ in fitness). Both reproductive advantage and genetic transmission are necessary for natural selection to act, and together they are sufficient. But because reproductive advantage and genetic transmission are logically and biologically independent of each other, the affirmation of one of them merely as a consequence of the truth of the other is a logical fallacy. This same point has been made in a variety of ways by Futuyma (1979), Gould (1980a, 1980b), and others.

It may be easier to see the kind of biological error that can result from this logical fallacy if the trait considered is a nonbehavioral one and, to further remove it from ourselves, a nonhuman one. To take a classical example, consider the trait "dry-weight yield" (total plant biomass) in Jesus' "Parable of the Sower":

> A sower went out to sow. And as he sowed, some seeds fell along the path, and the birds came and devoured them. Other seeds fell on rocky ground, where they had not much soil, and immediately they sprang up, since they had no depth of soil, but when the sun rose they were scorched; and since they had no root they withered away. Other seeds fell upon thorns, and the thorns grew up and choked them. Other seeds fell on good soil and brought forth grain [Matthew 13:3–8].

Plants from seed that fell on good soil had the highest dry-weight yield and the greatest reproductive success (highest seed set). Plants from seed that fell among thorns produced a smaller dry-weight yield and set fewer seeds, whereas plants from seed that fell elsewhere produced still smaller yields and still fewer seeds. Clearly, there was a correlation between phenotype (dry weight) and reproductive success.

Does it follow that the differences in dry-weight yield among the four "treatments" is genetically transmitted? Does it follow from the data presented that natural selection has acted, or is acting in the current generation of plants in the parable, to increase dry-weight yield? Assuming (as the parable clearly intends) that the seeds falling in any patch of ground were an unbiased sample of what the Sower had in his sack, the answer to both questions is a definite "No." The differences in the trait (dry-weight yield), among treatments, were purely a consequence of differences in environment, not genotype, so natural selection is inoperative. (Natural selection could, however, be operating among the seeds that fell on a given soil type, but that is another matter.)

With apologies to the Sower and his contemporaries, who, one supposes, knew perfectly well that it was soil conditions that mattered, we shall find it convenient to refer in the following sections to the "Sower's Fallacy": the error committed in assuming that a correlation between phenotype and reproductive success necessarily implies the operation of natural selection on genetically transmitted differences.

Traits that Follow Lineages

To clarify further the pitfalls of the inferential approach to evolutionary adaptation as it applies to human behavior, we need to introduce the concept of a "lineage-following trait." In a family pedigree, human geneticists would refer to such a trait as "familial" or "vertically transmitted" (with no presumption of genetic transmission), but we can broaden the idea to lineages of any order, from a nuclear family to a population, species, or a higher taxon (e.g., the primates). Genetically transmitted traits always follow lineages, though they need not be universal within a lineage. For example, feathers and the capacity for flight are lineage-following traits in the lineage Aves (the birds), but feathers are universal, whereas the capacity for flight is not (ostriches, penguins, and others are flightless). Likewise, the Rh blood factor and the capacity for learning and using a language are lineage-following traits in the lineage *Homo sapiens,* but language capacity is universal, whereas the Rh factor is not.

Because they may be transmitted between nonrelatives, culturally transmitted traits need not follow lineages, but in fact they very often do so (Cavalli-Sforza & Feldman, 1981; Pulliam & Dunford 1980). Languages (say, English versus Chinese) are a good example, because children generally learn to speak from their parents or other close relatives. Among humans, differences in wealth, power, status, control of resources, diet, religious practices, sexual practices, and sex-differentiated behavior are clearly lineage-following traits, and there is evidence that to some degree temperament, perceptual skill, psychomotor skill, memory, sexual preference, extroversion–introversion, and mental disorders tend to follow lineages as well (E. O. Wilson, 1978). It is logically sound to suppose that the transmission of any or all of these traits of human behavior is entirely cultural. Like the Sower's seeds, which had different fates simply because they landed in different environments, different people are born into different psychological, social, and cultural environments. (If the Sower had planted a perennial crop that drops its own seeds near the parent plant, the analogy would be even more complete.)

On the other hand, the transmission of differences in every lineage-following trait is potentially genetic. No one, to our knowledge, claims that differences in the language (e.g., English versus Chinese) people speak are in any way genetically transmitted, but many of the other aspects of human behavior listed in the preceding paragraph are much more controversial in this regard, and have figured in arguments for the genetic transmission of differences among contemporary humans, or among our ancestors. (The sociobiology references given earlier provide abundant examples of such arguments.)

Why are certain lineage-following traits candidates for evolutionary arguments based on genetic transmission? Because differences in these characteristics can be plausibly asserted, and sometimes demonstrated, to be correlated with differences in number of offspring (Barkow, 1977; Irons, 1980). However, a difference in reproductive success, associated with differences in a lineage-following behavioral trait, does not in itself imply that the trait is subject to natural selection—that is the Sower's Fallacy. Especially in humans, a difference in reproductive success may well be the result of a culturally transmitted trait. Cultural practices roughly constrain the transmission of social status, and control over resources, to biological lineages (families, bands, tribes, or states), and in difficult times this phenomenon may result in differential reproduction, with no genetic basis for these differences (Barkow, 1977; Cvalli-Sforza & Feldman, 1981; Williams, 1980).

Consider a trait that tends to follow human lineages entirely through cultural (as opposed to genetic) transmission; some relatively uncontroversial examples would be particular languages and dialects, or regional and class accents. Let us also assume that, in principle, the trait can be transmitted to any normal human being, regardless of genotype. Again, languages are a good example: Any young child can learn any language, if heard from infancy. If human lineages differ in some such characteristic, and coincidentally in reproductive success, then the relative frequency of the different states of the characteristic (e.g., different languages) will change with time, simply because reproductively more successful lineages come to form an increasing proportion of the total population, and the trait is passed primarily within lineages. Inevitably, the *genes* of more productive lineages, which may differ from the genes of less productive lineages for incidental reasons, also increase in relative frequency. However, the simultaneous increase in relative frequency of the trait and of particular genes is coincidental; the increase in frequency of the trait *cannot* be seen as the result of Darwinian natural selection favoring it, because its transmission is cultural.

This line of reasoning seems clear enough for traits, like specific languages, that have no obvious connection with differential reproduction itself. What about a trait, again with purely cultural transmission, that *itself* affects reproduction? Such a trait is scarcely different from the previous kind; as long as transmission and acquisition do not depend on genotype, changes in the frequency of the trait *cannot* be seen as the result of natural selection on the trait, even if accompanied by incidental changes in gene frequency. But in this case, there is one important difference. Unlike reproductively unrelated traits, the *nature* of the trait becomes important: Lineage-following cultural traits that *increase* relative reproduction will inevitably be more common, at any given time, than otherwise comparable traits that *decrease* relative reproductive success. *The similarity with evolution by natural selection is strictly allegorical.* However, this similarity has led some authors (e.g., Alexander, 1979, and earlier) to infer that the commonness of human behavioral traits or cultural practices that are associated with increased relative reproductive success (''inclusive-fitness maximizing'' traits), compared to the supposed rarity of traits associated with decreased success, is compelling evidence that the former traits evolved by genetic natural selection. However, as the foregoing reasoning indicates, purely cultural transmission, with genotype entirely irrelevant, would produce the same effect. With ''seed output'' substituted for ''dry-weight yield,'' this is the Sower's Fallacy writ large. Therefore, simple inference alone is incapable of distinguishing among the alternatives of purely cultural, purely genetic, or mixed transmission. For some traits however, contrasting predictions of alternative models of transmission may lead to testable hypotheses (Pulliam, 1982; Werren & Pulliam, 1981), and quantitative models of cultural and mixed transmission (Cavalli-Sforza & Feldman, 1981; Pulliam & Dunford, 1980) have begun to provide a basis for rigorous empirical studies.

Cultural Factors Can Cause Genetic Evolution

There is, however, a paradox: If there are any differences in gene frequencies among lineages within a population, and some lineages have a higher reproductive output, then genetic evolution (a change in population-wide gene frequency) will occur—even if the differences in reproduction are entirely the result of lineage-following *cultural* factors. In this way, a culturally based trait responsible for conferring greater reproduction on a lineage increases in relative frequency in the population and drags behind it completely unrelated changes in gene frequency. The genes whose frequencies are increased in this way have not been favored by natural selection. They may even be harmful: In this case, if a deleterious allele is present by chance in a lineage whose reproductive success is favored by a cultural trait, that allele will increase in frequency until the negative effect of the allele on reproduction or survival comes to equilibrium with the positive effect of the cultural trait (Cavalli-Sforza & Feldman, 1981).

On the other hand, culturally transmitted behavioral patterns may act as selective agents on human gene frequencies. For example, contemporary differences among human populations in the capacity of adults to digest lactose in milk are genetically transmitted, whereas the husbandry of milk-producing animals and the use of their milk is culturally transmitted. Livingstone (1980) makes a case for the historical primacy of culture as a selective force in this and certain other examples.

Criteria for Genetic Transmission of Variable Traits

To restate the argument in its strongest form: Differential reproduction by the bearer of a behavioral trait by no means proves that the trait is genetically transmitted. Even if descendants or collateral relations also bear the trait, although it is rare or absent in other lineages in the same population, and even if bona fide genetic differences exist among lineages, the genetic differences need not be the basis of the behavior. Because purely cultural traits may also follow lineages, and often affect reproduction, unambiguous inference concerning genetic transmission is impossible in such cases without: (1) evidence of a causal mechanism of gene action directly relating genetic differences to behavioral differences; or (2) good fit of a genetic model based on pedigree analysis, and the statistical rejection of environmental models. (We discuss some methods of genetic analysis later.) To behaviorists, these may seem to be the labors of

Hercules, but in fact they now represent the standard criteria among human geneticists (Vogel & Motulsky, 1979).

To be consistent, one should insist that the very same criteria be met in the study of lineage-following behavioral traits in nonhuman animal studies, and indeed, they often are (Beach, 1978; DeFries, 1980; Ehrman & Parsons, 1976). However, as we proceed from the social arthropods to social vertebrates, and then from fish to birds and mammals, the likelihood that a purely inferential approach to genetic adaptation will fall victim to the Sower's Fallacy increases steadily (Mainardi, 1980). The irony is that the difficulty of doing the experiments and/or observations to confirm genetic transmission of variable traits also increases. When we proceed from nonhuman mammals to humans, there is an abrupt and enormous increase in the likelihood of cultural transmission, the likelihood of committing the Sower's Fallacy, and the difficulty in obtaining evidence to reject it.

Why should the rigid criteria we suggest be imposed in the study of variable human characteristics suspected to have a genetic component? There are two reasons. First, experience in the rigorous study of human genetics has shown that environmental effects, including cultural practices, so often mimic genetic transmission that simple inferences from pedigrees are very risky. For example, lung cancer is very frequent in some families. The pattern of lung cancer occurrence in some families is consistent with genetic transmission. However, on further study, what is usually transmitted in these families is the smoking habit. Therefore, the reason the characteristic, "lung cancer," clusters in families is overwhelmingly environmental—the cultural inheritance of a behavior that has direct physiological consequences for individuals who engage in it (Higginson & Keller, 1975). In nonhuman species, the degree to which cultural transmission is likely to mimic genetic transmission is much less. Proponents of evolutionary homology in comparisons of human and nonhuman animal behavior would do well to take it as their task to falsify the hypothesis of cultural transmission within human lineages, even if cultural transmission can be excluded for other species under consideration (Beach, 1978; Caplan, 1980; Grene, 1978; Simon, 1980).

The second reason for demanding such rigor in the study of human characteristics is that in proclaiming any trait or condition to be genetically transmitted, we may impose severe social and psychological burdens on individuals exhibiting the trait, and on their relatives. For example, elucidation of the genetic basis of certain anemias based on anomalous forms of hemoglobin or on a deficiency of the enzyme glucose-6-phosphate dehydrogenase led to social ostracism of healthy carriers and their families in the villages where the condition was studied (Stamatoyannopoulos, 1974). Genetic counselors describe similar problems facing their clients who are carriers or potential carriers of genetically caused conditions (Kelly, 1976). In the field of behavioral conditions or disorders, these burdens can be particularly acute. On a societal scale, recent American and

European history provides us with all too familiar examples of oppression, exclusion, and even extermination of groups of people defined by their genetic ancestry on the grounds that genetic ancestry predisposed these people to undesirable behavioral characteristics (Allen, 1975; Graham, 1977; Kamin, 1974). Public ignorance of the complexities of gene–environment interaction and of the distinction between genetic predisposition and genetic predestination seems to be no less now than in previous decades (Montague, 1980; "Why You Do What You Do. Sociobiology: A New Theory of Behavior," 1977). However lamentable, this public ignorance places an extra burden on scientists. In our view, that the burden is undeserved does not justify ignoring it.

Universal Traits

Having discussed traits that vary among humans, we turn to the question of "universal" traits—human behavioral patterns believed to be manifest in all human cultures. But before considering human behavior, it would be well to look at the significance of "universal" traits in biology in general.

If all members of a lineage share some characteristic, we may speak of the characteristic as "universal" in the lineage (Futuyma, [1979], provides a useful analysis of this problem). For example, all verebrates have bony support structures; in all mammals, the female suckles the young, and so on. "Shared, derived" characteristics such as these are the basis of evolutionary taxonomy, inasmuch as the concordant appearance of many such traits strongly suggests that they became "fixed" (universal) in some ancestral population, which then gave rise to the lineage in question.

For example, *Homo sapiens* is classified with the great apes, rather than with some other group of primates, on the basis of a set of traits not present in ancestral or collateral lineages ("derived" traits) that are universal ("shared") among all humans and all great apes. There is another concordant set of traits, universal among humans, absent among the great apes, that distinguishes us from them. This is the method that taxonomists use to classify any biological species. Any trait that is universal in a lineage, of course, is a "lineage-following trait," in our terminology, but the traits used to make taxonomic judgements are almost entirely morphological and, more recently, biochemical, rather than behavioral. They are assumed by their nature (e.g., amino acid sequence in proteins, DNA sequences) or their relative lack of variation and critical functional role (e.g., pelvic or cranial structure) to be genetically "canalized"—expressed in much the same way, regardless of environment.

Further, it is assumed in evolutionary biology in general that any trait now invariant in a lineage was once subject to genetic variation, but became fixed by natural selection. This must almost always be an assumption, not susceptible to experimental test. The assumption of historic variation and natural selection for currently universal traits is most convincing when an adaptive function can be demonstrated for the trait, and artificial variation created, to test the proposed mechanism of selection. For example, Kettlewell (1973), in his famous series of experiments on industrial melanism, showed that avian predation on experimentally introduced, nonmelanic moths was greater than on cryptic, melanic moths that are now virtually "universal" in populations inhabiting industrial regions of England.

What about human behavior? Are there universal characteristics of human behavior? Beyond such "basics" as infant suckling, there is much dispute over the very existence of human universals that are specific enough to be studied as distinct "traits." E. O. Wilson (1975, 1978), for example, believes there are many traits that are "distinctively ineluctably human" that "can be safely classified as genetically based" because of their supposed universality (E. O. Wilson, 1975). Yet the universality of nearly every one of Wilson's "universals" of human social behavior has been challenged with anthropological data by Harris (1980), Freeman (1980), Silverberg (1980), Leacock (1980), or Sahlins (1976). This dispute is beyond our expertise. Here, we concentrate instead on the methodological problems presented by the evolutionary analysis of "universal" characteristics of human behavior, on the assumption that some such characteristics do exist.

Proponents of a Darwinian view of human behavior (e.g., Alexander, 1979; Barash, 1978b; E. O. Wilson, 1978) believe that there are (or were, in hunter–gatherer societies) many universal human traits—for example, incest avoidance, male dominance, and polygamy. These are, indeed, lineage-following traits in at least the sense that human cultures tend to be somewhat isolated lineages. It is not hard to find some plausible adapative significance for each of these traits in terms of associated differential reproductive success relative to the success of hypothetical alternative behaviors. But, as we have discussed, culturally transmitted behaviors follow lineages too, *especially* if they are associated with reproductive success. Let us suppose that we can describe such a universal trait and that we can postulate some adaptive significance for it. Because it is universal, there can be, by definition, no phenotypic variation in the trait. Thus, the methods we detail in the following section—family studies and pedigree analysis—are of no use in distinguishing cultural transmission from genetic transmission.

Instead, the question becomes one of canalization itself (Futuyma, 1979): If the trait is indeed universal and genetically transmitted, then, of necessity, it must be tightly canalized, such that no normally occurring social or cultural environment produces a different result. (Canalization does not preclude a role for learning in the ontogeny of the trait.) On the other hand, a trait might be so critical to human social life that every human culture enforces its expression, regardless of genotype—a kind of "cultural canalization." Incest avoidance, a favorite example of the Darwinian approach to human behavior (Alexander,

1979; Barash, 1977; E. O. Wilson, 1975), is a case in point. The sociobiological argument runs like this: If there were once genetic variation in the tendency to avoid incest, alleles that produced more reliable avoidance would bring their bearers greater reproductive success, by lowering the frequency of deleterious recessive alleles in homozygous form among their offspring, or by increasing the frequency of superior heterozygotes. (By avoidance of *incest,* we refer here to behavior that specifically precludes sexual relations between parents and their children, or brothers and sisters—not simply a tendency to disperse from one's birthplace, or other nonspecific behaviors tending to increase outcrossing.)

Although the universality of incest avoidance in human cultures has been disputed (Harris, 1980; Livingstone, 1980), let us assume that the "trait" is indeed characteristic of all historic cultures. Universality of a trait that increases reproductive success (fitness) need not imply genetic transmission of the trait, any more than synchronic variation in a trait, positively correlated with variation in reproductive success, implies genetic transmission. As we pointed out earlier, the frequency of a culturally transmitted, lineage-following trait that improves reproductive success will increase over time, because its bearers become more numerous, and the trait may indeed eventually reach "fixation" (universality).

A great variety of purely cultural reasons for avoiding incest (or for promoting outcrossing) has been proposed (Livingstone, 1980), and in principle different reasons might operate in different lineages, with nothing in common but the enhanced biological fitness that incest avoidance itself produces, and without genetic transmission. Culturally transmitted traits can increase biological fitness. On the model of purely cultural transmission of incest avoidance, a brother and sister born into a culture without an incest taboo are like the seeds that the Sower casts among thorns, whereas a brother and sister born into a culture that teaches incest avoidance are the seeds cast on fertile soil.

The case of incest avoidance presents a special and quite interesting complication, as there is a positive feedback between outcrossing (of which incest avoidance is a special case) and inbreeding depression (the cost to fitness that results from inbreeding) (Haldane, 1940). Contrary to conventional wisdom (see, for example, Alexander, 1979), there are many, many species (mostly plants and invertebrates) that inbreed consistently—some even undergo obligate sibmating, or mother–son mating (Colwell, 1981; Hamilton, 1967; D. S. Wilson & Colwell, 1981). Regularly inbreeding species that have been tested have shown no measurable inbreeding depression whatsoever from "incestuous" matings (Hoy, 1977). The reason is simple: In inbreeding species deleterious recessive alleles are continually kept at low frequencies because they are quickly exposed to selection in homozygous condition (produced by inbreeding). The higher the level of outcrossing, the greater is the level of heterozygosity. With increased heterozygosity, a greater number of recessive alleles can "hide" from selection in heterozygous condition, and "incestuous" matings will have increasingly disastrous effects (Livingstone, 1980). Thus the avoidance of incest becomes all the more advantageous the more it is practiced, perhaps making what might have been originally a weak proscription progressively more rigid.

To conclude, for practically any human social behavior it is possible to conceive of both genetic (Darwinian) and cultural scenarios for its origin and maintenance. But if the trait is invariant, it is not possible, even in principle, to distinguish between the hypothesis that it is genetically canalized, appearing in the same form regardless of cultural environment, and the hypothesis that the trait is rigidly constrained by social forces common to all human cultures, so that genotype is irrelevant. Intermediate hypotheses—for example, that culture reinforces a weak innate tendency, or that the trait is canalized in most genotypes, but socially enforced in the rest—are, of course, equally indistinguishable. Unfortunately, choosing among these alternatives is not a matter of experiment or statistical inference and must be based on plausibility arguments, or possibly on developmental studies. It is no wonder that the meaning of "universal" traits is so controversial.

Genetic Assimilation

A different and subtle argument in the application of Darwinian principles to the study of human behavior is the following: consider a lineage-following, behavioral trait that enhances reproductive success. Initially, the trait is transmitted entirely by cultural means, with no genetic component. According to the theory of "genetic assimilation," after many generations the trait will be "assimilated" into the genotype of its bearers, because any genetically based tendency toward its more effective expression (canalization) will be differentially passed on. (E. O. Wilson, 1975, Barash, 1977, and Barkow, 1977, 1980, discuss genetic assimilation by name in the context of behavioral evolution, and Futuyma, 1979, provides a general treatment. Elsewhere [Alcock, 1975, pp. 454–455], the mechanism is invoked without labeling it.) As an argument in evidence of the genetic transmission of any particular trait, genetic assimilation suffers from a logical flaw; it assumes as a premise what it seeks to establish as a conclusion. Unless there exists genetically heritable variation in the expression of the trait to begin with, there will be no variation on which natural selection can act, and no genetic assimilation can occur.

"Genetic assimilation" is a misleading term for a legitimate evolutionary process: the increase in frequency, by natural selection, of preexisting genotypes that decrease variability in the expression of a trait (Futuyma, 1979). The important point for the case of human social behavior is to keep firmly in mind that natural selection operates on phenotypes. *A behavioral phenotype that is genetically transmitted, partially or entirely, has no selective advantage whatever over the same phenotype transmitted culturally.*

Arguments for genetic assimilation of culturally transmitted traits in human behavior require two assumptions. First, it is necessary to assume that cultural norms and sanctions are slower and less effective than genes at constraining the quality or effectiveness of the trait. Otherwise, there would be no selection in favor of particular genotypes. It is difficult to conceive of a straightforward way of testing this assumption for a given human behavioral trait, but proponents of genetic assimilation are obliged to confront it.

Second, it must be assumed that the relevant aspects of the cultural environment remain stable long enough for genetic assimilation of a culturally advantageous trait to occur. The maximum rate of change in culturally transmitted traits is immensely greater than the maximum rate of genetic change in human populations (Cavalli-Sforza & Feldman, 1981). It seems unlikely that any modern culture is sufficiently static to permit genetic assimilation. Just how far back in human history must we go to find sufficiently long periods of cultural stasis? Testing the assumptions of genetic assimilation presents a major challenge.

GENETIC ANALYSIS OF COMPLEX HUMAN TRAITS

The strongest forms of evidence that a particular behavioral trait is partially or wholly transmitted by a gene or genes are: (1) the elucidation of a causal mechanism of gene expression directly relating genetic differences to behavioral differences; and (2) statistical evidence for the existence of a gene or genes responsible for differences in the behavior among individuals. In fact, geneticists generally pursue these lines of evidence simultaneously. The demonstration of the genetic basis of mental retardation due to phenylketonuria (PKU), for example, was accomplished by both metabolic studies of genetic variants of phenylalanine hydroxylase and quantitative studies of families of PKU children (Stanbury, Wyngaarden, & Fredrickson, 1978). The same guidelines apply to the study of the determinants of nonpathological variation in human behavior.

Genetic analysis of human traits—behavioral or otherwise—generally focuses on the basis for differences in those traits among related individuals. The most famous—and most controversial—examples of this approach are studies of heritability of human behaviors in identical and fraternal twins. The basis of the twins approach in the study of human behavior is to estimate the degree of genetic transmission of a behavioral trait by comparing the behavioral similarity of identical twins with the behavioral similarity of fraternal twins of the same sex. To the extent that the identical twins are behaviorally more similar to each other than fraternal twins, the trait is said to be determined by the greater genetic similarity of identical twins.

The difficulty with the twin approach, particularly for the study of behavioral traits, is that identical twins may be more similar to each other than are fraternal twins of the same sex for any of several reasons. They may indeed be more similar behaviorally because they are more similar genetically, as the classical approach assumes. On the other hand, identical co-twins may be more similar behaviorally than fraternal co-twins because environmental and social influences on identical twins are more similar than on fraternal twins (Cavalli–Sforza & Bodmer, 1971; Haseman & Elston, 1970). The classical twin method cannot distinguish these two causes of similarity, and consequently, the genetic influence on a trait studied in this way is frequently overestimated. Furthermore, comparison of identical twins who shared a single chorion in utero, identical twins having separate chorions, and fraternal twins (who always have separate chorions) indicates that having shared a chorion can be more important than sharing identical genes in influencing the degree to which co-twins resemble each other (Corey, Kang, Christian, Norton, Harris, & Nance, 1976; Melnick, Myrianthopoulos, & Christian, 1978). Any twin study that does not account for this very early environmental effect (and most twin studies cannot) risks overestimating the degree of genetic influence on the characteristic of interest.

A number of modifications in the design of twin studies have been proposed to circumvent these difficulties. Alternative estimates of the degree of genetic influence can be used, but each of these can over- or underestimate genetic contribution, depending on the particular nature of the twin pairs and the trait studied (Feinleib, Garrison, Fabsitz, Christian, Hrubec, Borhani, Kannel, Rosenman, Schwartz, & Wagner, 1977). It may also be possible to disentangle genetic and environmental influences on specific traits by studying both genetic and environmental reasons for differences between co-twins. If we can identify the environmental influences and how important they are, then we can adjust the differences in the trait for each twin pair to control for environmental influences. Only then would we compare identical twins with fraternal twins to estimate the magnitude of genetic influence on the trait. This strategy has not yet been applied to twin studies, but could solve some problems in the traditional design. A third possible modification of twin design involves the children and spouses of identical twins, as well as the twins themselves (Nance & Corey, 1976). This approach has been applied to physiological traits, but not yet to behavioral questions (Nance, Corey, & Boughman, 1978).

The investigation of identical twins adopted into different families—and thus presumably raised apart—has historically been, and remains an appealing strategy for the study of human behaviors (Bouchard, Hefton, Eckert, Keyes, & Resnick, 1981; Rao, Morton & Yee, 1976). The rationale is that the degree to which such twins resemble each other more than their adoptive brothers and sisters must be due to the twins' shared genetic identity. However, the strategy suffers from at least two related, severe limitations. First, there exist very few pairs of identical twins adopted into different homes. Second, it has been recently argued that many such adoptions do not really lead to twins being "raised apart," but rather, that each twin may be adopted by a relative of the mother, or by neighbors in the same town, so that the twins in question share most of the same environmental and social influences, including interacting with each othe (Kamin, 1974). Therefore, the number of identical twins truly raised apart is even smaller.

Adopting parents, biological parents, and adopted children have been compared in a variety of ingenious ways in the study of schizophrenia. One design is to compare the frequency of schizophrenia and related disorders among: (1) biological relatives of adopted children who became schizophrenic; and (2) adopting "relatives" of these same children (Kety, Rosenthal, Wender, Schulsinger, & Jacobsen, 1975). A second design involves the comparison of: (1) biological children of a schizophrenic parent, adopted at birth into unaffected families; and (2) biological children of unaffected parents, likewise adopted at birth into unaffected families (Rosenthal, Wender, Kety, & Welner, 1971). Still another design is based on the comparison of: (1) biological children of a schizophrenic parent, adopted into unaffected families; and (2) biological children of unaffected parents, adopted into families in which an adopting parent developed schizophrenia (Wender, Rosenthal, Kety, Schulsinger, & Welner, 1974). Each of these studies indicates some degree of genetic transmission of risk of schizophrenia. However, because many identical twins of schizophrenics are not affected themselves, environmental factors also must influence the development of the disease (Fischer, Harvald, & Hauge, 1969). Genetic and environmental hypotheses suggested by these statistical studies can ultimately be confirmed only by discovering a biological mechanism for inheritance of susceptibility to schizophrenia, and by elucidating those environmental factors that trigger its development.

In the past few years, several related statistical methods have been refined that can be applied to the analysis of behavioral (or other) traits in families. The goal of these methods as a group is to distinguish the influences of: (1) individual genes with major effects; (2) many genes, each contributing in a minor way, or polygenic effects; (3) culturally inherited characteristics; and (4) common environmental exposures. To our knowledge, no one method identifies each of these influences, but various combinations of influences can be disentangled. For example, if information is available for a very large number of parents and their children, path analysis can be used to estimate the influences of identifiable environmental exposures, culturally inherited effects (which may not be directly identifiable characteristics), and polygenic influences (Rao & Morton, 1978). Even this complex model, however, does not take into account greater-than-random genetic similarities of parents, or synergistic effects of genes and environments (Rao, Morton, Gulbrandsen, Rhoads, Kagan, & Yee, 1979). Alternatively, it is possible to separate "major gene" effects, polygenic effects, identifiable environmental influences, and (indirectly) the influence of the common childhood environment of siblings (Lalouel & Morton, 1981).

Each of these statistical methods is intended to estimate the extent of genetic and environmental influence on behavioral or other traits. None of these methods can prove the existence of genetic transmission—only demonstrate the consistency of genetic models with observed traits in families. A powerful means of verifying a model specifying the strong influence of a single gene on a behavioral (or other) trait is through linkage analysis (Emery 1976). This strategy is based on the fact that any gene (including a hypothetical one influencing the behavior under study) is a physical entity, a length of DNA. Furthermore, every human gene must lie on one of the human chromosomes. Therefore, in principle, at least, it should be possible to find out where in the genome the hypothetical gene influencing this behavior lies, even though we don't know how the gene is expressed physiologically. The strategy of linkage analysis is to prove that a hypothetical gene of interest exists by locating it. A gene can be located by finding another, easily detected gene of known chromosome location, completely unrelated to the behavior of interest, that is inherited the same way in the same family.

Linkage analysis has contributed to understanding the genetic influence on several diseases and to locating the genes responsible (Fredrickson, Goldstein, & Brown, 1978; King, Go, Elston, Lynch, & Petrakis, 1980; Lawler & Sandler, 1954). As far as we know, the approach has never been applied to nonpathological human behaviors for which the influence of individually important genes has been postulated. If genes with substantial influence on sexual preference, temperament, special talents, and so on, really exist (E. O. Wilson, 1978), it should be possible to locate at least some of them by linkage studies in large families in which particular variants of these behaviors appear. Until rigorous genetic evidence appears for the existence of ''behavior'' genes with major influence, one must remain skeptical of genetic hypotheses. Cultural transmission remains a strong competing hypothesis for the inheritance of lineage-following behavioral traits in human beings, whatever the plausibility of genetic hypotheses for their origin and maintenance.

CONCLUSION

We introduced this chapter by posing three specific questions. We conclude it by summarizing the responses evolutionay genetics can currently provide for each:

1. At the level of proteins and genes, we are extremely similar to our closest primate relatives. However, at the level of anatomy and morphology, we are not nearly as similar. In other words, biochemical similarity does not necessarily indicate morphological similarity. Likewise, genetic similarity need not imply behavioral similarity. Evolution at different levels of biological organization can proceed at quite independent rates.

2. Darwinian principles of evolutionary change in gene frequency due to natural selection can often be successfully applied to the study of behavior

(including social behavior) of animals other than ourselves. However, because cultural evolution is a far more pervasive process in human society than among social animals, and can proceed so rapidly, it needs to be considered explicitly as a hypothesis for the origin and maintenance of specific human behaviors. Cultural transmission may account fully for variations in human behaviors, even those that have reproductive and evolutionary consequences. Cultural transmission may also account fully for "universal" human behaviors, even those that are very similar to genetically transmitted behavior of other species. There is no reason to expect a culturally transmitted trait to become genetically transmitted ("genetic assimilation").

3. In order to demonstrate conclusively that individual genes determine or modify a particular human behavior, it is necessary either to elucidate the biochemical steps involved in the determination or modification of the behavior by the gene, to determine the relative influence of genetic and environmental factors in families in which the behavior varies, or to locate the gene on a particular chromosome. These methods have been applied to the study of many human diseases, including several involving mental retardation or mental disorder. They very seldom have been applied to the study of normal variation in human behavior.

It may appear that these responses raise more problems than they resolve. They certainly do not provide a dogmatic answer to the question of the determinants of human behavior, or even a smooth path for their investigation. We believe that the study of the evolution of human behavior will be the most intriguing, productive, and responsible when all competing hypotheses are made explicit and appropriate tests for each are devised. To specify the alternative hypotheses for a given behavior and to devise a test for each hypothesis is likely to require integrating available evidence from psychology, sociology, anthropology, animal behavior, and genetics. The breadth of such an undertaking is both the frustration and fascination of questions in comparative behavior.

ACKNOWLEDGMENTS

Various drafts of this article were read and usefully criticized by D. B. Wake, D. S. Wilson, P. W. Sherman, D. J. Futuyma, G. F. Oster, R. L. Caldwell, T. E. Rowell, P. Kenmore, and S. C. Stearns. We very much appreciate their help and suggestions, and the suggestions of an anonymous reviewer and the editor.

REFERENCES

Adkins, E. K. Genes, hormones, sex, and gender. In G. W. Barlow & J. Silverberg (Eds.), *Sociobiology: Beyond nature/nurture?* Boulder, Colo.: Westview Press, 1980.

Alcock, J. *Animal behavior*. Sunderland, Mass.: Sinauer, 1975.

Alexander, R. D. *Darwinism and human affairs*. Seattle: University of Washington Press, 1979.

Allen, G. E. Genetics, eugenics, and the class struggle. *Genetics,* 1975, *79,* 29–45.

Barash, D. P. *Sociobiology and behavior*. New York: Elsevier, 1977.

Barash, D. P. Evolution as a paradigm for behavior. In M. S. Gregory, A. Silvers, & D. Sutch (Eds.), *Sociobiology and human nature*. San Francisco: Jossey–Bass, 1978. (a)

Barash, D. P. *The whisperings within*. San Francisco: Harper & Row, 1978. (b)

Barash, D. P. Predictive sociobiology: Mate selection in damselfishes and brood defense in white-crowned sparrows. In G. W. Barlow & J. Silverberg (Eds.), *Sociobiology: Beyond nature/nurture?* Boulder, Colo.: Westview Press, 1980.

Barkow, J. H. Conformity to ethos and reproductive success in two Hausa communities. *Ethos,* 1977, *5,* 409–425.

Barkow, J. H. Sociobiology: Is this the new theory of human nature? In A. Montagu (Ed.), *Sociobiology examined*. New York: Oxford University Press, 1980.

Beach, F. A. Sociobiology and interspecific comparisons of behavior. In M. S. Gregory, A. Silvers, & D. Sutch (Eds.), *Sociobiology and human nature*. San Francisco: Jossey–Bass, 1978.

Bouchard, T. J., Hefton, A., Eckert, E., Keyes, M., & Resnick, S. The Minnesota USA study of twins reared apart: Project description and sample results in the developmental domain. *Progress in Clinical and Biological Research,* 1981, *69,* 227–234.

Bourne, G. H. (Ed.), *The Chimpanzee*. New York: Karger, 1970.

Caplan, A. L. A critical examination of current sociobiological theory: Adequacy and implications. In G. W. Barlow & J. Silverberg (Eds.), *Sociobiology: Beyond nature/nurture?* Boulder, Colo: Westview Press, 1980.

Cavalli-Sforza, L. L., & Bodmer, W. F. *The genetics of human populations*. San Francisco: Freeman, 1971.

Cavalli-Sforza, L. L., & Feldman, M. W. *Cultural transmission and evolution: A quantitative approach*. Princeton, N.J.: Princeton University Press, 1981.

Cherry, L. M., Case, S. M., & Wilson, A. C. Frog perspective on the morphological difference between humans and chimpanzees. *Science,* 1978, *200,* 209–211.

Clutton-Brock, T. H., & Harvey, P. H. Comparison and adaptation. *Proceedings of the Royal Society of London,* 1979, *205,* 547–565.

Colwell, R. K. Group selection is implicated in the evolution of female-biased sex ratios. *Nature,* 1981, *290,* 401–404.

Corey, L. A., Kang, K. W., Christian, J. C., Norton, J. A., Harris, R. E., & Nance, W. E. Effects of chorion type on variation in cord blood cholesterol of monozygotic twins. *American Journal of Human Genetics* 1976, *28,* 433–441.

Dawkins, R. *The selfish gene*. New York: Oxford University Press, 1976.

Dawkins, R. Good strategy or evolutionarily stable strategy? In G. W. Barlow & J. Silverberg (Eds.), *Sociobiology: Beyond nature/nurture?* Boulder, Colo.: Westview Press, 1980.

DeFries, J. C. Genetics of animal and human behavior. In G. W. Barlow & J. Silverberg (Eds.), *Sociobiology: Beyond nature/nurture?* Boulder, Colo.: Westview Press, 1980.

Ehrman, L., & Parsons, P. A. *The genetics of behavior*. Sunderland, Mass.: Sinauer, 1976.

Emery, A. E. H. *Methodology in medical genetics*. New York: Churchill Livingstone, 1976.

Feinlieb, M., Garrison, R. J., Fabsitz, R., Christian, J. C., Hrubec, Z., Borhani, N. O., Kannel, W. B., Rosenman, R., Schwartz, J. T., & Wagner, J. O. The N. H. L. B. I. twin study of cardiovascular disease risk factors: Methodology and summary of results. *American Journal of Epidemiology,* 1977, *106,* 284–295.

Fischer, M., Harvald, B., & Hauge, M. A Danish twin study of schizophrenia. *British Journal of Psychiatry,* 1969, *115,* 981–990.

Fredrickson, D. S., Goldstein, J. L. & Brown, M. S. The familial hyperlipoproteinemias. In J. B. Stanbury, J. B. Wyngaarden, & D. S. Frederickson (Eds.), *The metabolic basis of inherited disease* (4th ed.). New York: McGraw–Hill, 1978.

Freeman, D. Sociobiology: The "antidiscipline" of anthropology. In A. Montague (Ed.), *Sociobiology examined*. New York: Oxford University Press, 1980.

Futuyma, D. J. *Evolutionary biology*. Sunderland, Mass.: Sinauer, 1979.

Goodall, J. V. *In the shadow of man*. Boston: Houghton Mifflin, 1971.

Gould, S. J. Sociobiology and human nature: A postpanglossian vision. In A. Montague (Ed.), *Sociobiology examined*. New York: Oxford University Press, 1980. (a)

Gould, S. J. Sociobiology and the theory of natural selection. In G. W. Barlow & J. Silverberg (Eds.), *Sociobiology: Beyond nature/nurture?* Boulder, Colo.: Westview Press, 1980. (b)

Graham, L. R. Political ideology and genetic theory: Russia and Germany in the 1920s. *Hastings Center Report*, 1977, *7*, 30–39.

Grene, M. Sociobiology and the human mind. In M. S. Gregory, A. Silvers, & D. Sutch (Eds.), *Sociobiology and human nature*. San Francisco: Jossey–Bass, 1978.

Haldane, J. B. S. The conflict between selection and mutation of harmful recessive genes. *Annals of Eugenics*, 1940, *10*, 417–422.

Hamilton, W. D. Extraordinary sex ratios. *Science*, 1967, *156*, 477–488.

Hamilton, W. D. Innate social aptitudes of man: An approach from evolutionary genetics. In R. Fox (Ed.), *Biosocial anthropology*. New York: Wiley, 1975.

Harris, M. Sociobiology and biological reductionism. In A. Montague (Ed.), *Sociobiology examined*. New York: Oxford University Press, 1980.

Haseman, J. K., & Elston, R. C. The estimation of genetic variance from twin data. *Behavior Genetics*, 1970, *1*, 11–19.

Higginson, M., & Keller, J. Familial occurrence of chronic respiratory disease and familial resemblance in ventilatory capacity. *Journal of Chronic Diseases*, 1975, *28*, 239–251.

Hoy, M. A. Inbreeding in the arrhenotokous predator *Metaseiulus occidentalis* (Nesbitt) (Acari: Phytoseiidae). *International Journal of Acarology*, 1977, *3*, 117–121.

Hull, D. Sociobiology: Another new synthesis. In G. W. Barlow & J. Silverberg (Eds.), *Sociobiology: Beyond nature/nurture?* Boulder, Colo.: Westview Press, 1980.

Irons, W. Is Yomut social behavior adaptive? In G. W. Barlow & J. Silverberg (Eds.), *Sociobiology: Beyond nature/nurture?* Boulder, Colo.: Westview Press, 1980.

Kamin, L. J. *The science and politics of IQ* Hillsdale, N.J.: Lawrence Erlbaum Associates, 1974.

Kelly, P. T. *Dealing with dilemma: A manual for genetic counselors*. New York: Springer, 1976.

Kettlewell, H. B. D. *The evolution of melanism*. Oxford: Clarendon, 1973.

Kety, S., Rosenthal, D., Wender, P. H., Schulsinger, F., & Jacobsen, B. Mental illness in the biological and adoptive families of adopted individuals who have become schizophrenic. In R. R. Friere, D. Rosenthal, & H. Brill (Eds.), *Genetic research in psychiatry*. Baltimore: Johns Hopkins Press, 1975.

King, J. C. The genetics of sociobiology. In A. Montague (Ed.), *Sociobiology examined*. New York: Oxford University Press, 1980.

King, M.-C., Go, R. C. P., Elston, R. C., Lynch, H. T., & Petrakis, N. L. Allele increasing susceptibility to human breast cancer may be linked to the GPT locus. *Science*, 1980, *208*, 406–408.

King, M.-C., & Wilson, A. C. Evolution at two levels in humans and chimpanzees. *Science*, 1975, *188*, 107–116.

Lalouel, J. M., & Morton, N. E. Complex segregation analysis with pointers. *Human Heredity*, 1981, *31*, 312–321.

Lawler, S. D., & Sandler, M. Data on linkage in man: Elliptocytosis and blood groups. *Annals of Eugenics*, 1954, *18*, 328–334.

Layzer, D. On the evolution of intelligence and social behavior. In A. Montague (Ed.), *Sociobiology examined*. New York: Oxford University Press, 1980.

Leacock, E. Social behavior, biology and the double standard. In G. W. Barlow & J. Silverberg (Eds.), *Sociobiology: Beyond nature/nurture?* Boulder, Colo.: Westview Press, 1980.

Lewontin, R. C. Adaptation. *Scientific American,* 1978, *239*(3), 156–169.

Lewontin, R. C. Sociobiology as an adaptationist program. *Behavioral Science,* 1979, *24,* 5–14.

Livingstone, F. B. Cultural causes of genetic change. In G. W. Barlow & J. Silverberg (Eds.), *Sociobiology: Beyond nature/nurture?* Boulder, Colo.: Westview Press, 1980.

Lumsden, C. J., & Wilson, E. O. *Genes, mind, and culture: The coevolutionary process.* Cambridge, Mass.: Harvard University Press, 1981.

Mainardi, D. Tradition and the social transmission of behavior in animals. In G. W. Barlow & J. Silverberg (Eds.), *Sociobiology: Beyond nature/nurture?* Boulder, Colo.: Westview Press, 1980.

Melnick, M., Myrianthopoulos, N. C., & Christian, J. C. The effects of chorion type on variation in IQ in the NCPP twin population. *American Journal of Human Genetics,* 1978, *30,* 425–433.

Midgley, M. Gene juggling. In A. Montague (Ed.), *Sociobiology examined.* New York: Oxford University Press, 1980. (a)

Midgley, M. Rival fatalisms: The hollowness of the sociobiology debate. In A. Montague (Ed.), *Sociobiology examined.* New York: Oxford University Press, 1980. (b)

Montagu, A. (Ed.). *Sociobiology examined.* New York: Oxford University Press, 1980.

Nance, W. E., & Corey, L. A. Genetic models for the analysis of data from families of identical twins. *Genetics,* 1976, *83,* 811–826.

Nance, W. E., Corey, L. A., & Boughman, J. A. Monozygotic twin kinships: A new design for genetic epidemiological research. In N. E. Morton & C. S. Chung (Eds.), *Genetic epidemiology.* New York: Academic Press, 1978.

Pulliam, H. R. A social learning model of the occurrence of conflicts in matrilocal versus patrilocal societies. *Human Ecology,* 1982, *10,* 363–373.

Pulliam, H. R., & Dunford, C. *Programmed to learn: An essay on the evolution of culture.* New York: Columbia University Press, 1980.

Rao, D. C., & Morton, N. E. IQ as a paradigm in genetic epidemiology. In N. E. Morton & C. S. Chung (Eds.), *Genetic epidemiology.* New York: Academic Press, 1978.

Rao, D. C., Morton, N. E., Gulbrandsen, C. L., Rhoads, G. G., Kagan, A., & Yee, S. Cultural and biological determinants of lipoprotein concentrations. *Annals of Human Genetics,* 1979, *42,* 467–477.

Rao, D. C., Morton, N. E., & Yee, S. Resolution of cultural and biological inheritance by path analysis. *American Journal of Human Genetics,* 1976, *28,* 228–242.

Rosenthal, D., Wender, P. H., Kety, S. S., & Welner, J. The adopted-away offspring of schizophrenics. *American Journal of Psychiatry,* 1971, *128,* 307–311.

Sahlins, M. D. *The use and abuse of biology: An anthropological critique of sociobiology.* Ann Arbor: University of Michigan Press, 1976.

Schopf, T. M., Raup, D. M., Gould, S. J., & Simberloff, D. S. Genomic versus morphologic rates of evolution: Influence of morphologic complexity. *Paleobiology,* 1975, *1,* 63–70.

Shields, S. A. Nineteenth-century evolutionary theory and male scientific bias. In G. W. Barlow & J. Silverberg (Eds.), *Sociobiology: Beyond nature/nurture?* Boulder, Colo.: Westview Press, 1980.

Silverberg, J. Sociobiology, the new synthesis? An anthropologist's perspective. In G. W. Barlow & J. Silverberg (Eds.), *Sociobiology: Beyond nature/nurture?* Boulder, Colo.: Westview Press, 1980.

Simon, M. A. Biology, sociobiology, and the understanding of human social behavior. In A. Montague (Ed.), *Sociobiology examined.* New York: Oxford University Press, 1980.

Simons, E. L. Primate evolution. New York: Macmillan, 1972.

Simpson, G. G. *Principles of animal taxonomy.* New York: Columbia University Press, 1961.

Simpson, G. G. The meaning of taxonomic statements. In S. L. Washburn (Ed.), *Classification and human evolution.* Chicago: Aldine, 1963.

Stamatoyannopoulos, G. Problems of screening and counselling in the hemoglobinopathies. In A. Motulsky & F. J. G. Ebling (Eds.), *Proceedings of the Fourth International Conference on Birth Defects.* Amsterdam: Excerpta Medica, 1974.

Stanbury, J. B., Wyngaarden, J. B., & Fredrickson, D. S. *The metabolic basis of inherited disease* (4th ed.). New York: McGraw–Hill, 1978.

Trivers, R. L. The evolution of reciprocal altruism. *Quarterly Review of Biology*, 1971, *46*, 35–57.

Vogel, F., & Motulsky, A. G. *Human genetics*. New York: Springer Verlag, 1979.

Washburn, S. L. (Ed.). *Classification and human evolution*. Chicago: Aldine, 1963.

Washburn, S. L. Human behavior and the behavior of other animals. *American Psychologist*, 1978, *33*, 405–418.

Wender, P. H., Rosenthal, D., Kety, S. S., Schulsinger, F. & Welner, J. Cross-fostering: A research strategy for clarifying the role of genetic and experiential factors in the etiology of schizophrenia. *Archives of General Psychology*, 1974, *30*, 121–128.

Werren, J. H., & Pulliam, H. R. An intergenerational model of the cultural evolution of helping behavior. *Human Ecology*, 1981, *9*, 465–483.

Why You do What You do. Sociobiology: A new theory of behavior, *Time*, August 1, 1977, 63, pp. 54–58.

Williams, B. J. Kin selection, fitness, and cultural evolution. In G. W. Barlow & J. Silverberg (Eds.), *Sociobiology: Beyond nature/nurture?* Boulder, Colo.: Westview Press, 1980.

Wilson, A. C. Evolutionary importance of gene regulation. *Stadler Genetics Symposia*, 1975, *7*, 117–134.

Wilson, A. C., Carlson, S. S., & White, T. J. Biochemical evolution. *Annual Review of Biochemistry*, 1977, *46*, 573–639.

Wilson, D. S., & Colwell, R. K. The evolution of sex ratio in structured demes. *Evolution*. 1981, *35*, 882–897.

Wilson, E. O. *Sociobiology*. Cambridge, Mass.: Belknap Press, 1975.

Wilson, E. O. *On human nature*. Cambridge, Mass.: Belknap Press, 1978.

Wilson, E. O. A consideration of the genetic foundation of human social behavior. In G. W. Barlow & J. Silverberg (Eds.), *Sociobiology: Beyond nature/nurture?* Boulder, Colo.: Westview Press, 1980.

[illegible]

Wynne-Edwards, V. C. [illegible]

[illegible]

Williams, G. C. [illegible]

[illegible]

Wilson, E. O. [illegible]

[illegible]

10 Homology, Genetics, and Behavior: Homology from a Behavior-Genetic Perspective

Stephen Zawistowski
Jerry Hirsch
University of Illinois at Urbana-Champaign

Homology describes similar structures with a common evolutionary history. For many individuals a discussion of homology may conjure up images of 19th-century naturalists discoursing about the shape and form of the bat's wing, the bird's wing, and man's arm. Such associations might cast doubt upon the need for the consideration of homology by those who study behavior. We believe a careful consideration of homology from a behavioral perspective to be quite important. For those unfamiliar with the term's history in biology, the discussion is informative. For others, it is an "update" with respect to the new uses to which the term homology is now being put. For those unaware, homology has been resurrected from the dusty-bone bins of museum basements. It has become a vital and important tool now used in ever more intricate constructions of phylogenetic trees. Like so much of modern biology, it has found its way into the laboratory of the molecular biologist. And for everyone—we offer our own perspective on the use of homology and its place in comparative research.

Conjecture arises as to the origin(s) of animal behavior study. Whatever its origins, by the mid-1900s it was represented by essentially two schools of thought, that of comparative psychology and that of ethology. Although fundamentally different in theory and method (Jaynes, 1969; Lockard, 1971), they do bear some strong similarities to one another. Each claimed to study the true causes of behavior. Both recoiled from and attempted to overcome the problem of the mind–body dichotomy. Each relied on often unstated but implicit homologies to justify interpretations. Thorndike, who stands at the roots of American experimental psychology, after his analysis of learned behavior in several species suggested a similar underlying learning process in each, a view shared by Pavlov (Bitterman, 1975). With the aid of microcircuitry and computers, psychology has advanced beyond salivating dogs, and cats in a puzzle box. Although Seligman and Hager (1972) argue otherwise, the message, now couched in operational terms, is often the same: There is a continuity of the learning process across species and phyla, an inferred homology (Razran, 1971). Ethology, on the other hand, emphasized its link to evolution and natural selection. Each discipline unflinchingly applied its theories to the explanation of animal and human behavior, and to the manipulation of Man's fate. For the S–R school this was epitomized by behavior-modification therapy and Skinner's *Walden II* (Skinner, 1948). Ethology's stigma is probably best represented by Konrad Lorenz's endorsement of Nazi race policies based on his ethological perspective. Kalikow (1980) indicates that Lorenz:

> (1) saw *changes* in the instinctive behavior patterns of domesticated animals *as* symptoms of *decline*. (2) He assumed a *homology* between domesticated animals and civilized human beings, i.e., he assumed there must be similar causes for effects assumed to be similar, and he further believed that civilization was in a process of 'decline and fall.' Finally, (3) he connected the preceding concerns to racial policies and other features of the Nazi program (p. 1).

Although his interpretations were not shared by all ethologists (notably Niko Tinbergen), they were consistent with ethology's biological approach to behavior. Behavior study is now faced with a more subtle and seductive use of inferred homology, sociobiology. E. O. Wilson's massive text, *Sociobiology: The New Synthesis* (Wilson, 1975) has been much debated (e.g., Caplan, 1978; Montagu, 1980; Ruse, 1979; Hirsch, 1976, and accompanying discussions by others). Yet of its nearly 700 pages, it is the last chapter of 30 pages that has aroused the most heated discussion. For it is here that assumptions of homology prompt Wilson to theorize and make predictions about human social behavior, the impact of which can be devastating to society. It is for this reason that we wish to discuss homology, its history, usage, and relevance to behavior study. We should like to pay special attention to the rigorous analyses employed in the identification and uses of a homology. One should note the loss of resolution as our discussion moves from organic molecules to proteins, chromosomes, structures, and finally to behavior.

The concept of homology has long been used in mathematics (back to the 17th century) to describe a correspondence of ratios or relative values among the parts of structures, such as corresponding sides in similar geometric figures (Oxford English Dictionary). In chemistry, homology has been applied to a series of compounds differing in composition successively by a constant amount of certain constituents, for example, in the paraffin series compounds, expressed by the general formula C_2H_{2n+2}, show a regular difference of one carbon and two hydrogen atoms, CH_2: CH_4 (methane); C_2H_6 (ethane); C_3H_8 (propane); C_4H_{10} (butane); C_5H_{12} (pentane); and so on. A similar relation exists in the series of alcohols beginning CH_3OH (methylalcohol); C_2H_5OH (ethylalcohol); C_3H_7OH (propylalcohol); etc. (International Encyclopedia of Chemical Sciences).

As used in biology, the term *homology* was introduced by Richard Owen (1843). He defined "homologue" (Owen, 1866) to be: "the same part or organ in different animals under every variety of form and function [p. xii]." According to Heslop-Harrison (1962): "The concept of homology had for Darwin a perfectly clear meaning, well conveyed in the statement in *Fertilization of Orchids:* all homologous parts or organs, however much they may be diversified, are modifications of the same ancestral organ [1862, p. 233]. . . . The use of anatomical evidence in tracing homologies, whilst not unknown before Darwin, was carried by him to a new level of precision, and to this extent he must be regarded as the originator of this particular tool of interpretive morphology [p. 283]."

In 1943, Boyden reevaluated the homology concept and reconciled Owen's original usage with current practice. Boyden wrote: "Homology *is* a genetic phenomenon, and, in the case of special and serial homology, the mechanism is that of the orderly interaction of genes and cytoplasms with each other and with the environment [p. 243]." Boyden presented an elegant discussion at once clarifying homology and retaining its original spirit by integrating it with the burgeoning data from Mendelian genetics, especially as presented by H. J. Muller. He dismissed the idea of one gene–one organ and stressed the interaction of many genes with the cytoplasm. Overall, in regard to homologies he criticized the lack of "precision and clarity of meaning which is essential to their effective application in comparative zoology."

THE BIOLOGICAL BASIS OF HOMOLOGY

Once again, there is a need to reassert Boyden's message emphasizing the fundamentally genetic nature of homology. In cytogenetics the homology concept has proven most useful. In fact, its application represents an important milestone in the experimental study of both heredity and evolution. When chromosomes duplicate at meiosis, the two resulting products are *homologues*—exact replicas of one another, at that instant. In diploid species (those having pairs of chromosomes), only homologous chromosomes come together to form a pair during cell division in meiosis. Both parents contribute a haploid chromosome set to form the diploid chromosome complement of the offspring (that is, each parent, at random, contributes one member of each pair). The diploid set consists of heterologous duos, each being a homologous pair. The members of each pair "recognize" one another and come together at the appropriate stage in the cell cycle. Over generations of chromosome replication, changes do occur in the constitution and structure of a chromosome. Genes undergo mutation from one allelic form to another and chromosomes break, exchange parts, and recombine so that inversions (rotation of chromosomal segments), crossing over (exchanges between members of a pair), or translocations (exchange between members of heterologous pairs) occur with differing and unpredictable frequencies in "homologues" descended from a single ancestral chromosome. Over enough generations, therefore, sufficient changes might occur so that initially identical homologues might evolve through stages of partial homology to some ultimate stage (separate species) at which two chromosomes, no longer homologous, have become as unlike as are the members across heterologous chromosome pairs in an individual.

Where it has been possible to hybridize species and to study the chromosome complement of the hybrid, it has been found that one or more sections of a chromosome from one species will pair with part(s) of a chromosome from the other species, revealing their homology; whereas other sections of the same chromosomes will not pair, revealing a lack of homology between the unpaired sections. Studies of *Drosophila melanogaster* and *D. simulans* have shown much

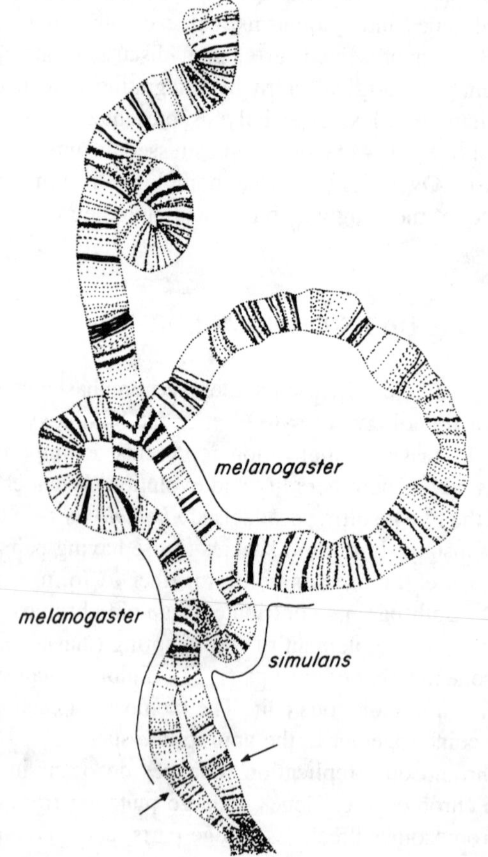

FIG. 10.1. Chromosomes of a *D. melanogaster/simulans* hybrid. Homologous sections exhibit pairing while nonhomologous sections do not pair. (Reprinted with permission of W. W. Norton and Company, Inc., from *Chromosomes, Giant Molecules and Evolution* by Bruce Wallace, 1966).

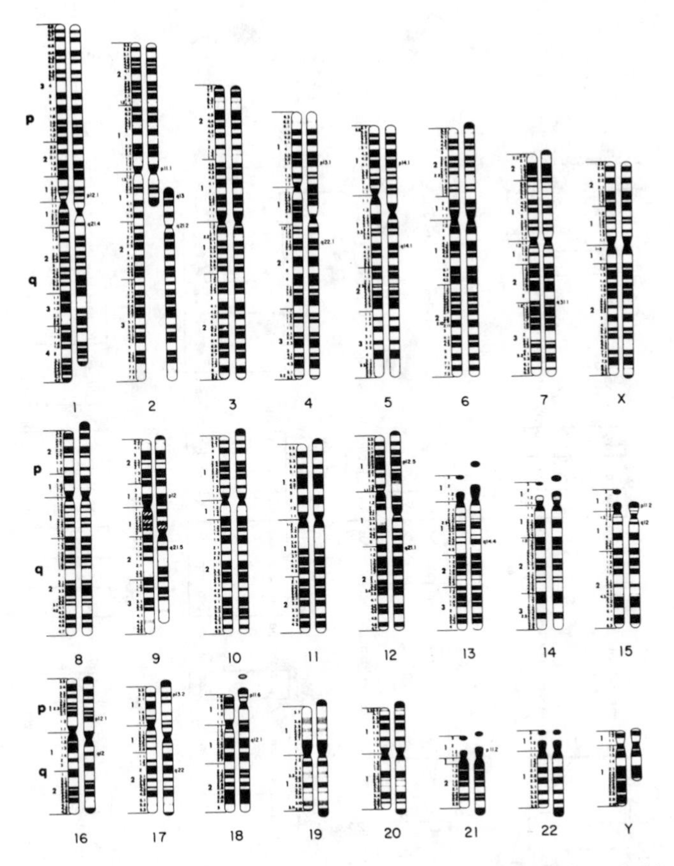

FIG. 10.2. Schemata of photomicrographs showing human (left) and chimpanzee (right) G-banded chromosomes. (Reprinted from Yunis, Sawyer, and Dunham, 1980; copyright 1980 by The American Association for the Advancement of Science).

of their chromosome complement to be homologous (Fig. 10.1). In other instances where it is not possible to hybridize, one can align photomicrographs of chromosomes from each species. Such an analysis led Yunis, Sawyer, and Dunham (1980) to wonder at the origin of the phenotypic differences between man and chimpanzee in light of the apparent homology of chromosomal structure (Fig. 10.2).

Because homology is fundamentally genetic, we consider its expression in primary gene products. Genes consist of sequences of DNA that code for assembling proteins from amino acids. The resulting proteins in several species may be similar in sequence and structure. Figure 10.3 presents the amino acid sequences for human alpha- and beta-hemoglobin. There are a number of portions that have

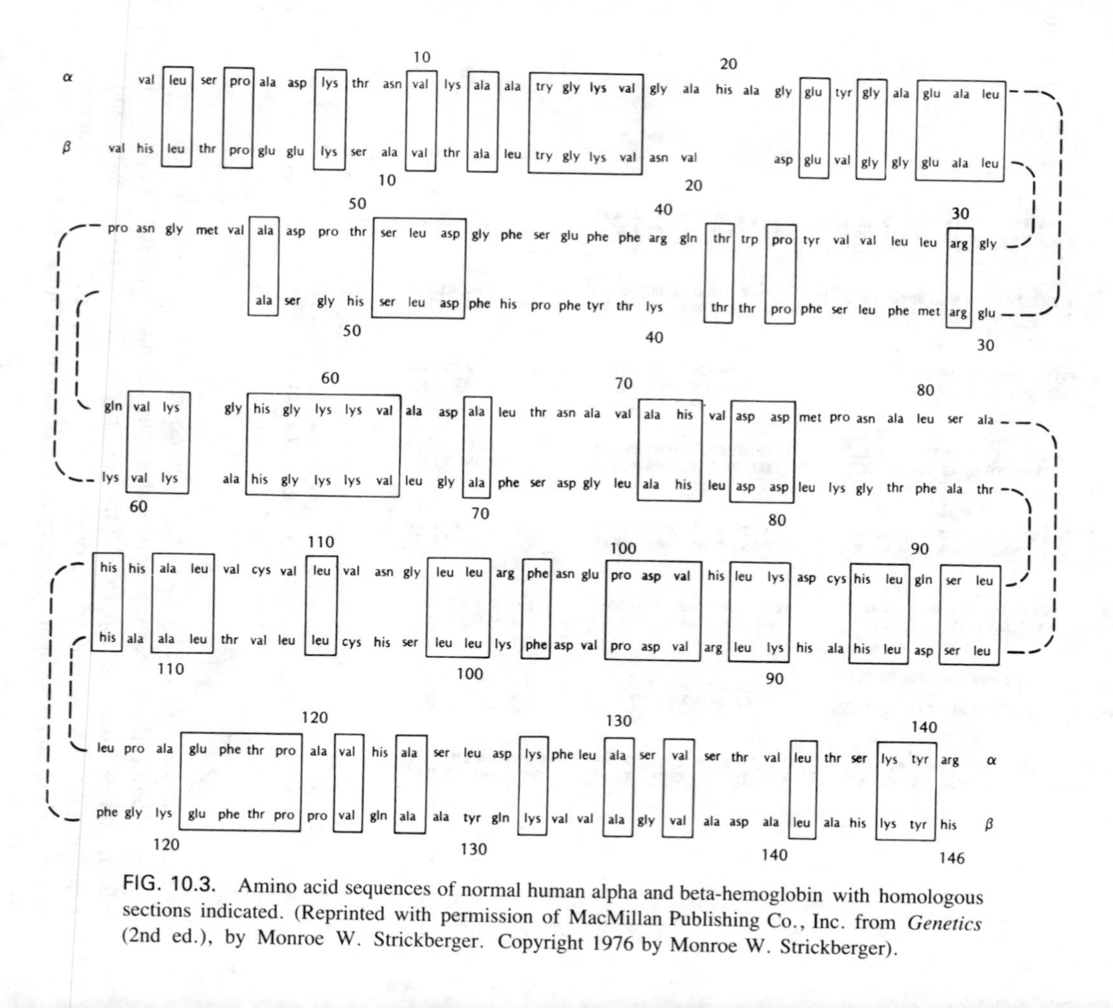

FIG. 10.3. Amino acid sequences of normal human alpha and beta-hemoglobin with homologous sections indicated. (Reprinted with permission of MacMillan Publishing Co., Inc. from *Genetics* (2nd ed.), by Monroe W. Strickberger. Copyright 1976 by Monroe W. Strickberger).

identical components and sequences. Their similarity is such that they might be considered homologous. Goodman and Lasker (1974) state that:

> by broad definition of genetic homology all vertebrate myoglobin and alpha and beta-hemoglobin chains are homologs related by their descent from a common gene ancestor which presumably existed in the first vertebrates. However, by a more restricted definition of genetic homology employed for constructing animal phylogenetic trees, the alpha chains in mammalian hemoglobin are a set of homologs; the beta chains are another set [p. 7].

Fitch (1966, 1970) and Margoliash and Fitch (1970) present a scheme to analyze biochemical homologies. First, one must ask whether the observed similarities between the two molocules are greater than would be expected by chance. One can compute such a probability by determining the number of base pair substitutions (in the original DNA strand) needed to convert one molecule to the other. This ''objective'' manner of determining a homology agrees favorably with instances where the homology is strong based on other data (e.g., hemoglobin and myoglobin). Margoliash and Fitch (1970) make a crucial addition to the scheme: Is the observed similarity a result of common ancestry or convergent evolution? Strictly speaking, the former would be homology, the latter analogy. Doolittle (1981) provides an excellent summary of current research on the similarity of amino acid sequences. He indicates that: ''It is altogether likely that the overwhelming majority of extant proteins—and certainly most enzymes—have evolved from a very small number of archetypal proteins [p. 149].''

BEHAVIOR AND HOMOLOGY: DESCRIPTION AND DEFINITION

Both the study of behavior and its use in species comparisons (homologies and analogies) have been complicated by problems with the definitions and the reliability of the behavioral observations. For example, what are aggression, appeasement, and courtship? In one of the early classics of ethology and comparative psychology, *The Expression of the Emotions in Man and Animals,* Charles Darwin (1872/1965) emphasized the description and interpretation of behavior. In it he gave detailed descriptions and anecdotal accounts of animal (and human) behavior. Lacking the traditions of experimental analysis and of the descriptive paradigms later to be developed in experimental psychology and ethology, respectively, Darwin relied on common sense and a little logic. An animal was doing what it appeared to be doing. For example, because people find it enjoyable to touch and embrace those they love, when a pet dog or cat rubs against its master's leg, the act is interpreted as a sign of affection (Darwin 1872/1965, p. 213). One can also make inferences based on an animal's behavior in two different situations. Darwin (1872/1965, p. 45) described the actions of his pet terrier upon receiving a biscuit. The dog played with the food using the same actions as when it captured a mouse. By inferring that the actions with the biscuit were intended to add ''relish'' to the food, Darwin joined a list of many ethologists who interpret certain behaviors as what have come to be called ''displacement activity''. Thorpe (1951) calls this: ''The performance of a behavior pattern out of the particular context of behavior to which it is normally related [p. 4].''

Perhaps the best known anecdotalist was a junior contemporary of Darwin, George Romanes. He collected and solicited stories of animal behavior and used them to begin a ''systematic'' analysis of animal intelligence and behavior (Romanes, 1882/1970). He also engaged in some rather interesting comparative interpretations:

> I may point out the remarkable change which has been produced in the domestic dog as compared with wild dogs, with reference to the enduring of pain. A wolf or a fox will sustain the severest kinds of physical suffering without giving utterance to a sound, while a dog will scream when any one accidentally treads upon its toes. This contrast is strikingly analogous to that which obtains between savage and civilized man: the North American Indian, and even the Hindoo, will endure without a moan an amount of physical pain—or at least bodily injury which would produce vehement expression of suffering from a European. And doubtless the explanation is in both cases the same—namely, that refinement of life engenders refinement of nervous organization, which renders nervous lesions more intolerable [p. 441].

Possibly to avoid the subjectivity of such description and interpretation, psychology borrowed from physics and adopted Bridgeman's ''operationism.'' Bridgeman (1950) related that he did not set out to start an ''ism'' but merely to describe what he saw his fellow physicists doing. Bridgeman (1927) stated that: ''In general, we mean by any concept nothing more than a set of operations; the concept is synonymous with the corresponding set of operations [p. 5].'' These operations are generally meant to be objectively specifiable and this served to place theoretical constructs on a physical level where they are amenable to experimentation. This meshed well with the growing tradition of behaviorism in psychology. If an experimenter were analyzing the conditioning of bar press behavior, a higher rate of responding would mean better conditioning.

Operationism was an important development for psychology. Pratt (1945) saw it to be a solution to the mind–body problem. As a methodology, it identified most clearly with the school of behaviorism. Skinner (1945) stated that ''behaviorism has been (at least to most behaviorists) nothing more than a thoroughgoing operational analysis of traditional mentalistic concepts.'' Operational thought was a part of early behaviorism even before Bridgeman fully articulated it. What Bridgeman saw his fellow physicists doing was experimentation with the possibility of replication. Two individuals could study the same concept if they used the same methods—operations. For the study of behavior to advance beyond the anecdotal stage, it ought to be possible to observe the phenomena repeatedly. Using operational principles, it would be possible to reproduce what was to be studied at the observer's convenience rather than relying on chance and the whims of nature for it to appear. A critical difficulty however was the stability (reproducibility) of the substrate of observation—the individual subjects. Chemistry and physics use well-defined substrates, elements and molecules, with environmental controls (STP: Standard temperature and pressure). The critical "contribution" made by the founder of behaviorism, John B. Watson, came through his misinterpretation of the Johannsen experiments on pure lines of beans (Hirsch, 1967). Johannsen studied pure lines of princess beans and measured seed weight. For geneticists his results demonstrated the effects of both heredity and environment—that variance between lines was genetic and variance within a line was environmental. Watson focused on the within-line results—only part of the experiment—and crystalized his concept of "environmental determinism." By reasserting the *tabula rasa* concept attributing all variance to experience, he provided the replicate individual substrate needed for the rigorous application of operationism to psychology. Not only could one repeatedly use the same operations to define a concept, but one could also use the "same organism" every time, in the words of Watson's (1959) well-known boast:

> Give me a dozen healthy infants, well-formed, and my own specified world to bring them up in and I'll guarantee to take any one at random and train him to become any type of specialist I might select — doctor, lawyer, artist, merchant-chief and, yes, even beggar-man and thief, regardless of his talents, penchants, tendencies, abilities, vocations, [and] race of his ancestors. I am going beyond my facts and I admit it, but so have the advocates of the contrary and they have been doing it for many thousands of years. [p. 104].

In other words, there are no differences among individuals, prior to the effects of experience, because they are all born alike (i.e., healthy infants).

The study of behavior eventually developed along two diverging paths, the distinctness of which has become most evident with the recent convergence of perspectives from laboratory experimental psychology and from ethology. One can describe behavior as either its actions (head nodding, tail wagging, etc.) or its consequences (e.g., courtship is behavior leading to mating.) It is this issue that Purton (1978) addresses from an analytic–philosophic perspective, calling attention to the problems it creates. Confusion may arise when a concept can be defined as both function (result) and form (actions), especially if one analyzes the form of a behavior experimentally when the objective is to understand its function.

As behavior geneticists we wish to analyze the genetic correlates of behaviors. To perform a genetic analysis it is necessary to be able reliably and accurately to categorize and/or to measure individual differences in phenotypic expression. It is through phenotypic variation that we gain access to the genetic system. By selecting phenotypes for breeding, we infer the possibility of reorganizing the distribution of the genetic correlates of the trait in which we are interested. For this phenotypic assessment it is then necessary to be able to operationalize a behavior so that we can measure it. This however does not necessarily mean that we can not later step away from our operationally defined behavior and assess its correspondence to a theoretical construct, such is the nature of validation. An example of our own work will be discussed in a later section.

At this point it is useful to reevaluate behavior as having two dimensions, form and function. Behavior as form can show constancy. Head nodding might look much the same in several species of birds, as might the cumulative response curves from operant behavior studies of bar pressing by rats and key pecking by pigeons. Function, however, is context dependent. Purton provides an example showing how, in the sandwich tern a "nodding" behavior, which displays the black top of the head against a white background, is a threat display whereas the same movement in the night heron, showing white feathers against a black background, is an appeasement display. Here one behavior as form has two functions in different species.

It is important for behavior geneticists to realize whether they are manipulating form or function. Manning (1961) selectively bred lines of *D. melanogaster* for slow and for rapid mating. The resulting lines differed significantly in latency to copulation following introduction of males and females into a mating chamber. Manning's further analysis showed the males of the slow-mating line to be very active and to engage in more irrelevant (noncopulatory-directed) behavior before courting. Bastock (1956) studied the mating of wild-type and of mutant yellow-bodied males. She found them to differ in components of the courtship sequence. Manning selected for latency to copulate based on results irrespective of the form of the behavior preceding copulation. It could be construed that he produced a change in the function of the preceding behavior. Slow-line males when placed in the chamber did not court. Bastock, however, observed a difference in form. Yellow-bodied males differed from wild-type males in the amounts of licking, wing flicking, and other courtship components. They attempted to court but their behavior differed from that of wild-type. In both the yellow-bodied males and the slow-mating-line males, a change respectively in form and function led to a similar result—a loss in mating efficiency.

We do not advise that behavior henceforth be described only as form or only as function. We agree with Purton, however, that models based on form need to be tested by experiments on form and those based on function tested on function.

We believe that behavioral homologies should be studied as form, not function. Recall our earlier discussion of hemoglobins and myoglobins. Their homologous nature is confirmed as a result of a similarity in sequence, composition, and structure. A functional perspective alone can be misled by convergent evolution. Willey (1911) discusses the homologous nature of the air bladder in teleost fishes and the lungs of terrestrial vertebrates. There is no homology, however, between the piscine air bladder and the avian air sac, attached to or near a bird's lungs, even though these two structures may serve similar (analogous) functions of air pressure control. The problem of analogy can be especially troublesome for behavior study. According to Ebling (1981): "Whereas structural homology is unquestionably distinct from analogy in instances where functions have diverged, patterns of behaviors are selected for comparison precisely because they appear to have analogous functions [p. 198]."

Ritualization may alter a behavior's function but not its basic form. From a genetic perspective, gene duplication provides a species with superfluous genetic material upon which natural selection can experiment. If, as is sometimes hypothesized, the primordial life form was represented by a single gene, then the present genotypic diversities could be the result of the duplication, reduplication, and modification of that single gene (Dobzhansky 1970, p. 413). To pursue an extreme, one could say that all gene products are homologous because they stem from genes that are descended from the single primordial gene. Of course, this statement is tempered by the great diversity of gene products. As seen, however, some have maintained a significant similarity and it is these that are considered within the context of homologies. We detect a parallel relationship between gene duplication and the ritualization of a behavior. Each process involves the provision and modification of material for functions at times very different from that of the original form. This similarity leads us to suggest that certain terminology used in the discussion of genetic homologies might be employed when discussing behaviors. Margoliash and Fitch (1970) distinguish two types of homologous genes: orthologous genes whose products have similar functions (i.e., the hemoglobins of man and chimpanzee), and paralogous genes whose products have diverged in function (i.e., hemoglobin and myoglobin). In related species, courtship behaviors sometimes remain similar in function and form yet have diverged sufficiently to provide a reproductive isolating mechanism, an orthologous homology. On the other hand, the previously mentioned head nodding in terns and herons has maintained similarity in form but has diverged from the ancestral function such that it serves different functions in the two species, paralogous homology. The distinction between paralogous and orthologous might prove to be quite useful for those who study behavior. It could help in the classification of those behaviors that through ritualization have maintained similarity in form yet have diverged in function. Inasmuch as we have restricted homology to form, one might still examine behaviors of different functions within the context of paralogous homology.

BEHAVIOR AND HOMOLOGY: CONFIRMATION AND INTERPRETATION

Konrad Lorenz (1956) has described how the great similarity of behaviors among more or less closely related species is a fact that has played a central role in the founding and development of ethology:

> Whitman and Heinroth, independently from each other . . . discovered the fact that behaviour patterns could be used as taxonomic characters, as characters that are not only characteristic of a species, but often of a genus, an order or even of one of the largest taxonomical group categories. The classical example of such a behavior pattern is the drinking movement of the Columbidae. Pigeons and sand grouse (Pteroclidae) have been put as two sub-orders into the one order of Columbidae by naturalists who were guided by morphology alone and not interested in behavior. This classification finds a convincing support in the movement in question, which characterizes the group better than any single morphological detail. When you read the zoological diagnosis of Columbidae in modern textbook, it is quite a paragraph and yet not an absolutely satisfying definition of the group. But if you say they drink water by sucking it up with a peculiar peristaltic movement of the oesophagous you have fully characterized the order.
>
> The invariability of such movements by far exceeds what Prof. Hebb has aptly termed species predictability. They are predictable, not only for a species, but for a genus, an order, or even a larger taxonomic unit. It is, for instance, predictable with supreme certainty that any new species belonging to the order of Columbidae will drink in the manner just described or that any species of Anatidae will take oil from the oilgland by rubbing its head on it in a rotatory movement.
>
> This remarkable distribution of motor patterns throughout the whole zoological system was the superlatively unexpected positive discovery whose tremendous inferences caused research to take that particular direction which ethology has pursued ever since [p. 52].

It is however one thing to recognize similarities and quite another to ascertain their origin. As we have seen, it is possible to quantify relatedness statistically at the molecular level. As complexity increases at the chromosomal level, determination of homology becomes more qualitative. Let us now consider both the experimental nature of confirmation and the usage of behavioral homologues.

It has been possible to carry on some behavioral studies on interspecific hybrids. Von Schilcher and Manning (1975) studied courtship song and mating speed in *D. simulans/melanogaster* hybrids. In their statement concerning interpulse intervals (ipi's) they make explicit an implicit component of many such studies. "it could equally well be that *D. simulans* ipi-determining genes are dominant over the homologous *D. melanogaster* genes [p. 402]." Their assumption is that the similar behaviors in the two different species are under the control of shared gene correlates. If such were the case, the data would meet our criterion of the basic genetic nature of homology. There have been other such studies of behaviors using hybrids: Bentley and Hoy (1972, 1974) with crickets, Sharpe and Johnsgard (1966) with ducks, Dilger (1962a, 1962b) with lovebirds, and Hinde (1956) with finches.

The studies have several shortcomings. It is often difficult to get the parental types to breed and, when they do reproduce, the yield of progeny is often small compared to the sample sizes required statistically to test genetic hypotheses. Furthermore, the F_1s are usually sterile or have highly reduced fecundity and viability. Hinde (1956) suggests that to determine homology it is necessary to observe a possible breakdown in the F_2 generation of the stimulus–response relationships. This is quite difficult considering the generally small numbers of animals (e.g., 11 male F_2 progeny for Sharpe and Johnsgard) available for study, even when the F_1 are not sterile. Thus this represents an interesting Catch-22 dilemma: A meaningful statistical analysis of F_2 progeny requires a rather large number of individuals. On the other hand, if the F_1 is fertile enough to produce a sufficient number of F_2 progeny, the populations being studied fail, by this very breeding test, to qualify as separate species. Whereas correctly performed hybrid analyses can provide substantial information about the gene correlates of similar behaviors in related species, the restrictions (only closely related species) and difficulties (the aforementioned Catch-22, fecundity) tend to limit the practicality of such attempts to very special cases.

HOMOLOGY OF METHOD

Due to the difficulties inherent in confirming behavioral homologies, we do not recommend that it be the major focus of comparative study. We suggest that in addition to explanatory generalization, we should also consider generalization of method as a promising approach. This suggests the use of models to be tested on a convenient species at the inception of a research program.

Harrington (1975, 1983) has used such a strategy in his study of the question of test bias. He used different populations of rats, and sets of items from the Hebb–Williams, rat intelligence test maze to analyze the influence on test bias of level of representation (of a group, population, or race) in the test-standardization sample. In this apparently culture-free environment, he found that groups in the numerical minority in the test-standardization sample were discriminated against by the test thus constructed. Harrington's insightful analysis does not assume homology between rat "intelligence" and human "intelligence." He used the animal behavior to study properties of a method, psychometric test construction. Animal models can also be used to study systematically the effects of violating assumptions of a method (e.g., random mating). Although it is possible to perform many such operations by computer simulation, such work nevertheless depends on what the researcher writes into the computer program (i.e., his assumptions [recall the operators' maxim, GIGO: garbage in, garbage out]). If one employs an animal model, the ancillary conditions are provided by the animal population. A living system can, and often does reveal unanticipated properties, which the experimenter would not have written into his computer program.

In part, our own work is an attempt to develop a general method for the behavior-genetic analysis of a species. By studying the feeding behavior of the blow fly we are learning how to analyze the genetic correlates of behavior of the species. This is consistent with the pattern of the history of genetics. Even though Mendel worked with peas, the classic crosses he developed have been used time and time again to analyze new and different diploid species. The work of Morgan, Sturtevant, and others on mapping the X chromosome of *Drosophila* did not prompt other geneticists to infer the same genes on the human X chromosome, but it did give them a method of analysis. The later success of these subsequent analyses led to McKusick's (1980) assertion that an "absolute homology for gene content appears to exist for the X chromosomes of all placental and marsupial mammals [p. 379]."

BEHAVIOR-GENETIC ANALYSIS

We have recently completed a hybrid-correlational analysis of Central Excitatory State (CES) and Conditionability in the blow fly (*Phormia regina*), combining two analyses: (1) a hybrid analysis (F_1, F_2, and backcrosses) of two pure breeding lines "fixed" for high and for low expression, respectively of CES; and (2) a correlational analysis comparing (correlating) conditioning and CES performance in the same samples of individuals. Combining the correlational and hybrid analyses yields information about: (1) relations between the two behavioral measures (phenotypes); (2) relations between each behavior and its genetic correlates; and (3) relations among the genetic correlates of the two behaviors. CES and conditionability are operationally defined hypothetical constructs, each of which is inferred from the same behavior, the proboscis extension reflex (PER) used in feeding by the blow fly. The blow fly possesses chemosensitive receptors on its proboscis (mouth parts) and tarsi (legs). Stimulation of these receptors can elicit a PER (Fig. 10.4). Dethier, Solomon, and Turner (1965) described CES as an: "induced responsiveness to water [that] reflects a perseverating central nervous system excitatory state (CES)." They found that a water-satiated fly will not respond to water. If, however, water stimulation is preceded by sucrose stimulation a response to water can be elicited. The effective time frame between sucrose and water stimulations is a function of the sucrose concentration and food deprivation state of the fly. Using water and saline solutions as conditioned stimuli (CSs) presented to the fore-tarsi and a sucrose unconditioned stimulus (US) presented to the mouth parts of flies, Nelson (1971) was able to demonstrate classical conditioning of the PER. Water-satiated flies

FIG. 10.4. Blow fly (*Phormia regina*) mounted in plastic pipet tip and showing proboscis extension. (Photo by T. Tully)

that initially showed low probability of response to saline or water eventually showed higher probabilities of response after pairing of water and saline presentations with sucrose. At the conclusion of her 1971 paper, Margaret Nelson suggested that the observed conditioning may be an extension of the nonassociative CES. We have observed a phenotypic correlation for the two traits indicating a relationship and have followed it during a hybridization of lines fixed for high and low expression of CES (Tully,Zawistowski, and Hirsch, 1982). Our results show this correlation to persist in replicate F_2 generations. This would indicate that the genes correlated with the two traits are not assorting independently, suggesting that the respective correlates are tightly linked on the same chromosome or that some subset of genes may be common to both sets of correlates. Further analysis of this relationship may help to clarify the role of CES in conditioning.

In essence, we are using our *Phormia* PER system to explore the feasibility of using genetic techniques to analyze components of complex behaviors. By selection we can amplify expression of specific, identified components of conditioning and through hybridization analysis we can assess relationships based on segregation and recombination of gene correlates. It is also interesting that by our identification of individual differences (IDs) in behavior and subsequent bidirectional selection we have, in fact, been able to approach one of behaviorism's assumptions—homogeneity of subjects. This homogeneity or similarity of performance is of course restricted to the trait selected for, but a strain selected for high (or low) performance on a task does offer a unique sample space for behavioral analysis. Watson believed behavior was a result of environment. A broader perspective views behavior as a function of genotype–environment interactions. The *Phormia* system offers a valuable opportunity to manipulate both the genotypic and environmental aspects that influence "conditioned behavior." Experimental manipulation of S–R relationships would be considerably easier to evaluate if the individuals in the various treatment groups really do perform similarly under similar conditions. We caution, however, that this is distinct from analyses with inbred lines where essentially all variance (for all traits) is environmental. In our work with selected lines, the homogeneity is inferred (and observed) for only specific traits (that have undergone selection) and at specific levels of expression of those traits. We believe that this perspective has implications for comparative research that are important, even though subtle. Comparative psychology and learning theory have been successful without needing to claim that the behaviors studied are homologous. This is because conditioning and learning are defined by the operations employed and not necessarily the responses elicited. A Skinnerian observes an operant curve produced by subjects, not the response elicited. The acquisition, extinction, and spontaneous recovery curves produced by a rat pressing a bar or a pigeon pecking a key are methodologically homologous to one another only with respect to the method employed for their study. The Nelson paradigm has enjoyed unique success in regard to dipteran conditioning because it is in fact not unique. Nelson employed the classic controls and operations defined by learning theory to produce the type of data expected by the field dominated by those operations.

The genetic aspects of our work are in no way novel. Our breeding analyses are pre-Mendelian in their scatter-gun approach to phenotypic alteration. The hybridization of divergent forms to observe segregation and recombination (assortment) date to the genesis of genetics—Mendel, his peas, and the introduction of those two concepts of transmission genetics, segregation and independent assortment. The phenomenal growth of genetics in the 80 or so years since the rediscovery of Mendel's paper has been highlighted by an appreciation for the similarity of the genetic system in different species. It is interesting to note that our work should combine these two often polar disciplines and their methodologies, learning theory with its defined homologues of operations, and genetics with its underlying processes of meiosis, mitosis, and protein synthesis. We believe that our program blending these two disciplines will not only provide interesting data on the interaction of excitation, sensitization, habituation, and conditioning in the blowfly, but that it may also serve to develop a strategy for analyzing any species both behaviorally and genetically.

CONCLUSION

Within the course of our discussion it may appear that we have been overly detailed in our discussion of the basis of homology. Admittedly comparisons are fundamentally important to the speculation and modeling that may lead to the conceptualization of new and important ideas. Homology, however, implicates a more intimate relationship between structures, behaviors, and species. Even this relationship bears careful scrutiny. Take the case of the commonly recognized homology of the wing and forelimbs among vertebates. Upon examination of the wing, the anatomist does not interpret the forelimb as a rudimentary structure of flight, or alternatively the wing as a possibility for terrestrial locomotion. No, rather that differing niches and natural selection have altered homologous structures to differing functions. Why is such a commonsense analysis not applied to behaviors such as aggression and mate selection? Why is it that so much is so quickly read into behavioral similarities?

That as scientists we need to consider the nature of our speculation within the public domain is highlighted by an important example. In *Man and Beast: Comparative Social Behavior* (Eisenberg & Dillon, 1971) a footnote quotes Senator J. William Fulbright questioning Dr. Karl Menninger during the Smithsonian Man and Beast Symposium:

> What is important to those of us who happen to be in the Senate, in the Congress, is to feel that it is possible, or even probable, that we can influence the decisions which affect the future of this country. . . . If we assume that men generally are inherently aggressive in their tendency, evidenced by the historical experience of using military force, physical force, to solve problems, if this is inherent and man cannot be educated away from it, it certainly makes a great deal of difference in one's attitude toward current problems. . . . If we are inherently committed by nature to this aggressive tendency to fight, well then, I certainly would not be bothering about all this business of arms limitations or talks with the Russians [p. 373].

Given the nature of this question it becomes even more important when the volume indicates (p. 14) that Sen. Fulbright later met with Congressional colleagues to discuss the material covered at that symposium. This incident highlights the complex relationship of academic freedom, scientific rigor, and the public responsibility of the scientist. Homologies are potentially powerful analytic tools. It must be recognized however that they are part of the method, not the answer.

SUMMARY

We have discussed homology and comparative research from the somewhat conservative perspective of genetics and behavior-genetic analysis. Atz (1970), who is more favorably disposed towards "homologies" of function, provides a more general discussion of behavior and homology. We have been more strict in our acceptance only of homologies of form. In those instances where a similar function is served by dissimilar structures, we should accept the homology if there is a phyletic link between the species, the structures and the function.

It is hoped that our conservative concept of homology is not viewed as a criticism of comparative research. To the contrary, we encourage comparative research. The animal species vary over a rich variety of behaviors, some of which are homologous from one species to another, some of which are not. The primary goal of comparative research should not be to root out homologies. It can be just as useful in the development of a thorough analysis of animal behavior to study nonhomologous behaviors. However, if homologies are investigated and later used for further interpretive purposes we need carefully to guard against overzealous generalizations. The homologies we use can not be of the logical "What if . . .," "It is possible . . ." type. We conclude with Aristotle's advice in such circumstances: "We must not accept a general principle from logic only, but must prove its application to each fact; for it is in facts that we must seek general principles and these must always accord with the facts [Osborn, 1894]."

ACKNOWLEDGMENTS

This project was supported by Grants MH15173 for Research Training in Institutional Racism, from the National Institute of Mental Health and from the Koscuiszko Foundation.

REFERENCES

Atz, J. W. The application of the idea of homology to behavior. In L. R. Aronson, E. Tobach, D. S. Lehrman, & J. S. Rosenblatt (Eds.), *Development and evolution of behavior*. San Francisco: W. H. Freeman, 1970.

Bastock, M. A gene mutation which affects a behavior pattern. *Evolution,* 1956, *10,* 421–439.

Bentley, D. R., & Hoy, R. R. Genetic control of the neuronal network generating cricket (*Teleogryllus gryllus*) song patterns. *Animal Behaviour,* 1972, *20,* 478–492.

Bentley, D. R., & Hoy, R. R. The neurobiology of cricket song. *Scientific American,* 1974, *231,* 34–44.

Bitterman, M. E. The comparative analysis of learning. *Science,* 1975, *188,* 699–709.

Boyden, A. Homology and analogy: A century after the definitions of "homologue" and "analogue" of Richard Owen. *Quarterly Review of Biology,* 1943, *18,* 228–241.

Bridgman, P. W. *The logic of modern physics*. New York: MacMillan Company, 1927.

Bridgman, P. W. *Reflections of a physicist*. New York: Harper & Row, 1978.

Caplan, A. L. (Ed.). *The sociobiological debate*. New York: Harper & Row, 1978.

Darwin, C. *The expression of the emotions in man and animals*. Chicago: University of Chicago Press, 1965. (Originally published, 1872).

Dethier, V. G., Solomon, R. L., & Turner, L. H. Sensory input and central excitation and inhibition in the blowfly. *Journal of Comparative and Physiological Psychology,* 1965, *60,* 303–313.

Dilger, W. Behavior and genetics. In E. Bliss (Ed.), *Roots of behavior*. New York: Harper & Row, 1962. (a)

Dilger, W. The behavior of lovebirds. *Scientific American,* 1962, *206,* 88–98. (b)

Dobzhansky, T. G. *Genetics of the evolutionary process*. New York: Columbia University Press, 1970.

Doolittle, R. F. Similar amino acid sequences: Chance or common ancestry? *Science,* 1981, *241,* 149–159.

Ebling, F. J. G. Review of: *Methods of inference from animal to human behavior.* (Edited by M. von Cranach.) *Journal of Biological and Social Structures,* 1981, *4,* 198.

Eisenberg, J. F., & Dillon, W. S. (Eds.). *Man and beast: Comparative social behavior.* Washington, D.C.: Smithsonian Institution Press, 1971.

Fitch, W. M. An improved method of testing for evolutionary homology. *Journal of Molecular Biology,* 1966, *16,* 9–16.

Fitch, W. M. Further improvements in the method of testing for evolutionary homology among proteins. *Journal of Molecular Biology,* 1970, *49,* 1–14.

Goodman, M., & Lasker, G. W. Measurement of distance and propinquity in anthropological studies. In J. F. Crow & C. Denniston (Eds.), *Genetic distance.* New York, London: Plenum Press, 1974.

Harrington, G. M. Intelligence tests may favor the majority groups in a population. *Nature,* 1975, *258,* 708–709.

Harrington, G. M. An experimental model of bias in mental testing. In C. R. Reynolds & R. T. Brown (Eds.), *Perspectives on bias in mental testing.* New York: Plenum Press, 1983.

Heslop-Harrison, J. Darwin as a botanist. In S. A. Barnett (Ed.), *A century of Darwin.* London: Mercury Books, 1962.

Hinde, R. A. The behavior of certain Cardueline F_1 inter-species hybrids. *Behaviour,* 1956, *9,* 202–213.

Hirsch, J. Behavior-genetic, or "experimental," analysis: The challenge of science versus the lure of technology. *American Psychologist,* 1967, *22*(2), 118–130.

Hirsch, J. Review of E. O. Wilson, *Sociobiology: The new synthesis,* 1975. *Animal Behaviour,* 1976, *24,* 707–709.

Jaynes, J. The historical origins of "ethology" and "comparative psychology." *Animal Behaviour,* 1969, *17,* 601–606.

Kalikow, T. J. Die ethologische theorie von Konrad Lorenz: Erklärung und Ideologie, 1938–1943. [Konrad Lorenz's ethological theory: Explanation and ideology, 1938–1943.] In H. Mehrtens & S. Richter (Eds.), *Naturwissenschaft Technik und NS-Ideologie: Beitrage zur Wissenschaftsgeschichte des Dritten Reichs.* Frankfurt am Main: Suhrkamp, 1980. (English text kindly provided by T. J. Kalikow, In J. Hirsch Papers, University of Illinois Archives, No. 15/19/22.)

Lockard, R. B. Reflections on the fall of comparative psychology: Is there a message for us all? *American Psychologist,* 1971, *26,* 168–179.

Lorenz, K. The objectivist theory of instinct. In P. P. Grassé (Ed.), *L'Instinct dans le comportment des animaux et de l'homme,* Paris: Masson, 1956.

Manning, A. The effects of artificial selection for mating speed in *Drosophila melanogaster. Animal Behaviour,* 1961, *9,* 82–92.

Margoliash E., & Fitch, W. M. The evolutionary information content of protein amino acid sequences. In W. J. Whelan & J. Schultz (Eds.), *Homologies in enzymes and metabolic pathways/metabolic alterations in cancer.* Proceedings of Miami Winter Symposium, January 19–23, 1970. Amsterdam, London: North Holland Publishing Co., 1970.

McKusick, V. A. The anatomy of the human genome. *The Journal of Heredity,* 1980, *71,* 370–391.

Montagu, A. *Sociobiology examined.* New York: Oxford University Press, 1980.

Nelson, M. Classical conditioning in the blowfly (*Phormia regina*): Associative and excitatory factors. *Journal of Comparative and Physiological Psychology,* 1971, *77,* 353–368.

Osborn, H. F. *From the Greeks to Darwin.* New York: Charles Scribner's Sons, 1894.

Owen, R. *Lectures on the comparative anatomy and physiology of the invertebrate animals.* Delivered at the Royal College of Surgeons of England. London: Longman, Brown, Green, & Longmans, 1843.

Owen, R. *Comparative anatomy and physiology of vertebrates.* London: Longmans, Green, & Co., 1866.

Pratt, C. C. Operationism in psychology. *Psychological Review*, 1945, *52*, 262–269.

Purton, A. C. Ethological categories of behaviour and some consequences of their conflation. *Animal Behaviour*, 1978, *26*, 653–670.

Razran, G. *Mind in evolution*. Boston: Houghton Mifflin Company, 1971.

Romanes, G. J. *Animal intelligence*. London: Kegan Paul, Trench and Co., 1882. (Republished by Gregg International Publishers Limited, England, 1970).

Ruse, M. *Sociobiology: Sense or nonsense?* Dordrecht, Holland: D. Reidel Publishing Company, 1979.

Schilcher, F. von, & Manning, A. Courtship song and mating speed in hybrids between *Drosophila melanogaster* and *Drosophila simulans*. *Behavior Genetics*, 1975, *5*, 395–404.

Seligman, M. E. P., & Hager, J. L. *Biological boundaries of learning*. New York: Appleton–Century–Crofts, 1972.

Sharpe, R. S., & Johnsgard, P. A. Inheritance of behavioral characters in F_2 mallard × pintail (*Anas platyrhychos L.* × *Anas acuta* L.) hybrids. *Behaviour*, 1966, *27*, 259–272.

Skinner, B. F. The operational analysis of psychological terms. *Psychological Review*, 1945, *52*, 270–277.

Skinner, B. F. *Walden II*. New York: MacMillan, 1948.

Strickberger, M. W. *Genetics* (2nd ed.). New York: MacMillan, 1976.

Thorpe, W. H. The definition of some terms used in animal behavior studies. *Bulletin of Animal Behaviour*, 1951, *9*, 1–7.

Tully, T., Zawistowski, S., & Hirsch, J. Behavior-genetic analysis of *Phormia regina:* III. A phenotypic correlation between the central excitatory state (CES) and conditioning remains in replicated F_2 generations of hybrid crosses. *Behavior Genetics*, 1982, *12*, 181–191.

Wallace, B. *Chromosomes, giant molecules and evolution*. New York: W. W. Norton & Co., Inc., 1966.

Watson, J. B. *Behaviorism*. Chicago: Chicago University Press, 1959.

Willey, A. *Convergence in evolution*. New York: E. P. Dutton & Co., 1911.

Wilson, E. O. *Sociobiology: The new synthesis*. Boston, Mass.: Belknap Press, 1975.

Yunis, J. J., Sawyer, J. R., & Dunham, K. The striking resemblance of high resolution G-banded chromosomes of man and chimpanzee. *Science*, 1980, *208*, 1145–1148.

EDITOR'S CONCLUDING REMARKS

In thinking about all this work on comparing behavior I was reminded of a paper published by Harlow and associates a decade ago (Harlow, Gluck, & Suomi, 1972). They provided a pointed and witty appraisal of the state of the art of comparative psychology at that time. Harlow et al. ended their article with a pithy statement:

> Basically the problems of generalizations of behavioral data between species are simple—one cannot generalize, but one must. If the competent do not wish to generalize, the incompetent will fill the field [p. 716].

By now we are in a position to respond to the above conclusion and warning based on the contents of the book at hand. In the first place, it does not appear that the incompetent have filled the field. If the contributors to *Comparing behavior: Studying man studying animals* are any indication, the area is well stocked with energetic, critical, and articulate scientists and thinkers. It is especially interesting that workers from a variety of academic disciplines were willing to share their special views on the comparative approach.

Second, it continues to be the case that one (or at least some of us) must generalize. Despite the fact that critics periodically announce the demise of comparative psychology (cf. Beach, 1950; Lockard, 1971) research and theorizing go on. As an index of the vigor of the field one can point to the number of prominent journals that provide an outlet for such work. In fact, coincidental with this writing the American Psychological Association announced the creation of a new comparative psychology journal—*Journal of Comparative Psychology and Behavior*—to begin publishing in 1983. According to the above, the field of comparative work is in a position to say (as did Mark Twain, I believe) "the rumors of my death are greatly exaggerated."

Finally, it does *not* appear to be the case that one cannot generalize. Several of the chapters in this volume offer clear methodological and intellectual standards for making meaningful comparisons at various levels and in several ways. In saying this I do not mean that comparing behavior is any easier than previously. However, I think we are in a new era insofar as the establishment of valid generalizations is concerned. Detractors of, and nonbelievers in the comparative approach may still have their say occasionally, but if they wish to deny the relevance of a legitimate comparison they will have to shoulder the burden of proof.

REFERENCES

Beach, F. A. The snark was a boojum. *American Psychologist,* 1950, *5,* 115–124.

Harlow, H. F., Gluck, J. P., & Suomi, S. J. Generalization of behavioral data between nonhuman and human animals. *American Psychologist,* 1972, *27,* 709–716.

Lockard, R. B. Reflections on the fall of comparative psychology: Is there a message for us all? *American Psychologist,* 1971, *26,* 168–179.

Author Index

Numbers in *italics* denote pages with complete bibliographic information.

T

Subject Index

M

N

O

P

Q

R

S

T

U

V